The Philosophy of Scientific Experimentation

Edited by Hans Radder

The Philosophy of

Scientific Experimentation

UNIVERSITY OF PITTSBURGH PRESS

Published by the University of Pittsburgh Press, Pittsburgh, Pa., 15261
Copyright © 2003, University of Pittsburgh Press
Manufactured in the United States of America
Printed on acid-free paper
10 9 8 7 6 5 4 3 2 1

Library of Congress Cataloging-in-Publication Data
The philosophy of scientific experimentation / Hans Radder, editor.
    p. cm.
Includes bibliographical references and index.
  ISBN 0-8229-5795-7 (pbk. : alk. paper)
   1. Science—Experiments—Philosophy. 2. ScienceMethodology—Philosophy.
I. Radder, Hans.
  Q180.55M4 P45 2003
  507.8—dc21

2002014515

CONTENTS

Preface and Acknowledgments   vii

About the Contributors   ix

1 Toward a More Developed Philosophy of Scientific
Experimentation   1
Hans Radder

2 The Materiality of Instruments in a Metaphysics for Experiments   19
Rom Harré

3 Thing Knowledge: Outline of a Materialist Theory of Knowledge   39
Davis Baird

4 Physics, Experiments, and the Concept of Nature   68
Peter Kroes

5 Experimentation, Causal Inference, and Instrumental Realism   87
Jim Woodward

6 Technology as Basis and Object of Experimental Practices   119
Rainer Lange

7 Theory-Ladenness and Scientific Instruments in Experimentation   138
Michael Heidelberger

8 Technology and Theory in Experimental Science   152
Hans Radder

9 The Idols of Experiment: Transcending the "Etc. List"   174
Giora Hon

10 Models, Simulation, and "Computer Experiments"   198
Evelyn Fox Keller

11 Experiments without Material Intervention: Model Experiments,
Virtual Experiments, and Virtually Experiments   216
Mary S. Morgan

12 Designing Instruments and the Design of Nature   236
Daniel Rothbart

13 Varying the Cognitive Span: Experimentation, Visualization,
and Computation   255
David Gooding

References   285
Index   303

PREFACE AND ACKNOWLEDGMENTS

The first stage in the making of this book on the philosophy of scientific experimentation was a three-day workshop held at the faculty of philosophy of the Vrije Universiteit Amsterdam, June 15–17, 2000. The workshop was entitled "Towards a More Developed Philosophy of Scientific Experimentation." The participants were Davis Baird, Henk van den Belt, Bram Bos, Marcel Boumans, Willem B. Drees, Maarten Franssen, David Gooding, Francesco Guala, Rom Harré, Michael Heidelberger, Giora Hon, Evelyn Fox Keller, Peter Kirschenmann, Peter Kroes, Huib Looren de Jong, Joke Meheus, Mary S. Morgan, Margaret Morrison, Arthur Petersen, Toine Pieters, Hans Radder, Henk de Regt, Hans-Jörg Rheinberger, Daniel Rothbart, Joseph Rouse, Rein Vos, and Jim Woodward.

The present volume includes revised and extended versions of papers from the workshop. Rainer Lange, who was originally invited but had to cancel, was nevertheless willing to contribute a chapter. Furthermore, I added a first chapter, which reviews the state of the art in the philosophy of scientific experimentation and situates the following chapters within that state. The book includes contributions from both the United States and Europe. In this way, it aims to bring together and integrate the best features of the U.S. and European philosophical traditions.

It is a pleasure to thank the various people and institutions who have supported this project. The commentators at the workshop did a great job, and their contribution has clearly improved the quality of the separate chapters and of the book as a whole. The review report from the University of Pittsburgh Press was to the point and helpful.

Funds for the workshop were made available by the Netherlands Graduate School of Science, Technology and Modern Culture; the Nederlandse Organisatie voor Wetenschappelijk Onderzoek (Gebied Geesteswetenschappen); the Faculty of Philosophy of the Vrije Universiteit Amsterdam; and the Koninklijke Nederlandse Akademie van Wetenschappen. The Faculty of Philosophy of the Vrije Universiteit Amsterdam provided additional means for the production of the final manuscript. I would like to thank these organizations for their financial support.

On a more personal level, I am indebted to Anniek van der Schuit and, in particular, Wouter Radder for their helpful and conscientious assistance in preparing the manuscript for publication. The cooperation with director Cynthia Miller and others at the University of Pittsburgh Press has been pleasant and efficient. Final thanks should go to Sally Wyatt. Knowing the ups and downs of editing a multiauthored volume herself, she has been an invaluable source of linguistic, editorial, and personal advice.

ABOUT THE CONTRIBUTORS

Davis Baird is professor and chair of the Department of Philosophy at the University of South Carolina. He received his Ph.D. (1981) from Stanford University's program in philosophy of science, philosophy of language, and logic. In addition, he has a master's (1981) from Stanford and a bachelor of arts (1976) from Brandeis University. His research focuses on the history and philosophy of scientific instruments, particularly those developed for analytical chemistry during the twentieth century. Here he pursues a familial interest, being the son of a cofounder of an early developer of spectrographic instrumentation, Baird Associates. He is the author of *Thing Knowledge: A Philosophy of Scientific Instruments* (The University of California Press, forthcoming) and currently is working on a history of the firm his father founded. He edits *Techné: Journal of the Society for Philosophy and Technology* (http://scholar.lib.vt.edu/ejournals/SPT/).

David Gooding is professor of history and philosophy of science and director of the Science Studies Centre, University of Bath. He is interested in how science continually reshapes the interaction of intellect, technology, and human agency. In addition to a number of articles on experimentation and in particular the fine structure of Faraday's experimental work, he has authored *Experiment and the Making of Meaning* (Kluwer, 1990) and coedited *Faraday Rediscovered* (Macmillan/American Institute of Physics, 1985) and *The Uses of Experiment* (Cambridge University Press, 1989). He is currently working on the changing role of visualization in the sciences.

Rom Harré studied mathematics and physics and then philosophy and anthropology. His published work includes studies in the philosophy of the natural sciences, including *Varieties of Realism* (Blackwell, 1986) and *Great Scientific Experiments* (Oxford, 1983). He has been among the pioneers of the "discursive" approach in the human sciences. In *Social Being* (Blackwell, 1979), *Personal Being* (Blackwell, 1983), and *Physical Being* (Blackwell, 1991), he explored the role of rules and conventions in various aspects of human cognition, while in *Pronouns and People* (Blackwell, 1990), he and Peter Mühlhäusler developed the thesis that grammar and the sense of self are intimately related. His most recent work, *One*

Thousand Years of Philosophy (Blackwell, 2001), follows the philosophical enterprise in India, China, Islam, and Europe from 1000 A.D. He is emeritus fellow of Linacre College (Oxford), professor of psychology at Georgetown University, and adjunct professor of philosophy at American University, Washington, D.C. He holds honorary doctorates from the universities of Helsinki, Brussels, Aarhus, and Lima.

Michael Heidelberger recently moved from the Humboldt University in Berlin to the University of Tübingen, where he took the chair for logic and philosophy of science. He has a wide interest in topics related to causality and probability, measurement, and experiment, and he specializes in the history of the philosophy of science, especially of the late nineteenth and early twentieth century. He is the author of *Die innere Seite der Natur* (Klostermann, 1993), a book on the philosopher, physicist, and founder of psychophysics, Gustav Theodor Fechner, and his philosophy of nature, and he coedited, with Friedrich Steinle, *Experimental Essays—Versuche zum experiment* (Nomos Verlag, 1998).

Giora Hon teaches philosophy of science in the Department of Philosophy, University of Haifa. He read physics at the University of Tel-Aviv and holds a doctoral degree in history and philosophy of science from London University. His main interest is experiment and its pitfalls. Recent articles include "The Limits of Experimental Method: Experimenting on an Entangled System— The Case of Biophysics," in *Science at Century's End: Philosophical Questions on the Progress and Limits of Science* (University of Pittsburgh Press, 2000), edited by M. Carrier, G. J. Massey, and L. Ruetsche, and "From Propagation to Structure: The Experimental Technique of Bombardment as a Contributing Factor to the Emerging Quantum Physics," in *Physics in Perspective* (forthcoming). He is the coeditor with Sam S. Rakover of *Explanation: Theoretical Approaches and Applications* (Kluwer, 2001).

Evelyn Fox Keller received her Ph.D. in theoretical physics at Harvard University, worked for a number of years at the interface of physics and biology, and is now professor of history and philosophy of science in the program in science, technology, and society at MIT. She is the author of *A Feeling for the Organism: The Life and Work of Barbara McClintock* (Freeman, 1983); *Reflections on Gender and Science* (Yale University Press, 1985); *Secrets of Life, Secrets of Death* (Routledge, 1992); *Refiguring Life: Metaphors of Twentieth Century Biology* (Columbia University Press, 1995); and *The Century of the Gene* (Harvard University Press, 2000). Her most recent book is *Making Sense of Life: Explaining Biological Development with Models, Metaphors, and Machines* (Harvard University Press, 2002).

Peter Kroes studied physical engineering at the University of Technology Eindhoven (The Netherlands) and wrote a Ph.D. thesis on philosophical problems concerning the notion of time in modern physical theories (University of Nijmegen, The Netherlands, 1982). Since 1995 he has been a professor in philosophy—in particular, philosophy of technology—at the University of Technology Delft. His main areas of interest are philosophy of technology and philosophy of science. His principal publications are *Time: Its Structure and Role in Physical Theories* (Reidel, 1985); *Technological Development and Science in the Industrial Age* (Kluwer, 1992); *Ideaalbeelden van wetenschap* (Boom, 1996); and, coedited with A. Meijers, *The Empirical Turn in the Philosophy of Technology* (JAI/Elsevier, 2000).

Rainer Lange taught philosophy at the University of Marburg and is presently working for the German Science Council, a science policy advisory board. His main interests are the philosophy of biology, experimental practices, and the philosophy of action, especially cooperative agency. His publications include "Gibt es einen neuen Experimentalismus?" *Theory in Biosciences* (1997); *Experimentalwissenschaft Biologie* (Königshausen and Neumann, 1999); "Epistemology Culturalized" (with Dirk Hartmann), *Journal for General Philosophy of Science* (2000), and a number of articles and contributions in the philosophy of science, epistemology, and the philosophy of action.

Mary S. Morgan is professor of history of economics at the London School of Economics and professor of history and philosophy of economics at the University of Amsterdam. Her recent publications include *Models as Mediators,* coedited with Margaret Morrison (Cambridge University Press, 1999); *Methodology and Tacit Knowledge: Two Experiments in Econometrics,* with Jan R. Magnus (Wiley, 1999); and *Empirical Models and Policy-Making: Interaction and Institutions,* coedited with Frank den Butter (Routledge, 2000). She is currently working on a book on the history and philosophy of economics as a modeling discipline.

Hans Radder is associate professor in the Faculty of Philosophy at the Vrije Universiteit Amsterdam. He holds a master's degree in physics and a master's and Ph.D. degree in philosophy. Principal themes in his work are experimentation, scientific realism, and the normative and political significance of science and technology. In addition to a variety of articles, he published *The Material Realization of Science* (Van Gorcum, 1988) and *In and about the World* (State University of New York Press, 1996). His forthcoming book, *The World Observed/The World Conceived,* argues that concepts not only structure the world (through observation) but also abstract from it.

Daniel Rothbart teaches in the Department of Philosophy and Religious Studies, George Mason University, Fairfax. After receiving his Ph.D. in philosophy from Washington University in St. Louis, he was a visiting scholar at Dartmouth College and the University of Cambridge. He has published extensively on the philosophical aspects of scientific modeling and on the philosophy of experimentation. In addition to writing the book *Explaining the Growth of Scientific Knowledge* (Edwin Mellen Press, 1997) and editing *Science, Reason and Reality* (Wadsworth, 1997), his work appeared in anthologies, such as *Analogical Reasoning, Chemistry and Philosophy* and *Of Minds and Molecules*, and in journals, such as *Dialectica, Erkenntnis, Foundations of Chemistry,* and *Philosophy of Science*. He recently completed work on his next book, entitled *Philosophical Instruments: Minds and Tools at Work*.

Jim Woodward is J. O. and Juliette Koepfli Professor of the Humanities and executive officer for the humanities at the California Institute of Technology. His research interests include theories of causation and explanation, problems of inductive inference in experimental contexts, the evolution of cognition, and experimental investigations into moral reasoning and behavior. Recent papers include "Data, Phenomena, and Reliability," *PSA 1998*; "Independence, Invariance and the Causal Markov Condition" (coauthored with Dan Hausman), *British Journal for the Philosophy of Science* (1999); "Explanation and Invariance in the Special Sciences," *British Journal for the Philosophy of Science* (2000); and "Law and Explanation in Biology: Invariance Is the Kind of Stability that Matters," *Philosophy of Science* (2001). A forthcoming book, *Making Things Happen: A Theory of Causal Explanation,* will be published by Oxford University Press.

The Philosophy of Scientific Experimentation

Hans Radder

I / Toward a More Developed Philosophy of Scientific Experimentation

1. The Philosophy of Scientific Experimentation

The development of the philosophy of scientific experimentation over the past twenty years has two main features. After a rapid start in the 1980s (see Hacking 1989a), it seems to have lost much of this momentum during the next decade. At the very least, the expectation that the study of experiment would become a major issue within received traditions in philosophy of science has not been fulfilled. To verify this, it is enough to glance through the recent volumes of well-known journals, such as *Philosophy of Science*, *British Journal for the Philosophy of Science*, *Erkenntnis,* and the like. Alternatively, one may look at recent anthologies, which could be supposed to represent the core readings in present-day philosophy of science. For example, the six-volume set of collected papers in philosophy of science (Sklar 2000) contains no contributions that focus on experimentation. And in the voluminous *Companion to Philosophy of Science* (Newton-Smith 2000) the explicit analysis of experimentation is almost completely limited to one chapter. Thus, the fact that many scientists, perhaps even the majority of them, spend most of their time doing experiments of various kinds is not reflected in the basic literature in the philosophy of science.

In this respect, a strong contrast can be seen between philosophical and historical or social scientific studies of science. This contrast marks the second feature of the present state of the philosophy of scientific experimentation. A brief perusal of recent volumes of leading science studies journals confirms the claim.

Studies in History and Philosophy of Science and *Social Studies of Science,* for example, offer many detailed historical and social scientific articles on experimental practice. In addition, a major recent anthology (Biagioli 1998) includes many contributions that explicitly deal with empirical and theoretical issues of scientific experimentation.

Thus, the philosophy of experimentation is still underdeveloped, especially as compared to historical and social scientific approaches (Radder 1998). Given this state of affairs, many philosophers of experiment agree that the field needs a new impulse (see Lelas 2000, 203; Harré, this volume, chap. 2; Hon, this volume, chap. 9). In this spirit, in June 2000 a workshop was held in Amsterdam, also entitled "Towards a More Developed Philosophy of Scientific Experimentation." The following chapters are the reworked and expanded results of that workshop.

Having described the present state of the art in the philosophy of experimentation in these terms, two qualifications are in order. The first is that, of course, the noted characteristics of contemporary philosophy of experiment represent a trend, not an exceptionless regularity. Thus, the volumes edited by Buchwald (1995) and by Heidelberger and Steinle (1998) contain a number of philosophical chapters of experiment, in addition to primarily historical studies. A perhaps more significant and promising fact is that a surprisingly large number of the papers presented at the recent 2000 Biannual Meeting of the Philosophy of Science Association addressed issues in the philosophy of scientific experimentation.

A second qualification has to do with the fact that historical and social scientific work on experimentation is often relevant to, and sometimes contains explicit discussions of, philosophical issues. So, proponents of a science studies approach might say, What is the problem? A comprehensive answer to this question would require a discussion of methodical and substantive similarities and dissimilarities between a philosophical and a historical or sociological approach to the study of science and, in particular, scientific experimentation. I cannot address this question in its full generality here.[1] Instead, I shall illustrate the need for a more developed philosophy of scientific experimentation by way of example.

Consider, for instance, the notion of stability. Within the science studies approach, a major feature of experimental practice is claimed to be the emergence of an interactive stability between a variety of heterogeneous elements of experimental practice; for example, material procedures, models of the instruments, and models of the phenomena under study (Pickering 1989). In such accounts, "stability" functions as a descriptive term for a situation that displays certain constant features (at least for a relatively long period). But in fact, the notion of stability is richer than mere lack of change, and a more developed philosophy of scientific experimentation should exploit this surplus meaning. If being stable implies being robust against actual and possible disturbances, then further philo-sophical questions immediately suggest themselves: What kind of disturbances are involved? What characteristics of the stabilization procedures can explain this robustness? Are those characteristics only of a factual or also of a normative nature? Generally speaking, dealing with such questions requires the more the-oretical approach that is typical of philosophy.

As a second example, consider Latour's (1987) definition of an instrument as any setup that produces an inscription in a scientific text and, more generally, his interpretation of a laboratory as an "inscription factory." This definition and interpretation are discussed in the chapters by Rom Harré and Davis Baird. One of the criticisms put forward is that Latour's account of instruments is superficial because a comprehensive analysis of the role and function of scientific instru-ments is lacking. Hence, one of the goals of the present volume is to contribute to a more adequate account of the nature and role of instrumentation in exper-imental science (see also section 7). Another characteristic of Latour's approach is that it does not offer a conceptual account of the difference between an ade-quate and a useless instrument. Here we touch upon the issue of the evaluative and normative nature of philosophical accounts of science. Thus, in his contri-bution, Giora Hon argues that, for methodological reasons, there should be a clear distinction between the theories of the apparatus and material procedures, on the one hand, and the theoretical interpretation of the result of an experi-ment, on the other. Now, it is true that quite a few contemporary philosophers are naturalists who claim to be value-neutral and nonnormative. In spite of this, and rightly so, within philosophy the normativity issue is still alive and well, and here a further contrast with historical or sociological studies of science applies.

In this book, we focus on six central themes in the philosophy of scientific ex-perimentation, which run through the entire volume: the material realization of experiments; experimentation and causality; the science-technology relation-ship; the role of theory in experimentation; modeling and (computer) experi-ments; and the scientific and philosophical significance of instrumentation. Each chapter deals with some of those themes, while each theme is discussed by vari-ous authors. In part, the themes are approached from complementary perspec-tives, and, in part, authors address each other's accounts of a relevant theme. The latter means that sometimes they use the same results or endorse the same views (for example, the chapters by Harré and Rothbart and by Baird and Kroes) and sometimes they challenge each other's accounts (for instance, the chapters by Woodward and Lange and by Heidelberger and myself). Wherever appropriate, I will highlight those agreements and disagreements. In the following six sec-tions, I introduce and discuss our central themes. The final section briefly addresses some further issues that should be included in a mature philosophy of scientific experimentation.

To conclude, I would like to make one further observation. This book is a plea for a philosophy of experimentation as a subject in its own right. Yet, the philosophy of scientific experimentation should not degenerate into a philosophical "-ism"; in this case, "experimentalism." That is to say, taking full account of scientific experimentation does not commit one to the doctrine that all philosophical problems regarding science can be completely resolved on the basis of an analysis of experiment only.[2]

2. The Material Realization of Experiments and Its Philosophical Significance

In experiments we actively interfere with the material world. In one way or another, experimentation involves the material realization of an experimental process (the object[s] of study, the apparatus, and their interaction). The question, then, is this: What are the implications of this action and production character of scientific experimentation for philosophical debates on ontological, epistemological, and methodological issues about science?

A general ontological lesson appears to be the following. The action and production character of experimentation entails that the actual experimental objects and phenomena themselves are, at least in part, produced by human intervention. For this reason, if one does not want to endorse a full-fledged constructivism—according to which the experimental objects and processes are nothing but artificial, human creations—one needs to go beyond an actualist ontology and introduce more differentiated ontological categorizations.[3] This is precisely what is at issue in several chapters. Thus, Rom Harré argues that an adequate ontological interpretation of experimental science needs some kind of dispositional concepts, namely Gibsonian affordances. In the same spirit, Daniel Rothbart analyzes the practice of experimental design, the role of experimental reproducibility, and the conception of nature as a machine. He concludes that the use of pictorial symbols, the procedure of "virtual witnessing," and the role of the specimen in instrumentation entail the need to include nonactualist notions, such as possibility, capacity, and tendency, into the ontology of experimental science. Peter Kroes starts from a different problem context, namely the question of whether the distinction between natural and artificial objects and processes still makes sense for modern physical science. Kroes concludes that experimental interventions do create the actual, artificial "instances" of phenomena, but not the natural phenomena as such. Thus, by making such a distinction, he in fact presupposes a nonactualist ontology as well.

Next to ontological problems, the interventionist character of experimentation engenders epistemological questions as well. An important question is whether scientists, on the basis of artificial experimental intervention, can acquire knowledge of a human-independent nature. According to Harré, such back inferences, from the artificial laboratory systems to their natural counterparts, are possible in a number of cases, but their justification is different for different types of apparatus.

Another approach accepts the constructed nature of much experimental science but stresses the fact that its results acquire a certain endurance and autonomy with respect to both the context in which they have been realized in the first place and later developments. In this spirit, Davis Baird offers a neo-Popperian account of "objective thing knowledge," the knowledge encapsulated in material things. Illustrations of such knowledge are Watson and Crick's material double helix model, Davenport's rotary electromagnetic motor, and the indicator of Watt and Southern's steam engine.[4] Baird suggests analogues of the standard epistemological notions of truth, justification, and delocalization for the case of thing knowledge. On this basis, thing knowledge is claimed to be objective in the sense of transcending its context of creation. This idea of transcendence can be seen as complementary to Popper's account of objective propositional knowledge, according to which human ideas, problems, arguments, and the like can transcend their context of discovery toward an autonomous ontological domain.

A further epistemologically relevant feature of experimental science is the fact that scientific apparatus often works in the absence of an agreement on exactly how it does so. An example discussed by Baird is Faraday's electromotor in its early days. Thus, in scientific practice a significant distinction obtains between the working of apparatus and their theoretical accounts. More particularly, the claim is that variety and variability at the theoretical and ontological level may well go together with a considerable stability at the level of the material realization of experiments. Such claims can be used for philosophical purposes; for example, to vindicate an instrumental realism, as is done in Jim Woodward's contribution, or a referential realism, as I have proposed elsewhere (Radder 1988; 1996, 4.2).

Given the arguments and views set out so far, a natural question is whether they license or perhaps even entail a full-fledged materialism, in the ontological sense of that term. Although not all of the authors mentioned discuss this question in an explicit way, the answer is certainly not an unambivalent "yes." Baird, for one, explicitly leaves room for theoretical knowledge and abstract Popperian world-3 entities. Kroes emphasizes the indispensability of functions and their irreducibility to physical structures, and he argues that experimental and technological objects, being simultaneously physical and functional entities, possess a dual nature. The critical issue, then, is whether the methodologically indispensable notion of function should, or should not, be taken into account ontologically.

3. Experimentation and Causality

Theoretical and empirical studies of experimentation are preeminently suited to an investigation of the issue of causality. Conversely, the philosophy of scientific experimentation may fruitfully employ insights gained in the debate on causality (see, for example, Guala 1999b). In the following chapters, at least three different approaches can be found.

First, the role of causality in experimental processes and experimental practice may be analyzed. Both Rom Harré and Michael Heidelberger advocate a differentiated account of this role. Harré speaks of causally based instruments and distinguishes these from other types of apparatus. Following Cartwright (1983) and Hacking (1983a), Heidelberger intends to make a clear contrast between a causal and a theoretical level in scientific experimentation.

An issue that is relevant here is whether experimentation can be characterized as fully causal or whether free or intentional action is important as well. This issue is explicitly addressed in the contributions by Peter Kroes, Jim Woodward, and Rainer Lange. Interestingly, Niels Bohr, in his analyses of experimentation in atomic physics, has already dealt with this issue. He claimed that experimenters need a free choice of, first, where to put the necessary boundary between the instrument and the object under investigation and, second, which one of two complementary phenomena they decide to realize (Bohr 1958, 1963; Scheibe 1973, 25; Radder 1979, 427–428). In a different tradition, Peter Janich has emphasized the indispensability of free and intentional action—as contrasted with caused behavior—for an adequate account of scientific experimentation (Janich 1998, 102–107). The basic point, for both Bohr and Janich, is that in experimental interventions, we intentionally bring about certain states of affairs that would not have arisen without our interference, while we could have chosen to realize other states instead.

A second approach involves analyzing the role of experimentation in interpreting and testing causal claims. This is the approach chosen by Jim Woodward. Building on methodological literature on causation in the biomedical, behavioral, and social sciences, he introduces a specific version of the idea of (experimental) intervention. On this basis, he develops his view that causal inferences can only be justified through (possibly hypothetical) experimental interventions and not through "passive" observations. Basically, a causal claim is about what would happen if certain experiments were to be performed. Hence, this approach goes beyond the Humean regularity theory, in which the causal relation is reduced to a constant conjunction of two actually occurring events of a particular type. Woodward emphasizes that what he offers is a criterion of causality. Because the explanation of the notion of intervention itself employs the concept of causal processes, we do not have a reduction of causality to experimental intervention. From Woodward's perspective, a disadvantage of such a reduction would be that it makes causal processes in nature dependent on human action, and hence the resulting account would be anthropomorphic and subjectivist.

A third approach, however, just tries to do this: to explain the notion of causality on the basis of the notions of action and manipulation. This view is represented by Rainer Lange (see also Von Wright 1971; Janich 1998, 107–110). The central idea is to make use of a distinction between the intentional bringing about and the causal coming about of the states of experimental systems. Lange claims that this version of the manipulability account of causality need not be subjective or anthropomorphic, and that it contrasts with Woodward's version in being noncircular.

4. The Science-Technology Relationship

In my introductory section I pointed out that a decisive breakthrough of the topic of scientific experimentation within the philosophy of science has not yet been accomplished. The reason may be that many philosophers deem a topic significant to the extent that it contributes to reaching what they see as the aim, or the aims, of science. Traditionally, philosophers of science have defined the aim of science as, roughly, the generation of reliable knowledge of the world. Moreover, as a consequence of explicit or implicit empiricist influences, there has been a strong tendency to take the production of empirical knowledge for granted. From this perspective, then, the only interesting philosophical problems concern theoretical knowledge and its relationship to this taken-for-granted empirical base.

However, if we take a more empirical look at the sciences, both at their historical development and at their current condition, this perspective must be qualified as quite one-sided, to say the least. After all, from Archimedes' lever and pulley systems to the cloned sheep Dolly, the development of science has been intricately interwoven with the development of technology (see Tiles and Oberdiek 1995; Joerges and Shinn 2001). Hence, if one wants to attribute any aims to science, making a contribution to technology should certainly be one of them. From this alternative perspective, the relevance of experimentation for philosophy hardly needs any further justification. After all, experiments make essential use of (often specifically designed) technological devices and they often contribute to technological innovations. Moreover, there are substantial conceptual similarities between the realization of experimental and that of technological processes, most significantly the implied possibility and necessity of the manipulation and control of nature (see Radder 1987; 1996, chap. 6; Lee 1999, chap. 2).

In sum, if philosophers keep neglecting the technological dimension of science, experimentation will continue to be seen as a mere data provider for the evaluation of theories. If they start taking the science-technology relationship seriously, however, doing experiments can be studied as a topic in its own right, which poses—as we hope to show in this volume—many interesting and important philosophical questions.

One obvious way to study the role of technology in science is to focus on the instruments and equipment employed in laboratory experiments. Several chapters of this book take this route, and I will return to this approach in section 7. Here, I would like to focus on the general philosophical significance of the experiment-technology relationship. Quite a few philosophers who emphasize the relevance of technology for science endorse a "science-as-technology" account. That is to say, they advocate an overall interpretation in which the nature of science—not just experimental but also theoretical science—is seen as basically or primarily technological (see, for instance, Dingler 1928; Habermas 1978; Janich 1978; Latour 1987; Lelas 1993, 2000; Lee 1999).

Most chapters of this book, however, take a less radical view. As we have seen, Davis Baird argues for the importance of thing knowledge on a par with theoretical knowledge. Rainer Lange emphasizes the conceptual and historical proximity of (experimental) science and technology primarily through his notion of a reproducible experimental instruction; but he also argues that scientific laws cannot be reduced to technological operations. Michael Heidelberger distinguishes a theoretical level, where interpretation and representation take place, from a relatively independent causal level, where (technological) production and construction of phenomena prevail. In my own chapter, I take account of two essential aspects of scientific experimentation, its material realization and its theoretical interpretation. In particular, I offer an argument for the irreducibility of the theoretical meaning of replicable experimental results. Thus, while stressing the significance of the technological—or perhaps more precisely, the action and production—dimension of science, these views nevertheless see this dimension as complementary to a theoretical dimension.

5. Theory and Theoretical Knowledge in Experimental Practice

We are now led to a further central theme in the philosophy of scientific experimentation, namely the relationship between theory and experiment. The theme can be approached from two sides. First, one may study the role of existing theories, or theoretical knowledge, within experimental practices. This will include a discussion of the view of experiments as (mere) tests of theories. The overarching issue concerns the claimed (relative) autonomy of experimental science from theory.

The most far-reaching position is that, basically, experimentation is theory-free. The German school of "methodical culturalism" seems to come close to this position (Janich 1998; for a review, see Lange 1999, chap. 3). A more differentiated view is that, in important cases, theory-free experimentation is possible and occurs in scientific practice. Hacking (1983a) and Steinle (1998) make this claim primarily on the basis of a number of case studies from the history of experimental science. In his contribution, Michael Heidelberger aims at a more systematic underpinning of this view. He discusses the notions of theory ladenness put forward by Hanson, Duhem, and Kuhn and shows that they differ significantly. In particular, he argues that, for Hanson, theory ladenness primarily means "causality ladenness." Next, Heidelberger suggests that causal issues in experimentation can and should be distinguished from theoretical issues. The same distinction returns in his classification of scientific instruments. While experiments with "representative" instruments are theory-laden, the use of "productive," "constructive," or "imitative" instruments is causally based and claimed to be theory-free.

Still another view admits that not all concrete activities that can be discerned in scientific practice are guided by theories. Yet, according to this view, if certain activities are to count as a genuine experiment, they require a theoretical interpretation. This is the view I argue for in my own contribution (chapter 8), both on the basis of systematic philosophical arguments and on the basis of a criticism of cases of claimed theory-free experiments. Something like this view seems to be implied in Giora Hon's chapter. In the spirit of Francis Bacon, he proposes a systematic theory of the types of errors that may arise, and hence should be avoided, in performing and interpreting experiments (see also Hon

1989b, 1998a). This typology of errors is based on an account of scientific experimentation in which both theoretical knowledge and material realization play an indispensable role, and it is meant to illuminate the epistemic structure of experiments.

Two points, briefly discussed in my chapter, are crucial in settling the issue of the role of theory in scientific experiments. The first is that even posing the question of whether or not theory-free experiments are possible presupposes some notion of what we understand by "an experiment." Second, since nobody seems to deny that some kind of interpretation plays a role in performing and understanding experiments, the critical question is whether this interpretation is "theoretical" or not. Can we distinguish different kinds and levels of (theoretical) interpretation and, if so, what are the philosophical implications of such a "compartmentalization"? As I mentioned in section 1, Hon takes the view that the theories of the apparatus and material procedures can and should be distinguished from the theoretical interpretation of the result of the experiment. My own view is that such a compartmentalization of theories—if possible at all —may be helpful in dealing with some philosophical problems (in particular, the issue of circularity in testing theories) but not with others (primarily, the realism problem).

The second major approach to the experiment-theory relationship addresses the question of how theory may arise from material experimental practices, or, in Hon's terms, how to conceptualize the transition from the material process to propositional, theoretical knowledge. And, of course, even if experimental research is not merely a means to theoretical knowledge, experiment does play an epistemic role with respect to the formation of scientific theories. A balanced philosophical study of this issue may profit both from "relativist" science studies approaches (for example, Collins 1985; Gooding 1990) and from "rationalist" epistemological approaches (for example, Franklin 1986, 1990; Mayo 1996).

One aspect of the transition from experimental practices to theoretical knowledge is discussed in David Gooding's chapter. He argues that the mathematical nature of scientific theories is intrinsically connected to the possibility of quantitative measurement. Moreover, often the precision and repeatability required for quantification cannot be found in nature but has to be technologically manufactured.

My own approach to this issue—see chapter 8—is by means of a novel notion of abstraction. Abstraction, as a first major step toward the formation of theories, plays a significant role in experimental practice. It occurs whenever experimenters attempt to replicate experimental results by means of completely different processes. That is to say, a replicability claim entails an abstraction from the particular material circumstances and procedures through which the experiment has been realized so far.[5] I argue that such claims possess a nonlocal, theoretical meaning that cannot be reduced to the meaning of a fixed set of material realizations. By way of example, I briefly discuss the role of abstraction in the experiments that contributed to Edison's invention of the incandescent lamp.

6. Experiment, Modeling, and (Computer) Simulation

Over the past decades, the scientific significance of computer modeling and simulation has increased greatly. Many scientists nowadays are involved in what they call "computer experiments." Apart from its intrinsic interest, this development invites a philosophical discussion of what is meant by these computer experiments and how they relate to ordinary experiments.

The chapters by Evelyn Fox Keller and Mary Morgan deal with this topic. Both offer a classification of computer modeling and simulation. Keller proposes a historical typology, primarily derived from the development of the physical sciences during the last half century but with applications in the biological sciences. Within this development she distinguishes three different stages. The first uses of computer simulation were meant to provide an alternative to (cumbersome or unfeasible) mathematical methods. They are sometimes called "experimental" because of their nonanalytical and exploratory nature: they aim to solve certain problems in mathematical physics, which have proved to be intractable so far, by means of novel computational techniques. In the second stage we meet with "computer experiments" proper. Here it is the physical systems (for example, the molecular dynamics of gases or liquids) that are being simulated by theoretical models. The experiments, then, consist in varying certain parameters (for example, density or temperature), noting what happens in the model, and comparing the outcomes to observed features of the systems. The third stage tries to model phenomena for which no theory exists so far. Here Keller discusses "artificial life" studies, in which the modeled phenomena exhibit certain patterns that are similar to global processes of biological self-reproduction or evolution. Again, it is the opportunity for artificial manipulation of parameters of the model objects that motivates scientists to call this approach "experimental."

In section 2 I emphasized, with most of the authors of this volume, the significance of material realization for scientific experimentation. Scientists, however, often use the term "experiment" in a looser and more varied sense.[6] This is quite clear in the examples of model and computer experiments discussed in the chapters by Keller and Morgan. Here, the relevant models are conceptual or theoretical, in contrast to the physical or material models dealt with in Harré's and Baird's contributions. This raises the obvious question of the relationship between ordinary experiments and such model or computer experiments. This question is discussed in detail in Mary Morgan's contribution. In contrast to the historical approach by Keller, she offers a systematic typology of modeling and simulation experiments. It is based on a theoretical analysis in which the types are distinguished according to their kind of controls, their mode of demonstration, their degree of materiality, and their representational validity. Morgan discusses a number of experiments in mechanics, biology, and economics and classifies them on a continuum: from setups that materially intervene in a straightforward sense to the types of virtually, virtual, and model experiments. In this order, these types of experiment exhibit an ever decreasing amount of material intervention, while the ways in which they represent the world can be seen to vary as well.

Thus, in scientific practice we find various sorts of hybrids of material interventions, computer simulations, and theoretical and mathematical modeling techniques. Often, more traditional experimental approaches are challenged and replaced by approaches based fully or primarily on simulations or mathematical models (sometimes this replacement is based on budgetary considerations only). This development raises interesting questions for a philosophy of scientific experimentation. Prominently, there is the epistemological question of the reliability of the results of the new approaches. Mary Morgan suggests that experiments with a substantial material component should remain the standard because, generally speaking, they possess the greater epistemic power. Evelyn Fox Keller's assessment is more implicit, but she does seem to be wary of overly simplistic identifications of simulated reality with the real world, suggested by seductive computer imaging techniques (for instance, in the area of artificial life).

A further question raised by these chapters is metaphilosophical. How should the philosopher's notion of experiment relate to scientists' usages? Of course, this is just one example of a quite general hermeneutical issue: to what extent should scholars in the human sciences take into account the concepts and interpretations of the people who are being studied? When scientists use the notion of experiment in a broad sense—for example, by speaking of computer experiments —should philosophers follow them? Answers to these questions depend on the conception of philosophy one adheres to. Thus, if one aims at descriptive adequacy with respect to scientific practice, it is natural to be alert to actor uses and meanings and to the way these uses and meanings change over time (for example, Galison 1997). If one aims to uncover and evaluate general features or underlying principles of science, one will look for fruitful theoretical concepts, plausible generalizations, and reliable research standards. This approach is exemplified in Giora Hon's chapter, which proposes a general theory of experimental error based on a systematic and normative account of scientific experimentation. I myself see philosophy as primarily a theoretical, normative, and reflexive activity (Radder 1996, chap. 8). From this perspective, philosophy retains a relative autonomy vis-à-vis scientific practice (Radder 1997). Thus, as I argue in my own chapter, if we put forward philosophical claims about scientific experiments we need to make explicit what we understand by "experiment." In doing so, we draw on insights gained from descriptive studies of scientific practice, but we go beyond those studies by taking into account philosophical concerns and conceptions as well.

7. The Scientific and Philosophical Significance of Instruments

Both the older literature (for example, Gooding, Pinch, and Schaffer 1989) and the present collection of papers show that the study of scientific instruments is a rich source of insights for a philosophy of scientific experimentation. The chapters of this book exhibit a variety of features of the design, operation, and wider uses of instruments, and they discuss many of their philosophical implications. To give some idea of those features and implications I will briefly sum up the various descriptive and interpretive accounts, in as far as they have not yet come up in the previous sections.

Daniel Rothbart focuses on the design process. He points out the importance of schematic, pictorial symbols in designing scientific instruments, and he analyzes the perceptual and functional information that is being stored in those images. Philosophical themes of his chapter include the nature of visual perception, the relationship between thought and vision, the role of reproducibility as a norm for experimental research, and the ontological conception of nature as a nomological machine.

The contribution by David Gooding deals with the modes of representation of instrumentally mediated experimental outcomes. Gooding contrasts visual and verbal modes of representation with numerical and digital ones. His principal claim is that over the past centuries the former modes of representation seem to have been superseded by the latter, but a complete replacement of qualitative sensation and conceptual interpretation with quantitative measurement and formal calculation is neither possible nor desirable. A premise of this claim is the view that, ultimately, human beings are and will remain analogical by nature. Gooding illustrates his account with examples taken from experimental science and from research in artificial intelligence.[7]

Several authors propose classifications of scientific instruments or apparatus. As we have already seen, both Davis Baird and Michael Heidelberger suggest a typology of instruments with respect to their epistemic function. Baird distinguishes between instruments that generate material representations, instruments that present phenomena, and measuring instruments. Heidelberger identifies productive, constructive, imitative, and representative types of instruments. In drawing philosophical conclusions Heidelberger employs his typology of instrumentation to argue for the possibility of theory-free experiments, while Baird focuses on the notion of thing knowledge as a complement to propositional knowledge. Finally, Rom Harré's classification is based on distinct ontological relationships between laboratory equipment and the world. In his case, the epistemic functions of this equipment are derived from these ontological relationships. His prime distinction is that between "instruments" and "apparatus." Instruments are characterized by their causal relation to the (outside) world, and they enable a clear separation between the natural object and its measuring device. In contrast, apparatus are said to be "part of nature" because they either are inseparable from or (almost) identical to natural objects.

Thus, we have a variety of classifications of scientific instrumentation. They form, I think, an excellent starting point for investigating further questions. Do these classifications, taken together, exhaust the types and uses of scientific instruments? Where exactly do they overlap and where do they differ? To what extent are they compatible or complementary? And, last but not least, how plausible are the philosophical conclusions inferred from these classifications?

In concluding this section, I would like to add one point of comment. Surely, an analysis of instruments is indispensable for the philosophy of scientific experimentation; yet, an exclusive focus on the instruments as such may tend to ignore two things. First, an experimental setup often includes various "devices," such as a concrete wall to shield off dangerous radiation, a support to hold a thermometer, a spoon to stir a liquid, curtains to darken a room, and so on. Such devices are usually not called instruments, but they are equally crucial to a successful performance and interpretation of the experiment and hence should be taken into account. Second, a strong emphasis on instruments may lead to a neglect of the environment of the experimental system, especially of the requirement to realize a "closed system." I stress this point in my own chapter, in the account of Boyle's air-pump experiments. The point also arises in Rainer Lange's treatment of experimental disturbances and in Mary Morgan's discussion of experimental controls (see also Boumans 1999). In sum, a comprehensive view of scientific experimentation needs to go beyond an analysis of the instrument as such by taking full account of the specific setting in which the instrument needs to function.

8. Further Issues for a Mature Philosophy of Scientific Experimentation

Even if we have made an effort to address the most important themes from the philosophy of scientific experimentation, a volume like this cannot claim to offer a complete account (if such a thing exists). Further research is possible and desirable. In my view, a mature philosophy of scientific experimentation should systematically address, at least, these additional issues: the relationship between scientific observation and experiment; experimentation in the social and human sciences; and the various normative and social questions of scientific experimentation.

The study of the relation between experiment and observation may be pursued in several ways. First, we need to develop a philosophical account of how observations are realized in scientific practice and to what extent they differ from experiments. Here some work has been done already (see, for example, Pinch 1985; Gooding 1990). What has been shown as well is that, in actual practice, making scientific observations often includes doing genuine experiments. This is quite clear in the case of solar and stellar astrophysics (see Schaffer 1995; Hentschel 1998).

Next, the results of such studies should be used to develop a new conception of scientific experience, a conception that leaves behind all empiricist accounts in which experience is somehow seen as foundational and hence as philosophically unproblematic.[8] Such a novel view of experience should also be informed by knowledge of ordinary perception that has been developed within the cognitive sciences. Examples of such an approach can be found in Daniel Rothbart's use of Gibson's theory of perception and in David Gooding's discussion of the interaction between qualitative, ordinary experience and quantitative, technologically enhanced experience.

Finally, there is the question of whether or not significant epistemic differences exist between observation and experiment. With respect to causality, Jim Woodward's contribution affirms the epistemic importance of the observation-experiment contrast. He claims that causal inferences based on purely observational evidence often prove to be spurious. Something analogous has been done by Ian Hacking regarding the issue of scientific realism. In this case, it is claimed that observation alone cannot justify our belief in the reality of theoretical entities, while experimental manipulation can (Hacking 1989b). These claims certainly do not exhaust the range of views that can be taken with respect to this question, and further contributions are most welcome.

A second issue that merits more attention is the role of experimentation in the social and human sciences, such as economics, sociology, medicine, and psychology. Practitioners of those sciences often label substantial, or even large, parts of their activities as "experimental."[9] So far, this fact is not reflected in the philosophical literature on experimentation, which has primarily focused on the natural sciences. For example, the *Routledge Encyclopedia of Philosophy* (Craig 1998) has an entry on "Experiment" and one on "Experiments in Social Science," and the remarkable fact is that the two accounts appear to be almost totally unrelated. A sign of this is that their reference lists are completely different. Thus, one challenge for future research is to connect the primarily methodological literature on experimenting in medicine, psychology, economics, and sociology with the philosophy of science literature on experimentation. In the present volume, Mary Morgan has made a start with taking up this challenge, and Jim Woodward's chapter includes relevant material, but of course much more can be done.

One subject that will naturally arise in philosophical reflection upon the similarities and dissimilarities of natural and social or human sciences is the problem of the double hermeneutic. Although it is true that the nature of this problem has been transformed by the more recent philosophical accounts of the practices of the natural sciences (cf. Rouse 1987, chap. 6), the problem has by no means been resolved. Its point is this: in addition to the interpretations of the scientists, in experiments on human beings the experimental subjects will often have their own interpretation of what is going on in these experiments, and this interpretation may influence their responses over and above the behavior intended by the experimenters. As a methodological problem (of how to avoid "biased" responses) this is of course well-known to practitioners of the human and social sciences. However, from a broader philosophical or sociocultural perspective, the problem is not necessarily one of bias. It may also reflect a clash between a scientific and a life-world interpretation of human beings.[10] In case of such a clash, social and ethical issues are at stake, since the basic question is who is entitled to define the nature of human beings: the scientists or the people themselves? In this form, the methodological, ethical, and social problems of the double hermeneutic will continue to be a significant theme for the study of experimentation in the human and social sciences.

This brings us to our last issue. So far, within the philosophy of scientific experimentation the study of normative and social questions is clearly underdeveloped. This applies to the present volume as well. To be sure, the subject of epistemic normativity—primarily related to the proper functioning of instruments—is briefly mentioned in some of the chapters, but questions regarding the connections between epistemic and social or ethical normativity are hardly addressed. Posing such questions, however, is not at all far-fetched, and they often relate to ontological, epistemological, or methodological concerns quite directly (see Radder 1996, chaps. 6–7). Following are some examples.

First, those experiments that use animals or humans as experimental subjects are confronted with a variety of normative questions. In the case of humans, some of these came up in the above discussion of the double hermeneutic, but there are many more. By way of illustration, consider medical research, where experimental tests of therapies and drugs are increasingly carried out in developing countries. In addition to the question of the realizability of adequate testing conditions in those countries, this shift leads to many serious ethical problems resulting from the tension between the well-being of the subjects and the methodological requirements of the experimental trials (see, for example, Rothman 2000). The actual and potential conflicts arising from the increasingly intimate connections between medicine and the pharmaceutical industry constitute another area for future research. In these cases, the conflicts often involve clashes between commercial interests and methodological or ethical standards (see, for example, Horton 2001).

The issue of causality is socially and normatively relevant as well. Just think of the case, discussed in Woodward's chapter, of the social scientific claim that there is a link between being female and being discriminated against in hiring and salary. Here, it makes a social and normative difference whether this link is genuinely causal or a mere correlation due to an underlying common cause. More generally, proposed policies and interventions often seek a causal underpinning in order to be seen as really effective. A clear case is drug testing, where statistical results of observations gain credibility if they can be supported by causal accounts of experimental tests. Yet, in medicine an exclusive focus on objective, causal mechanisms is also being contested (for example, Richards 1991). One type of argument refers to the placebo effect, which, somewhat paradoxically, is one of the prime reasons for the practice of double-blind trials. Another argument stresses that causal knowledge of laboratory experiments does not automatically lead to successful therapeutic uses outside the laboratory.

Finally, consider the question of the contrast between the natural and the artificial, mentioned in section 2. This question is often discussed in environmental philosophy, and different answers to it may entail different environmental policies (see Lee 1999). More specifically, the issue is crucial to debates about patenting, in particular the patenting of genes and other parts of organisms. The reason is that discoveries of natural phenomena are not patentable, while inventions of artificial phenomena are (see Sterckx 2000).

Although philosophers of experiment cannot be expected to solve all of those broader social and normative problems, they may be legitimately asked to contribute to the debate on possible approaches and solutions. In this respect, the philosophy of scientific experimentation could profit from its kinship to the philosophy of technology, which has always shown a keen sensitivity to the interconnectedness between technical and social or normative issues (see, for example, Mitcham 1994).

NOTES

1. For a detailed discussion of this issue, see Radder (1996, chap. 8).

2. See, with respect to the issue of experimental realism, Radder (1996, 75–76).

3. On experimentation and the ontology of actualism, see Bhaskar (1978), Harré (1986), and Radder (1996, chap. 4).

4. According to Baird, the idea of "reading" an instrument points to a hermeneutics of material (in contrast to textual) representation. This subject has been discussed in more detail in Heelan (1983) and Ihde (1990).

5. Thus, this notion of abstraction differs from the usual one, which is mostly defined as the inference of a universal concept from its particular instantiations. This kind of abstraction and its role in scientific practice is discussed and assessed in Gooding's chapter.

6. Such broader usages are understandable enough, given the fact that, in ordinary language, experimenting often has the general meaning of trying out something new.

7. His overall position is congenial to Patrick Heelan's view of the anthropological primacy of hyperbolic over Euclidean vision. See Heelan (1983, especially chaps. 14 and 15).

8. In this respect, Van Fraassen's constructive empiricism is typical in that a substantial account of experience and observation is conspicuously absent (see Van Fraassen 1980).

9. Cf. Dehue (1997) on the rise of comparative randomized experiments in the life sciences, psychology, and the social sciences.

10. See, e.g., Feenberg (1995, chap. 5). This chapter, "On Being a Human Subject: AIDS and the Crisis of Experimental Medicine," describes a case of AIDS patients who challenged the established methodological and ethical separation of cure and care.

Rom Harré

2 / The Materiality of Instruments in
a Metaphysics for Experiments

1. Introduction

There is no doubt that the philosophical study of experiments and the apparatus
and instruments with which they are conducted has been neglected. In recent
years, the topic has been broached, but there is still much to say. The invisibility
of the experiment in the period during which logicism dominated the philosophy of science will serve as a starting point for these investigations. The recent
inroads that have been made into the philosophy of science from the sociology
of science have brought experimental activity to the fore and have led to some
attention being paid to laboratory equipment. However, there is still a tendency
to see science wholly in terms of the discourse of scientific communities. In that
light, concrete experimental procedures and the instruments and apparatus
used to perform them are still almost invisible. If they are attended to, they are
given a constructionist interpretation as allies in the argumentative discourses
of scientific communities. This chapter is about laboratory equipment, in all its
materiality as part of the material world.

2. Some Philosophical Views of Experiments

An experiment is the manipulation of apparatus, which is an arrangement of
material stuff integrated into the material world in a number of different ways.

I shall refer to such integrated wholes as apparatus-world complexes. In the course of the manipulation some process of interest is made to occur in an apparatus-world complex. The process more often than not results in discernible transformations of the apparatus. A close analysis of the kinds of equipment in use in the sciences will lead us to see that there are two very different ways of interpreting these transformations. In those cases in which an apparatus is serving as a working model of some natural system, the changes brought about by experimental manipulations must be interpreted as analogous to states of the natural system being modeled. In those cases in which an instrument is causally affected by some natural process, the changes in the instrument are effects of the relevant state of the material world. These effects must be interpreted in terms of the causal relation presumed to hold between the process in nature and the state of the instrument. It will be convenient to use the word "instrument" for that species of equipment which registers an effect of some state of the material environment, such as a thermometer, and the word "apparatus" for that species of equipment which is a model of some naturally occurring structure or process, such as the use of a calorimeter to study the effect of salt on the freezing point of water or in vitro fertilization. Science has advanced so far now that there are pieces of apparatus in which processes are made to occur that have no analogues in nature, for instance the equipment used for cloning large animals.

The manipulation of material stuff extends beyond the experiment into the past and into the future. The apparatus has to be created, designed, and built by technicians from available material. As Toulmin (1967) pointed out, the materials used must be subjected to processes of purification. This demand covers not only the reagents used in chemistry but also the material of which an apparatus is made. There are all sorts of presumptions, then, about the past of an instrument. There are also presumptions about the future of an instrument. Though a specific piece of equipment need not survive in order for an experiment to be replicated, it must be possible to reproduce versions of the material structures of the apparatus in the future. The replicability of experiments depends on that condition. Even something like a space probe that is burnt up in the Jovian atmosphere should, in principle, be able to be duplicated and the experiment done again. If an instrument does survive to be used again it must be presumed that no radical changes have occurred during the course of the original experimental procedure.

Whether the equipment is an apparatus built to be a working model of some feature of the world or an instrument causally transformed by processes in the world, the result of an experiment is a reading of the end state of a process. Much can happen to equipment after the moment at which the experiment ends. However, the moment at which the experiment is taken to have ended, and so which state is taken as the end state of the experiment, is not an arbitrary cut in a continuous causal process. It is determined by the project for which the experiment was undertaken.

2.1. Experiments in Logicist Philosophy of Science

The long-standing idea that philosophy of science is a branch of logic has two aspects. There is, or might be, an inductive logic that legitimizes generalizing the results of observation and experiment. These generalizations are often presented as laws of nature. Scientific knowledge, so obtained, is presented in writings organized according to the principles of deductive logic. The laws of nature and other generalizations are used as axioms in the deductive process. The result of working within this framework has been the privileging of the proposition as the entity around which discussion and debate as to the acceptability of putative contributions to scientific knowledge must center.

For the most part, the discourse of "experiments" was shaped by a standard format. According to logicism, the relation between the material world and the discourses of science was taken to be entirely captured by the relation between the propositional forms of an updated version of the Aristotelian Square of Opposition. Experiments and observations were reported in the "Some A are B" or "Some A are not B" forms, and general laws or lawlike propositions, of the "All A are B" or "If x is A then x is B" forms. Occasionally the "No A are B" propositional form was required. Philosophy of science was thought to start where the propositional universe began. All this seemed entirely natural during the domination of logicism in philosophy of science.

Experimental results were just there to be described, brute facts, so to say. Observations were usually mentioned in the same breath as experiments. Even when the first intimations of a more subtle account of how states of the world were captured in propositions appeared, in the thesis of the theory-ladenness of descriptions, the role of the apparatus as the locus of the genesis of phenomena was ignored.

Logicist philosophy of science is the legacy of two philosophers above all, Mill and Mach. Mill's methods (Mill 1872) were the result of his attempt to identify the logical forms of patterns of reasoning that led from particular propositions to general propositions. His general propositions were causal laws in the Humean sense; that is, they reported exceptionless correlations between types of observable states of the world. Mach's contribution was twofold. His "sensationalist" metaphysics confirmed a phenomenalist strand in philosophy of science, having its source in British empiricism. Science was concerned with discovering the relations between sensations. His account of laws of nature was Humean. They were nothing but summaries of instances of correlations between types of sensations, serving as mnemonics to recover any instance at will. The Mach-Mill analysis of scientific knowledge confirmed the project of the search for scientific rationality as the development of an inductive logic (Mach 1883).

The once popular Popperian fallibilism involved an inversion of the general schema of inductive logic: from true statements describing the results of experiments or of observations one can infer a true general statement that can be judged to be worthy of being accepted as knowledge. The logic of scientific discovery for Popper could be expressed as a general schema from deductive logic: from a true statement describing the result of an experiment or observation that contradicts a prediction deductively drawn from a general hypothesis, one can infer the falsity of the hypothesis, which must be judged worthy of being rejected as knowledge. Both the inductivist and the fallibilist principles are offered as rules governing scientific discourse. Popper's only discussion of the sources of descriptive propositions is confined to a few comments on the conventionalist origins of descriptive predicates and the need to take certain basic statements to be true by convention (Popper 1965).

From a logicist point of view, experiments are simply the unproblematic sources of descriptive propositions in the Aristotelian I and O forms, "Some A are B" or "Some A are not B." The only feature of experiments that matters in logicism is the restriction of their spatial and temporal span to the here and now. This means that descriptions of empirical studies in science can exhibit only the I and O forms. It is also worth noting that experiments have traditionally been lumped in with observations in the frequently recurring phrase "observations and experiments." Both are cited as sources of I and O propositions.

The upshot is a complete neglect of the apparatus itself. Since it was never attended to, the variety of kinds of apparatus was never discussed.

2.2. Experiments in Constructionist Philosophy of Science

Latour (1987) and others *have* discussed apparatus and experimental manipulations. I shall use his account as a stalking horse.

We can use technically crafted things to do science. The relation between apparatus and nature needs to be analyzed. Its value to science as the best means for obtaining representations of aspects of nature needs to be justified. Latour's move—which, as we shall see, sidelines this issue—is to delete the idea of nature as an independent being from the account of science. In his treatment, apparatus is wholly a humanly created artifact that brings other artifacts into being. These determine what nature is for the scientific community.

When we use experimental equipment of either kind, apparatus or instruments, according to Latour, we are taking up material things into discourse. Here is Latour's interpretation of an experiment: "The guinea pig alone would not have been able to tell us anything about the similarity of endorphin to morphine; it was not mobilizable into a text and would not help convince us. Only a part of the gut, tied up in a glass chamber and hooked up to a physiograph, can be mobilized in the text and add to our conviction" (Latour 1987, 67). The observation that Latour uses to push one toward the textual interpretation of experimental activity is his account of an instrument. It is deliberately severed from its place as a material thing integrated into a material system that includes the world. The material entity, though ultimately of natural origin, is detached from its place in nature. According to Latour, it is "mobilized in the text." The laboratory, he says, is "a set of new [rhetorical] resources devised in such a way as to provide the literature with its most powerful tool: the visual display" (Latour 1987, 68). "I will call an instrument (or inscription device) any set-up, no matter what its size, nature or cost, that provides a visual display of any sort in a scientific text. . . . What is behind a scientific text? Inscriptions. How are these inscriptions obtained? By setting up instruments" (Latour 1987, 68, 69). Just as in the logicist neglect of the instrument that makes it invisible, in this rendering of its role it becomes relevant only in so far as it produces inscriptions. Only these are relevant to and define its importance for the enterprise of science. The triad "world/apparatus/inscription" is replaced by a dyad, "apparatus/inscription." An apparatus is something that produces inscriptions, like a ticker-tape machine.

There are two obvious difficulties with this account. First, it does not seem to me to differ in any substantial way from the logicist account. Experiments are relevant only in so far as they are the source of inscriptions to be added to the text. Second, if inscription producing were all that apparatus is good for, then surely any material setup that produced inscriptions would do! By what criteria do we reject De La Warr's boxes[1] and accept Wilson's cloud chamber as genuine scientific equipment?

"At the beginning of its definition the 'thing' is a score list for a series of trials" (Latour 1987, 89). The "things" behind the scientific texts are all defined by their performances and are similar to the heroes of folktales. Thus the thing is identified or picked out by its performance: that is a cloud chamber and it makes droplets, and that is a Geiger counter and it makes clicks. The thing is defined by what it does.

That is not yet all. As an ally in a scientific dispute, the cloud chamber of metal and glass dissolves not into lines of droplets, but into a visual display of subatomic particles—in short, a kind of statement. Similarly, the Geiger counter, having dissolved into clicks, dissolves even further into part of an auditory display of a wavelike distribution of particles. This raises a nice point, since the cloud chamber favors a particle interpretation and an array of Geiger counters favors the wave picture. What is the relationship between the displays? It cannot just be that one is an auditory display and the other visual. Why should they have anything to do with one another?

That this is a constructionist account is confirmed by Latour's third rule of method (Latour 1987, 99): "since the settlement of a controversy is the cause of Nature's representation not the consequence, we can never use the outcome—Nature—to explain how and why a controversy has been settled." "Nature's representation," something propositional and abstract, is equated with "nature," something concrete and material. This follows inexorably if we are willing to accept that it is the use of apparatus that creates nature by displaying readable representations of nature.

The final step in assimilating the material and the propositional is Latour's special notion of the "black box." The story of Herr Diesel and his engine (Latour 1987, 105) provides the setting. For the use of the engine to spread through the world, it must be possible to treat it wholly in terms of its performance as a source of motive power. Those who use the engine need not be forever worrying about the fine details of how it works. However, to get these fine details right required the recruitment of a great many "allies" to the project. At times these allies fell away and the engine's fate was uncertain. Only as a unified whole with all the components working properly did it make its way into the practical world. According to Latour, a "black box" exists when many elements are brought together to act as one.

This step having been taken, the reconstitution of the formerly material thing, the apparatus, as text seems to be a natural step. As Latour says, "no distinction has been made [by him] between what is called a 'scientific' fact [a proposition] and what is called a 'technical' object or artifact" (Latour 1987, 131).

There is only one difference between colleagues as allies and machines as allies in settling scientific controversies, according to Latour. It is easier to see that the gathered resources are made to act as one unified whole in the case of machines and pieces of hardware than in the case of colleagues and communities and runs of learned journals.

Latour has attended to the importance of practical skills in such matters as "making the equipment work." However, these appear as bargaining counters in social competition for community hegemony. Such skills are defined in a quite complex way with respect to the material nature of experimental equipment and the tasks of technicians in using it to display what it is expected to display, and sometimes does not. Failures are not always due to incompetent manipulations but to the intransigence of nature, which, as integrated with this apparatus, is other than the scientific community thought it to be. In Latour's treatment, it is hard to see that any distinction between these kinds of failures can be made.

No doubt we get the idea of what nature is like from experiments. However, it is not that idea that is integrated with a bit of itself in a laboratory apparatus. Despite attention to the "real world of the laboratory," Latour (and others) do not clearly distinguish the role of people talking and writing and so producing "science," in the sense of a community discourse, from the role of people making and manipulating material things and using them. The former are manipulating an idea or representation of nature, the latter are manipulating nature. It is evident that Latour does not treat these as different. He is led by the illuminating power of his metaphors, especially "ally," to assume an identity. Nevertheless, the fact that a metaphor "sticks" does not justify taking the similarity of source and subject that the metaphor makes visible to be an identity.

3. What Is Apparatus?

3.1. The Instrumentarium

I owe this useful phrase to Robert Ackermann (1985). Each laboratory has its characteristic instrumentarium, the actual equipment available to an experimenter. Depending on the generosity of the budget, the instrumentarium will consist not only of what is in the storeroom but also what is to be found for purchase in the catalogs of instrument makers.

There seem to me to be two main groups of philosophical questions that the use of apparatus raises once logicism and constructionism are set aside:

1. What is the ontological status of a piece of laboratory equipment? Is it part of the material world, or can it be treated as if it were detached from or outside of the material world, a probe that is affected by but does not affect that which it samples?

2. What is the epistemological status of the induced states of an apparatus? In particular, what can we "back infer" about nature from those states?

I will show that the answers to question 2 depend on the category of the equipment, apparatus, or instrument. To that end, I will develop a classification scheme in some detail.

It is unlikely that a comprehensive survey and analysis of all that is called "apparatus," "instruments," or "equipment" would reveal a common essence. What follows is a preliminary effort at a taxonomy setting up some generic categories, based on an analysis of laboratory equipment/world relationships.

The broadest distinction, which gives us distinct families of laboratory equipment, is between instruments, some relevant states of which are causally related to some feature of the world in a reliable way, and apparatus that is not so related because it is serving as a working model of some part of the world. Causal relations relevant to apparatus are within the model system. Causal relations relevant to instruments link the equipment to the world.

Among the most important genera of apparatus are working, bench-top models of natural processes and the material systems in which they occur. Allied to these are computer-generated models of natural systems.

However, there is another genus of apparatus of great interest in contemporary physics. Apparatus, as conceived by Niels Bohr, does not model the production of naturally occurring phenomena as a discharge tube might. It creates phenomena that do not occur in nature in the absence of the apparatus. I will offer a detailed analysis of Bohrian apparatus below.

3.2. Apparatus as Models of the Systems in the World

Material Models as Domesticated Versions of Natural Systems

An apparatus of this genus is a domesticated and simplified version of a material setup, which has two main features of interest to us:

> 1. It is found in the wild; that is, it occurs in nature in the absence of human beings and their constructions and interventions. For example, a model of a ferro-concrete structure would not fall into this genus, however useful it might prove in architecture.

> 2. The feral setup is such that certain phenomena can be perceived, seen, heard, tasted, and so on.

The apparatus is a material model of the naturally occurring material setup. I shall use the metaphor of "domestication" to explore the relation between apparatus as model and that of which it is a model. The history of experimental science offers us a rich catalog of apparatus as domesticated material systems. Let me illustrate this genus of models with some examples.

> Example 1: Theodoric of Friborg set up a rack of water-filled spherical flasks as an apparatus to study the formation and geometry of the rainbow. It is not too fanciful to think of his rack as a domesticated version of the curtain of raindrops implicated in the coming to be of a rainbow. The drops replace each other sufficiently quickly in the falling shower that they can be considered as if they were a fixed array. If certain conditions on manipulability of the domestic version are met—for example, finding a moveable light source to simulate the sun—the whole setup makes possible an experimental laboratory study of the rainbow. The rack of water-filled

flasks is a curtain of spherical drops, which is like the curtain of spherical drops in the naturally occurring shower of rain.

Example 2: A drosophila colony is a domesticated version of an orchard replete with a breeding population of fruit flies that display variation by selection. If certain conditions on manipulability of the laboratory colony are met, it makes possible the experimental, laboratory study of inheritance.

Example 3: An Atwood's machine is a domesticated version of a cliff that a stone falls down or a leaning tower from which objects can be dropped. The machine consists of a graduated vertical column with various movable attachments, allowing for the releasing of standard weights from different heights, and the determination of the locations of the falling masses at different times.

Example 4: A tokamak is a domesticated version of a star. A powerful magnetic field confines hydrogen atoms in a small volume, fusion to helium being ignited by an external energy source. The process set going is a domesticated version of stellar fusion.

A useful image with which to explore the metaphor of apparatus as domesticated versions of feral originals could be farmyard creatures. A cow is a domesticated version of the auroc, which was found in the wild and which did give milk. However, cows are more tractable than aurocs. One notes that life on the farm is simpler than in the wild. There is abundant fodder and there are no predators. Such a life is lived with more regular and less extreme mental and bodily states than life in the wild. Hunger and fear are rare in the farmyard. Not only is a cow spared the anxieties of feral living; it has been bred to be docile. Therefore, unlike its feral relatives, it is easily manipulable; for example, it will stand patiently to be milked.

Domesticated versions of material setups and processes that occur in feral form in nature, versions that we know as experimental apparatus and procedures, are, relative to their feral ancestors, simpler, more regular, and more manipulable. The drosophila colony in the laboratory is a simpler biosystem, with more regular life patterns and more manipulability than the swarms of flies in the apple orchard.

Things can happen in the domesticated model world that do not happen in the region of nature that is the source of the model. For example, strange variants of the insect appear and can even be maintained as living examples of mutations that would either not occur at all in the wild or be immediately eliminated.

Back Inference from "Domesticated" Models to the World

Domestication permits strong back inference to the wild, since the same kind of material systems and phenomena occur in the wild and in domestication. An apparatus of this sort is a piece of nature in the laboratory. Of course, the richness of back inference will depend on how relations of similarity and difference are weighted by the interests of the researcher in performing the experimental manipulations.

There is no ontological disparity between apparatus and the natural setup. The choice of apparatus and procedure guarantees this identity, since the apparatus is a version of the naturally occurring phenomenon and the material setup in which it occurs. Theodoric's apparatus brings a rainbow to be by refraction and internal reflection in spherical volumes of water, just as the curtain of raindrops does. So, whatever can be learned about the paths of rays of light in the laboratory can be back inferred to the wild, to nature.

A weight falls in an Atwood's machine just as a stone falls from a cliff or an iron ball falls from the Tower of Pisa. What we study in the laboratory by the use of this apparatus is not the effect of a causal relation between a state of nature and the corresponding state of an apparatus. It is a simplified version of the phenomenon itself.

Galileo's inclined plane experiment is, in some respects, an intermediate case. In order to use the results of times and distance of descent as a test of his hypothesis of free fall, he could not treat the ball rolling down the plane as just a domesticated version of free fall. He had to perform a mental operation on his results, effectively resolving the inclined motion into a vertical and a horizontal component. The back inference required this intermediate step.

Apparatus-World Complexes and the Production of Phenomena

The analysis offered above does not seem to fit the case of the Wilson Cloud Chamber, the Stern-Gerlach apparatus, and many other well-known pieces of laboratory equipment. At first sight, it might seem that these belong in the causal family. We might be tempted to say that the lines of droplets in the cloud chamber are the effects of the ionization produced by the passage of electrons. Yet, are there electrons as moving particles in the absence of this kind of apparatus? Are there electrons as interfering wave fronts in the absence of double slits and photosensitive screens? There is molecular motion in the absence of thermometers and the roughly global earth exists in the absence of cartographic surveys.

An apparatus is a piece of junk until it is integrated into a unitary entity by fusion with nature. A retort exhibited in a museum is not an apparatus. Let us call the apparatus/world complexes that scientists, engineers, gardeners, and cooks bring into being Bohrian artifacts. Properly manipulated they bring into existence phenomena that do not exist as such in the wild—that is, in nature. In general, there is no material structure in nature like the apparatus. Ice cream does not occur in nature, only in kitchens with freezers.

In the famous Bohr-Einstein debate around the EPR paradox, it is possible to see the outlines of Bohr's account of experimental physics. While Einstein is insisting that for every distinct symbol in a theoretical discourse there must be a corresponding state in the world (logical atomism under another name) Bohr (1958) is concerned with the concrete apparatus and its relation to the world as part of the world. An apparatus is not something transcendent to the world, outside it, interacting causally with nature. That is the role of the instrument. The apparatus and the neighboring part of the world in which it is embedded constitute one thing.

Bohr realized that the seeming ontological paradoxes of subatomic physics could be resolved by taking a right view of experimental apparatus. It is possible to see, for example, how nature can yield both particles and waves by treating particle phenomena and wave phenomena as products of the running of different apparatus/world complexes. Particles and waves are phenomena that occur in such complexes. They do not occur in nature.

Bohr's philosophy of experimentation was misconstrued as some kind of positivism. It was never Bohr's intention to argue, as Mach had argued, that science was the study of the properties of apparatus. In an experiment, what was "run" was not just the apparatus. Nor was Bohr advocating a straightforward realist interpretation. Physicists could not treat the apparatus in this class of experiments as a transparent window through which to see the world as it would have existed had the apparatus never been constructed and switched on. Science was the study of apparatus/world complexes. Neither component could be detached from the reality that produced phenomena.

The laboratory is full of equipment, apparatus, drawn from the local instrumentarium. People are setting it up and making it work, and so bringing phenomena into being. In some cases, the apparatus is a materially independent entity, with all relevant causal processes entirely internal to it—for instance, the experimental drosophila colony. Its relation to the world is analogy. In Bohrian experiments, the apparatus is indissolubly melded with the world. In that case, the phenomena are properties of the apparatus/world complex. It is materially part of the world. The question of whether there are natural setups like the Bohrian complexes we have constructed, such as tokamaks, is germane to this class of apparatus, as it is to the simpler, materially independent models such as Theodoric's flasks. The flasks contain spherical masses of water. That is what raindrops are.

Therefore, we have two matters to examine. There is the issue of the nature of the apparatus as a constructed material object in relation to a naturally occurring material system. We must also examine the nature of the phenomena created by using it. These may be states of the apparatus that come into being as effects of causal processes in the world. On the other hand, they may be phenomena that are brought into being by running the apparatus as a model of some material system. If the apparatus is a model of something in the world, we can ask what is the relation between the phenomena we produce in the apparatus and those that occur naturally. It is well to bear in mind that the apparatus has properties before it is switched on, heated up, or otherwise manipulated.

In thinking through the meaning of the products of experimental activity it is important to keep in mind that the phenomena generated by experimenting with Bohr type apparatus are properties of a complex unity, the apparatus/world entity. Bohr was driven to this insight by the duality of types of quantum phenomena, but the point is quite general. In this chapter, I shall use Humphrey Davy's isolation of sodium in the metallic state by electrolysis as my prime exemplar of Bohrian experimentation. As far as I know, there is no setup similar to Davy's equipment anywhere in the universe. Free metallic sodium exists only by virtue of the apparatus/world complex Davy built. Humphrey Davy used electrolysis on molten common salt in a crucible to bring metallic sodium to light. Think of how much is presupposed in describing this experiment as the "discovery of sodium" or as the "extraction of sodium." There is no metallic sodium in the universe to the best of my belief. Sodium-as-a-metal is a Bohrian phenomenon.

This experiment contrasts sharply with Faraday's use of a tube of rarefied gas to study discharge phenomena. A similar setup to the apparatus existed "in nature," in the electron wind in the rarefied upper atmosphere. Therefore, we can understand the glow in the laboratory tube as an analogue, in a domesticated version of the upper atmosphere of the aurora borealis.

Just as the cow and the auroc can serve as a metaphor for the relation of apparatus to the world, so the homely image of a loaf of bread can serve as a metaphor for the Bohrian apparatus/world complex. A loaf is brought into existence from wheat and other ingredients by the use of material structures that do not exist in the wild, such as flour mills and ovens. Loaves do not appear spontaneously in nature.

At the Cern laboratories a huge apparatus/world complex brings certain tracks into existence, which, simplifying, we could imagine are recorded in photosensitive plates. One such set of tracks was lauded as the "discovery of the W particle." There are probably no free W particles in the universe now. They are exchange particles, intermediate vector bosons, postulated in quantum field theory. They are wrenched into momentary isolation at Cern. The pattern of reasoning that lies behind the "discovery of the W particle" seems to have been something like this: photons can be studied in the propagation of light, and they also play a role as exchange or virtual particles in quantum field theory. So we have the idea of the free version of the virtual exchange particle. The W particle was introduced into physics as the virtual exchange particle for a certain class of interactions. By parity of reasoning there should be a free W particle analogous to the free photon.

Is there an analogy between the discovery of the W particle and the isolation of metallic sodium? I think that few chemists would interpret the Na atom as a virtual constituent of common salt. So Davy's experiment brought metallic sodium to light by aggregating enough preexisting Na atoms. Reflecting on the possible analogy of this case to that of the W particle we can see that what would be at issue if the analogy were to be taken seriously is the ontological status of virtual particles in quantum field theory. This is a deep issue that cannot be gone into in this discussion, except to point out that the virtual W particle is a representation of just one of the exchange modes possible in any relevant particle interaction. It is hard to make a case for a preexisting particle in any exchange process (Brown and Harré 1988).

Back Inference from Phenomena to Nature

Back inference from phenomena created in Bohrian artifacts is problematic, since there is an ontological question to be solved. What is the standing of the apparatus/world complex in relation to the world, to nature?

The general form of this question seems to be whether it is legitimate to analyze the situation in Aristotelian terms (cf. Wallace 1996)—that is, in the principle: An actual phenomenon produced in an apparatus is the manifestation of a potentiality in the world. This would allow a back inference that would simply ascribe a natural propensity or potentiality for whatever occupies some region of nature in contact or causal connection with the apparatus to appear as the experimental phenomenon. This sounds as if we could say "an apparatus makes actual in the laboratory that which is potential in nature." This still treats apparatus as a kind of window on the world.

It ignores the contribution of the apparatus to the form and qualities of the phenomenon. Reflecting on this issue takes us deeper into the Bohr interpretation. The Bohrian phenomena are neither properties of the apparatus nor properties of the world that are elicited by the apparatus. They are properties of a novel kind of entity, an indissoluble union of apparatus and world, the apparatus/world complex.

This makes the question "In what form does metallic sodium exist before the electrolysis begins?" illegitimate. Nature, in conjunction with Davy's apparatus, affords metallic sodium, just as nature, in conjunction with Wilson's apparatus, affords tracks and thereby affords electrons as particles. By parity of reasoning, the question "In what form do electrons as particles exist before the cloud chamber is activated?" is equally illegitimate.

To follow this line of analysis further would take us into the metaphysics of powers, dispositions, and affordances, the neo-Aristotelian metaphysics of physics implicit in the writings of Nancy Cartwright (1989) and explicit in the recent work of William Wallace (1996). I will return to this issue in the final section.

3.3. Instruments in Causal Relation to the World

The distinction between apparatus and instruments is vital to an understanding of how knowledge is acquired in laboratories and what sort of knowledge it is. In many discussions of the nature of experiments it is simply assumed that the state of an instrument is an effect of an independently existing state of the world. In the ideal experiment the producing of the effect in equipment, the instrument, does not change the state of the world of which the state of the instrument is an effect. Sometimes the thermometer requires so much heat to expand the mercury that the liquid being studied cools down substantially. Skilled experimenters know how to compensate for these exceptions. Usually the pressure in the car tires is not significantly reduced by the amount needed to activate the tire gauge.

Kinds of Instruments

There are two main genera of "causal" instruments. They can be differentiated by the use of an extended version of the old but useful distinction between primary and secondary qualities. Ideas of primary qualities are those, which, as apprehended by a sensitive organism, resemble the state of the material entity that caused the experience. For example, "shape" is a property of a material thing whether or not it is being observed. It is experienced by an observer as a shape, according to some rule of projection. Ideas of secondary qualities are those conscious states of an organism that do not resemble the states of the material entity that induce them. Rapidity of molecular motion is experienced not as motion but as heat.

A distinction somewhat similar to that between primary and secondary qualities and the corresponding experiences or "ideas" can be used to identify the two main relations that the states of an instrument can bear to the states of the world that caused them. Thus, the image on a photograph of a spiral nebula is a spiral shape, a projection of the shape of the nebula. Shape is a primary quality, since the representation in the photograph is analogous to shape as experienced by a conscious being and both are of the same nature as the spatial structure of the nebula itself. The energy of molecular motion in a material entity results in a change in the length of a mercury column in a thermometer. The length of a mercury column in thermometers is something like an "idea of a secondary quality," representing the energy of motion that caused it but not resembling that motion.

Some instruments are calibrated so that they yield numerical measures of qualities that can occur in different degrees. Other instruments simply detect the presence of some natural state, process, or entity. For example, litmus paper detects the presence of acidity, free hydrogen ions, without indicating the pH or strength of the acidity.

Back Inferences from States of Causally Based Instruments

It may seem at first sight that inferences from "primary" qualities of instruments to the qualities that cause them via various manipulations must be based on a different principle from the back inferences from "secondary" quality detectors and measurers. I would contend, however, that the principles are essentially the same. Both kinds of instruments depend on the reliability and verisimilitude of the relevant causal relations, the evidence for which can be found somewhere else in physics and chemistry. Change in internal energy of a sample of gas is a change in the root mean square velocity of the constituent molecules, which causes a change in the motion of the molecules in the mercury column. This motion causes the column to expand. This is a fact in physics, if it is a fact at all. A similar but longer chain of causes links an infection conceived as the invasion of an organism by viruses to the presence of antibodies, which can be chemically identified. This is a fact of biochemistry, if it is a fact at all.

3.4. A Taxonomy for Laboratory Equipment

The Basic Classification

Laid out as a tree diagram the classification of the items in an instrumentarium might look something like Fig. 2.1.

Two Kinds of Back Inference Schemata

The schemata by means of which legitimate back inferences can be made from the state of laboratory equipment to the relevant state of Nature are quite dif-

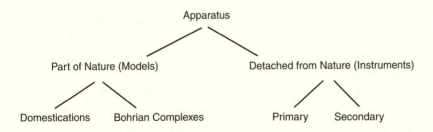

FIG. 2.1

ferent in the two families of equipment: apparatus as models (be they domestications or Bohrian) and instruments.

For apparatus as models, the back inference schema is not causal. It is ontological. The relevant state of the model is not caused by some state of nature. The relation of model to nature is analogy within the framework of a qualified ontological identity. Showers of rain and racks of flasks are both subtypes of the ontological supertype "curtains of spherical water drops." A Bohrian apparatus is a part of the material world with its characteristic products. So there is no question of seeking a ground for back inferences elsewhere in physics or chemistry.

For instruments, as I have outlined above, back inferences are based simply and directly on some causal laws already established with whatever degree of certitude the relevant branches of physics and chemistry admit. These may include quite theoretical matters with their own foundations in models. For instance, there is a causal law linking molecular motion in a sample to the extension or expansion of the substance used in the detector. There is a law linking the vibrations of electrons in hot atoms to the spectral lines detected in a spectrometer.

4. Toward an Ontology for a Bohrian Theory of Experimentation

A superficial glance at the situation would suggest that attributions based on experimenting with apparatus/world complexes would be conditional: If such and such a material setup is created and activated, then it will display certain phenomena. The Bohrian interpretation of experimental physics suggests an ontology of dispositions. There is something right about this intuition, but it is too simplistic to do duty to the subtlety of Bohrian experimentation. To prepare the way for an understanding of what Bohrian experimentation can disclose, I will begin by taking a general look at the metaphysical foundations of physics and chemistry.

4.1. The Metaphysics of Physics

The success of the corpuscularian metaphysics in the modern period, particularly in inorganic chemistry and the physics of gases, great triumphs of the nineteenth century, supported heroic efforts to generalize this ontology and the grammar in which it was immanent, across the board. There were held to be myriad material things, strictly located in space, with continuous temporal existence and clusters of essential properties. In different versions, Descartes and Newton shared this metaphysical vision. This metaphysics was naturally realized in the use of subject/predicate grammar to describe the natural world.

However, there were important dissidents, especially Leibniz, Boscovich, Kant, and Faraday. They espoused versions of an ontology of forces and active powers, distributed throughout space and structured around charges and poles taken as the sources of the field. This was the metaphysical basis of the field ontology. Trying to work with the presuppositions of the substance-attribute conceptual framework in setting up the science of field physics around strains in a substantival ether came to seem unnatural. Instead, we have intimations of a new ontology, of potentials at space-time points related to charges and poles. These are not some strange new kind of entity. There are no material entities in this metaphysics. Fields are distributions of dispositions through which the powers of charges and poles are manifested.

Which ontology shall we choose?

There are the familiar problems with justifying claims about imperceptible substances with imperceptible properties that are characteristic of the ontology of the traditional mode of theorizing by invoking hidden mechanisms constituted of localizable and enduring or quasi-enduring particles. How are alternative accounts of the hidden realm to be decided among? There are difficulties in understanding the role of apparatus in Bohrian apparatus/world complexes in the genesis of knowledge if we persist in thinking in terms of substances and their occurrent properties. It seems quite natural to think in this way about the metaphysics implicit in the use of instruments and for many of the model worlds realized in laboratory apparatus.

4.2. The Structure of Field Thinking

Starting with the metaphysics common to field theories from Gilbert's *De Magnete* to Faraday's *Experimental Researches,* the story has been something like this distribution of explanatory formats:

TABLE 2.1. Distribution of Explanatory Formats

Task	Ontology
Descriptions of phenomena	Eclectic
Explanations of observable phenomena	Dispositions
Explanations of dispositions, etc.	Causal powers

Philosophers and logicians have put a huge effort into developing only one of the possible options available in the eclectic grammars of ordinary life, namely substances and their occurrent properties. This privileges subject/predicate sentence structures. After Descartes, the attributes in focus became more or less restricted to extensive magnitudes. Physics, however, continued to develop its own version of the dispositions and powers scheme implicit in the mathematical analyses of the field concept. I contend that the adoption of that scheme will help reveal the role of apparatus in a much clearer way than trying to re-jig the substance/occurrent property scheme yet again. The exigencies of field physics demand conditional sentence forms.

It is worth remarking that the substantive content of a dispositional attribution, expressed in the usual conditional form, consists of observables only. There are various truth functions of observables, for example "p & q," "p v q," and so on. The conditional form by means of which we express simple dispositional attributions, "p ? q," is just another one. Dispositional ascriptions have no explanatory value. However, it is here that the distinction between occurrent and nonoccurrent properties gets its first outing. A disposition is not an occurrent property.[2] A substance can be properly said to have a certain disposition when it is not manifesting it. Thus we say aspirin is an analgesic when the tablet is still in the bottle.

Powers and the Concept of Energy

Dynamicist ontologies, that is metaphysical schemes based on activity concepts such as "force," "energy," and so on, as used by Leibniz, Boscovich, Kant, and Faraday and modern field theorists, offer explanations of the same general form as corpuscularian explanations. They invoke unobservables to account for observable phenomena. However, unlike the corpuscularian unobservables, which are potential perceptibles, the unobservables of the dynamicist ontology are imperceptibles. Forces are manifested in their effects. They are not observable as such. The visualizability constraints in theoretical thinking that seem so natural within the corpuscularian tradition are simply not relevant for dynamicists.[3]

The simplest dynamicist layout uses imperceptible energy transformations to explain perceptible accelerations via the relevant field potentials—that is, as described dispositionally. The structure is tightly bound together conceptually since the dispositional attributions ascribe tendencies to accelerate to test bodies located at specific points and moments in the field in question. The content of a dispositional attribution is wholly observable, indeed perceptible, but as properties, dispositions are not occurrent. For example, a gravitational field is described at the level of observables by the spatiotemporal distribution of dispositions of test bodies to accelerate, while it is described at the level of unobservables in terms of the distribution of potential energy. In mechanics, energy is neither a substance nor a disposition, but a power that accounts for the existence of dispositions. Energy is the power to do work.

I have shown elsewhere how the dynamicist scheme can be generalized to quantum field theory (Harré 1988). However, to do that a new kind of dispositional concept must be introduced. This is the affordance, which will prove indispensable in making sense of the role of Bohrian apparatus.

Affordances

Gibson's concept of affordance, which he deployed to such effect in his account of the dynamics of perception (Gibson 1986), can serve a similar purpose in the analysis of experiments. An experiment shows people what a particular apparatus/world complex can afford people who are skilled enough to use it. Of course, any piece of matter can afford all sorts of things. Ice affords walking, skating, cooling drinks, and so on. Which affordances are invoked depends on what human beings are proposing to do.

Gibson introduced the concept of an affordance to distinguish a certain class of dispositions. The general form of the ascription of a dispositional property to something is conditional: "If certain conditions obtain then a certain phenomenon will (probably) occur." In many cases the outcome of activating a disposition does not depend on any particular human situation, interest, or construction. However, in some cases the phenomenon has a specifically human aspect. Compare the generic outcome that ice of a certain thickness can bear a certain weight per unit area, expressed in a generic disposition, with the claim that ice of that thickness affords walking for a person.

Generalizing the notion of an affordance we can say that an apparatus/world complex can afford *things*. For instance, wheat, yeast, and a stove can afford loaves of bread. A lathe can afford chair legs, and a discharge tube can afford gamma rays.

An apparatus/world complex can also afford *activities*. For instance, some rapids can afford whitewater rafting. A reamer can afford boring, and a chemistry laboratory can afford gravimetric analyses.

In both groups of cases, what is afforded would not have existed without human action to bring it into being. Thus ice affords walking only if there are actual and potential walkers. The heterogeneity of the first group shows up clearly, in that while there are no loaves of bread or perfectly regular cylindrical lengths of wood in nature, there are gamma rays. However, even that is not quite right, since there is fermentation and there are fairly regular cylindrical branches on some trees. Perhaps the gamma rays in the Cavendish laboratory were not quite the same in type as those from a sunburst.

Back inference from Bohrian artifacts to the attribution of affordances never escapes wholly from the human domain of material constructions. However, it is not confined to that domain. The Bohrian artifact is a hybrid being, part construction and part nature. The phenomena that are produced in activated Bohrian material systems are the manifestations of affordances. These are dispositions that bring together two sets of causal powers that cannot be disentangled. There are the powers of the material stuff organized as an apparatus and the powers of the world realized in the phenomena. What can we say? Bohrian apparatus displays the affordances of the world, relative to that apparatus. At the deeper layers of scientific work there are no transparent windows on the world as it would be did the apparatus not exist.

5. Conclusion

It has been important to distinguish two broad families of laboratory equipment. The first family consists of domesticated versions of material setups in nature and Bohrian artifacts. The second family consists of instruments that yield knowledge of nature by virtue of causal relations with states of the world.

The first family of equipment, apparatus as part of nature but isolated in the laboratory, can serve as a springboard for back inferences from the laboratory to the world outside the laboratory. Simple modeling depends on and requires only simple ontological identity of subtypes under a common supertype. We are still in an extension of the everyday world of substances and their occurrent properties. In Bohrian modeling, the apparatus is also part of nature. However, unlike simple models, a Bohrian apparatus cannot be isolated from nature, since it is indissolubly melded with it. The interpretation of the states of Bohrian apparatus, phenomena, requires a shift in ontology from Newtonian occurrent properties to dispositions and powers. These dispositions are affordances and permit limited inferences from what is displayed in Bohrian artifacts to the causal powers of nature.

The various pieces of equipment that one finds in the second family, instruments in causal interaction with nature, may throw up problems of the justification of back inferences from the state of the instrument to the presumed state of the world. However, these are in principle resolvable by doing more physics and chemistry, taking for granted the verisimilitude of the reactions of the instruments used for that subsidiary work.

NOTES

1. De La Warr produced diagnostic "machines" that were made of odds and ends. They were shown to be worthless.

2. For an excellent account of dispositional concepts, see Mumford (1998).

3. In general philosophy of science, the realism debate takes a quite different form when the candidate realist ontology is dynamicist.

Davis Baird

3 / Thing Knowledge

OUTLINE OF A MATERIALIST THEORY OF KNOWLEDGE

1. Need for a Materialist Epistemology

1.1. Thing Knowledge

I present here in brief a materialist theory of knowledge, which I call "thing knowledge." This is an epistemology where the things we make bear our knowledge of the world, on a par with the words we speak. I oppose an epistemological picture where the things we make are simply instrumental to the articulation and justification of knowledge expressed in words or equations. Our things do this, but they do more. They bear knowledge themselves, and, frequently enough, the words we speak serve instrumentally in the articulation and justification of knowledge borne by things.

This is important for a variety of reasons. I do not think we can fully understand experiment without a materialist epistemology along the lines I sketch. Nor do I think we can fully understand the relationship between industry and science without a materialist epistemology. And, indeed, a detailed understanding of tacit knowledge will remain elusive without a materialist epistemology.

Scientific instruments provide a particularly useful class of things to examine. Given our current understanding of science one cannot contest the claim that instruments are essential to the development of scientific knowledge. Yet in many instances instrument development has preceded theory development. We may know that an instrument works; we may know how to develop, refine, and use it. But we may have either an erroneous understanding or no theoretical understanding of it. Here is a prima facie argument for thing knowledge based on the common enough autonomy of its development. While autonomy is useful for philosophical argument, I acknowledge that it is much more common to see instruments and theories develop in tandem, each supporting our understanding of the other.

There is a second reason instruments are a useful class of things to examine. Thing knowledge, as I articulate it below, is not a unitary epistemology. Different things, and even different aspects of the same thing, operate epistemologically in different ways. There are many different kinds of instruments, and in this variety we can find different instruments that present in different pure forms different kinds of thing knowledge, different epistemologies. Focusing on instruments provides a particularly useful venue for the analysis and articulation of these different epistemologies, of the varieties of thing knowledge.

While I focus on scientific instruments, I urge a more general understanding of thing knowledge. Thing knowledge embraces genetically modified life forms, wonder drugs, expert systems, and generally the material products of science and technology. Thing knowledge is an epistemology of the knowledge borne by the things we make, scientific, technological, arts and crafts, or otherwise.

1.2. Things and Theories

Making is different from saying, and made things bear a different kind of knowledge than expressed sentences. Thomas Davenport, a Vermont blacksmith with little schooling and no training in electromagnetism, made a rotary electromagnetic motor after seeing a demonstration of Joseph Henry's electromagnet. Patenting his motor in 1837, Davenport is a claimant to being the first to make such a thing. Davenport's motor worked well enough to power a printing press. One major obstacle to making a commercial success of his invention was the lack of good batteries or other sources of electric power. Davenport's intellectual achievement was a commercial failure. (See fig. 3.1.)

The invention bears some of Davenport's knowledge of electromagnetism. Certain aspects of the relations between electricity and magnetism can be expressed in literary terms with words and equations. Other aspects can be expressed in material terms with wire, iron, and wood. This was Davenport's way. Davenport was able to see relationships in the material terms in which they were presented in Henry's electromagnet. He could manipulate these relationships in his mind's eye and ultimately manually to make something new. He was not working with equations or propositions. He was working with materials. What he learned from Henry's electromagnet he learned from the

FIG. 3.1. Davenport's 1837 model motor that accompanied his patent application *(Davenport 1929, 144)*

material object. He then worked with what he learned, with his hands and his blacksmith shop's materials, to produce his motor. This was his contribution to our knowledge of electromagnetism. It was a material contribution, not a theoretical contribution.

Davenport's case demands the following epistemic analogy. Theoretical knowledge—the kind we are used to (and used to arguing about as philosophers) —is borne by a curious collection of "marks," typically (or now proverbially) on paper. These marks have content because of linguistic conventions that allow us "to read" them. Thing knowledge works analogously, but in place of marks on paper typically (or perhaps proverbially) we have crafted objects placed in various physical—spatiotemporal—relations to each other. These objects have content and "can be read"—as Davenport "read" Henry's electromagnet. The nature of the content borne materially is not the same (with the significant exception of models, but see below) as the nature of the content borne linguistically. "Material conventions" for "reading" things differ from linguistic conventions for reading texts. And, finally, the knowledge borne in these two different media differs in kind. But there can be little doubt that Henry's material thing, his electromagnet, "spoke volumes" to Davenport—as material things in all ages, cultures, and times have "spoken" to craftspersons, engineers, experimental scientists, and "ordinary people."

This is a kind of content and knowledge, borne in material things, that has been virtually invisible to philosophers. Consider the graphic (fig. 3.2) from Steve Woolgar and Bruno Latour (1979). Here is the function of the laboratory. Animals, chemicals, mail, telephone, and energy go in; articles go out. The picture Latour and Woolgar present of science is thoroughly literary. "Nature," with the help of "inscription devices" (that is, instruments), produces literary outputs for scientists; scientists use these outputs, plus other literary resources (mail, telephone, preprints, etc.), to produce their own literary outputs. The material product the scientists happened to be investigating in Latour and Woolgar's study—a substance called "TRF"—becomes, on their reading, merely an instrumental good, "just one more of the many tools utilized as part of long research programmes" (Latour and Woolgar 1979, 148).

This picture of the function of a laboratory is a travesty. There is a long history of scientists sharing material other than words. William Thompson sent electric coils to colleagues as part of his measurement of the ohm. Henry Rowland's fame rests on the gratings he ruled and sent to colleagues. Chemists share chemicals. Biologists share biologically active chemicals—enzymes, etc.—as well as prepared animals for experiments. When it is hard to share devices, scientists with the relevant expertise are shared; such is the manner in which E. O. Lawrence's cyclotron moved beyond Berkeley. Laboratories do not simply produce words. And Davenport's fame does not rest on the words he produced.

Much needs to be done to articulate how this material epistemology, this thing knowledge, works. What are the "conventions," what is it to "read" things, and what kind of knowledge is borne in things? It is the task of this chapter to take some steps toward answering these questions.

Once we recognize and articulate how things and their making bear knowledge we will be in a better position to understand experiment, the epistemological relations between industry and science, and tacit knowledge. Beyond common usage, why should we treat "know-how" as a species of knowledge? A materialist epistemology can answer this question. How does knowledge flow back and forth between industry and science? Words help, of course, but it was not words that taught Thomas Davenport and it was not the lens maker's words that made the telescope, which in Galileo's hands transformed astronomy. Henry Rowland was one of the first great American physicists. But the "experimental results" for which he is best remembered—contra Latour and Woolgar—are things, not words. Rowland's diffraction gratings, distributed the world over,

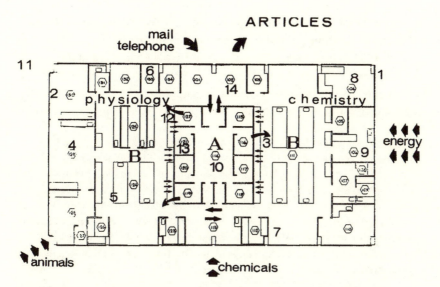

FIG. 3.2. Latour and Woolgar's "architectural take" on the function of a laboratory (*Latour and Woolgar 1979, 46*)

transformed spectrographic analysis of the heavens and of the earth; they transformed the experimenter's laboratory.

Philosophers and historians express themselves in words, not things, and so it is not surprising—although it is unfortunate—that those who hold a virtual monopoly over saying (words!) what scientific knowledge is, characterize it in terms of the kind of knowledge with which they are familiar—words. This strikes me as the wordsmith's prejudice. Derek de Solla Price captured my reaction exactly when he noted, "It is unfortunate that so many historians of science and virtually all of the philosophers of science are born-again theoreticians instead of bench scientists" (Price 1980, 75). Epistemologists interested in science and engineering should attend to what, during the 1980s, was a credo of MIT's Media Lab, not publish or perish, but demo or die.

1.3. Subjective and Objective

Before I come to the heart of the matter—an articulation of the kinds of knowledge borne by instruments—I need to make a first pass on the locus of thing knowledge. Thing knowledge is in things, not minds. Instruments, not beliefs, are the carriers of thing knowledge.

Consider a question Larry Bucciarelli asks at the beginning of *Designing Engineers* (1994): Do you know how your telephone works? A speaker at a conference on technological literacy noted with alarm that less than 20 percent of

Americans knew how their telephone worked. But, Bucciarelli notes, the question is ambiguous. Some people (although perhaps fewer than 20 percent) may have an inkling of how sound waves can move a diaphragm and drive a coil back and forth in a magnetic field to create an electric current. But there is more to telephony than such simple physics. Bucciarelli wonders whether the conference speaker knows how his phone works:

> Does he know about the heuristics used to achieve optimum routing for long-distance calls? Does he know about the intricacies of the algorithms used for echo and noise suppression? Does he know how a signal is transmitted to and retrieved from a satellite in orbit? Does he know how AT&T, MCI, and the local phone companies are able to use the same network simultaneously? Does he know how many operators are needed to keep the system working, or what these repair people actually do when they climb a telephone pole? Does he know about corporate financing, capital investment strategies, or the role of regulation in the functioning of this expansive and sophisticated communication system? (Bucciarelli 1994, 3)

Indeed, Bucciarelli concludes, "Does *anyone* know how their telephone works?" (Bucciarelli 1994, 3).

Here, following the conference speaker, Bucciarelli uses the word *know* in a subjective sense. He makes a persuasive case that, in this sense, no one knows how his or her phone works. In the first place, the phone system is too big to be comprehended by a single "subjective knower." In the second place, the people who developed pieces of the hardware and software that constitute the phone system may have moved onto other concerns and forgotten the hows and whys of the pieces they developed. Their "subjective knowledge" is lost, while the objective artifacts they helped bring into existence still work. In the third place, complicated systems with many interacting parts do not always behave in ways we can predict in detail. Despite having created them, programmers cannot always predict, and in this sense do not "subjectively know," how their complicated computer programs will behave.

It is, of course, well and proper to engage in what might be called subjective epistemology. This is the attempt to understand that aspect of knowledge that is a species of subjective belief. But if we want to understand scientific and technological knowledge, this is the wrong place to look. Bucciarelli's telephones teach us why. If no one—subjectively—knows how the phone system works, the situation with all scientific and technological knowledge is radically worse. The epistemological world of science and technology is too big for a single person to comprehend. People change the focus of their research and forget. Expert knowledge systems transcend their makers. Yet it still makes sense to speak of the objective knowledge that comprises our technology and science. This objective knowledge includes material published in journals, but also crafted things, and especially experimental apparatus. This objective knowledge may be subjectively understood by no one.

2. Three Kinds of Instrument

2.1. Model Knowledge

Some of the things we make, we make to represent the world. Such things work like theories, but they use material bits of the world, not words, to represent. For this reason, they provide a different entry for our cognitive apparatus. Conceptual manipulation provides one entry; material manipulation—"hand-eye manipulation"—provides a second entry.

Watson and Crick's model of DNA provides a fine example of material representational knowledge.[1] Their remarkable collection of metal plates and rods represents the structure of DNA (fig. 3.3). Watson and Crick's DNA model provides a two-part argument for understanding material models as knowledge. There is a negative part: It makes little sense to think of the Watson-Crick model in other terms. They did not use the model as a pedagogic device. They did not simply extract information from it. The model was not part of some intervention in nature. It was not a part of an experiment.

There also is a positive part: Watson and Crick's model performs theoretical functions with contrived bits of the material world instead of words. Their model has the standard theoretical virtues. It has been used to make explanations and predictions. It was confirmed by evidence—X ray and other—and it could have been refuted by evidence, for example, if DNA had been found with markedly different quantities of adenine and thymine. While made of metal, not words, there can be little doubt that Watson and Crick's model of DNA is knowledge.

There remains a somewhat stickling business of sorting out the relationship between material models, such as Watson and Crick's, and theories understood in usual propositional terms. This is best done with the semantic view of theories, according to which a theory does not directly describe the world.[2] Instead, it describes a "model" or class of "models." With a successful theory, the model or class of models represents a portion of the world. This use of "model" is more general than that covering my use here. But the material models I am concerned with are instances of this general concept of model.[3]

R. I. G. Hughes provides a three-part account for how models represent the world. He calls it the "DDI account," referring to its three parts: denotation,

FIG. 3.3. James Watson, *left,* and Francis Crick, *right,* with their 1953 model of DNA
(Watson 1981, 125)

demonstration, and interpretation (Hughes 1997). In the first place, models *denote* some part of the world, and to properly understand a model one must know which parts denote what. Thus, the sticks in Watson and Crick's model denote bond lengths, not rigid metallic connections. The plates denote "base groups" —adenine, thymine, guanine, and cytosine—that is, groupings of atoms that can be treated as units in their own right.

Models also allow for *demonstrations*. With material models these demonstrations can be manual. Such is the magic moment when Jim Watson discovered base-pair bonding in DNA. He played around with cardboard cutout versions of the bases and their bonding distances. Through these manual manipulations he found that one of the two purine-bases (adenine) hydrogen bonds with one of the two pyrimidine-bases (thymine) while the other purine (guanine) hydrogen bonds with the other pyrimidine (cytosine), A with T and G with C. The bonding distances were such that bases could bond on the interior of the DNA chain while a two-strand backbone twisted in a double helix without distortion.

Models then are *interpreted*. One immediate result of the Watson-Crick model for DNA was the requirement that there be the same amounts of adenine as thymine, and of guanine as cytosine. Chemical analysis bears this out. The most significant interpretation, of course, is the possibility it presents for genetic replication.

2.2. Working Knowledge

Material models are not too difficult to recognize as bearing knowledge, for they do so in a way that is very similar to words, through representation. This is not how Thomas Davenport "read" Joseph Henry's electromagnet, however. Here we have a different kind of interaction with the material and a different kind of knowledge borne in materials. I call it "working knowledge."[4] In coining this neologism I call on our use of "working" to describe an instrument or machine that performs regularly and reliably. I also draw on the phrase "to have a working knowledge." Someone with a working knowledge of something has knowledge that is sufficient to do something. My neologism, "working knowledge," should draw attention to the connection between knowledge and effective action.

Faraday's 1821 electromagnetic motor provides a good example (see fig. 3.4). When he first made the motor, there was considerable disagreement over the explanation of the phenomenon it exhibited. Yet, no one contested what the apparatus did: It exhibited a kind of rotary motion as a consequence of a suitable combination of electric and magnetic elements. It exhibited a phenomenon for which, at the time, there was no adequate theoretical language.[5]

Despite the disagreement over how it worked when he first made the motor, and in the face of the fact that today many people do not know the explanation for how it works, no one denies *that* it works. When Faraday built it, this phenomenon was striking and proved to be very important for the future development of science. Any explanation offered for the nature of "electromagnetical motions" would now have to recognize the rotations Faraday produced. We do not need a load of theory (or indeed any "real" theory) to learn something from the construction and demonstration of Faraday's device.

It is because Faraday and his scientific contemporaries could reliably create, re-create, and manipulate this phenomenon, despite their lack of an agreed-upon theoretical language, that I refer to the motor itself as "working knowledge." Indeed, Faraday's motor worked reliably enough that, lacking any theoretical explanation, other researchers could learn from the motor itself—as Davenport learned from Henry's electromagnet. Six months after Faraday made his device, Peter Barlow produced a variant (Faraday 1971, 133). (See fig. 3.5.) Current runs

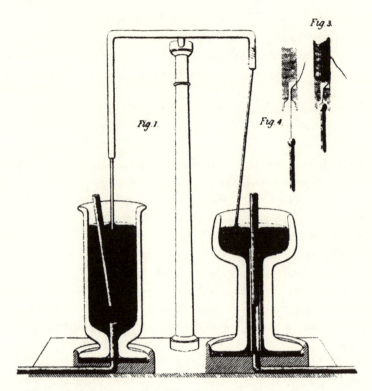

FIG. 3.4. Two versions of Michael Faraday's 1821 electromagnetic motor
(Faraday 1844, plate 4)

FIG. 3.5. Barlow's Star, Peter Barlow's
1821 variant of Faraday's motor
(Faraday 1971, 133)

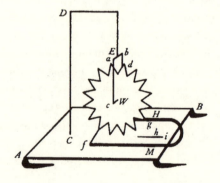

from one "voltaic pole" to the star's suspension *(abcd)* through the star to the mercury bath *(fg)* and thence to the other voltaic pole. A strong horseshoe magnet *(HM)* surrounds the mercury bath and, as Barlow put it in a letter to Faraday, "the wheel begins to rotate, with an astonishing velocity, and thus exhibits a very pretty appearance" (Faraday 1971, 133, letter dated March 14, 1822).

It is another step to figure out how to create such rotary motion without the use of mercury. Then we might have something useful. There is a significant story here, a story not primarily about the evolution of our words and equations, but a story about material manipulations. The story involves many players, including Thomas Davenport, and a full telling would not serve here (see King 1963; Gee 1991). Nonetheless, the material creation, re-creation, and manipulation of Faraday's motor shows that the scientists, engineers, and tinkerers involved had a working knowledge of Faraday's phenomenon and the successor phenomena they developed from it. The things they made bore this working knowledge.

With his motor, Faraday contributed to science a compact instrumentally framed fact, a technological certainty in a sea of theoretical confusion, and the foundation for further instrumental development. Faraday's motor is an instance of an instrument—a contrived part of the material world—that creates and exhibits an element of material agency. While there was no consensus on the right sequence of words to describe this agency, it, the agency itself, could not be denied. Charles Sanders Peirce used these words to describe such a situation: "When an experimentalist speaks of a *phenomenon*, such as 'Hall's phenomenon,' 'Zeeman's phenomenon' . . . he does not mean any particular event that did happen to somebody in the dead past, but what *surely will* happen to everybody in the living future who shall fulfill certain conditions. The phenomenon consists in the fact that when an experimenter shall come to *act* according to a certain scheme that he has in mind, then will something else happen, and shatter the doubts of skeptics, like the celestial fire upon the altar of Elijah" (Peirce 1934, vol. 5, para. 425). Ideally a phenomenon has the striking and persuasive quality of the divine blaze by which Elijah embarrassed the 450 prophets of Baal, but it must also be constant and reliable, a permanent fixture of the living future.

The ability to make, manipulate, adapt, and develop material agency as Faraday and his followers did with his motor is ample proof of knowledge of the agency. In the subjective sense, the people involved possess the necessary "know-how"—working knowledge—to produce a reliable permanent fixture of the living future. In the objective sense, there are devices, made by humans, exhibiting particular phenomena over which we have substantial material—if not linguistic—control. They are the material bearers of working knowledge.

2.3. Measurement

Measuring instruments tell us something about a "specimen"; they measure it in some manner. To succeed at this, their makers must synthesize in the material form of the instrument both model knowledge and working knowledge. At a fundamental level, measurement requires control over material agency. In a mercury thermometer the (approximately) linear expansion of mercury with temperature is a material phenomenon over which—after many decades of work —we developed control. Here is our working knowledge built into a mercury thermometer. But it is the instrument maker's selection of a field of possibilities that drives the choice of signal generated and the transformations that are made to the signal as it is rendered "a measurement." The instrument maker builds a representation of this field of possibilities into the material form of the instrument. So we have on the thermometer's glass tube a scale that displays the field of possibilities that we embrace with the thermometer. Here is a representation of the structure of temperature. It is model knowledge. When both kinds of material knowledge are integrated, the instrument appears to extract information from nature, and we have encapsulated into the material form of the instrument model and working knowledge.

A different example can clarify my meaning. The Indicator Diagram—or indicator, for short—is an instrument attached to the working cylinder of a heat engine. It produces a simultaneous trace of the pressure and volume inside the cylinder as the engine runs through its cycle. It can be used to show how much work is being extracted from the heat energy run through the engine (see fig. 3.6). James Watt and his assistant John Southern invented the indicator probably during the early months of 1796 (Baird 1989; Hills 1989, 92). It played significant roles both in improving steam engine performance and in developing the modern understanding of thermodynamics.[6]

As we see things now, the indicator gives a graphic display of the amount of work generated on each stroke of a heat engine. The area enclosed by the Indicator Diagram is the amount of work done on a cycle of the steam engine (see fig. 3.7).[7] When the piston compresses the gas in a cylinder, work is done on the gas; when the gas expands in the cylinder, the gas does work on the piston. If more work is extracted when the gas expands than is spent compressing the gas, then the cycle yields a net production of usable work, or motive power. The area under the pressure-volume curve in the Indicator Diagram during the compression of the gas (under F-K-C in fig. 3.7) measures the work required to compress the gas. The area under the curve during the expansion of the gas (under C-E-F, in fig. 3.7) measures the work done by the gas on the piston. The difference, which is the area enclosed by the Indicator Diagram curve, is the usable work extracted in this cycle of the engine.

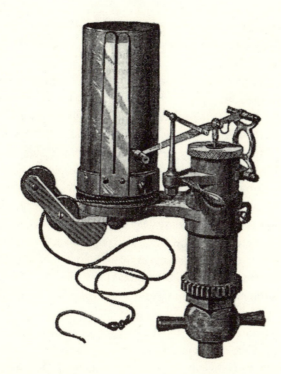

FIG. 3.6. The Indicator Diagram, used to produce a simultaneous trace of pressure and volume as a steam engine cycles *(Roper 1885, 251)*

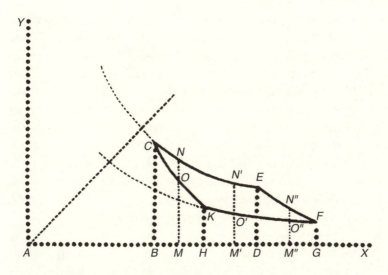

FIG. 3.7. Emile Clapeyron's 1834 Indicator Diagram schematic *(Mendoza 1960, 75)*

Both model and material knowledge are synthetically joined in the indicator. In the first place it presents a phenomenon. The instrument harnesses an instance of material agency, the behavior of pressure and volume in a steam engine cylinder as it goes through its cycle. The ability of Watt and Southern and subsequent steam engineers to contrive this device, to make it work regularly and reliably is how the instrument bears their working knowledge of this material agency. At the same time, the indicator presents information. A field of possibilities is constructed in terms of the pressure-volume graph on which the indicator "writes." Here is model knowledge—a representation of the possibilities for steam engine performance. The two-dimensional graph on which the indicator "writes" presents a space in which engine performance can be "read." It is the same with Watson and Crick's work, where there was a space of possibilities defined by the geometry of the chemical subgroups (the bases and backbones, etc.) and the particular geometry that Watson and Crick found in this space— *the* structure of DNA. With the indicator we have the space of possibilities given by the two-dimensional graph and the particular performance shown in the particular graph provided.

How the representational space of the indicator is understood has changed. Now we understand the output of the indicator in terms of work. The area enclosed by the diagram is the amount of work done by the engine on a single cycle. When they developed the indicator, Watt and Southern did not have our contemporary concepts of "work," "force," and the like with which to understand what it measured. They used the indicator to measure the "steam economy" or "duty" of their engines. By measuring the pressure at several points in the expansion part of the cycle, they could compute the average pressure exerted in the cycle. Watt and Southern wanted to know how much "average pressure" they could get from how little steam. By varying the time when steam was no longer admitted to the cylinder and measuring the average pressure on the piston with the indicator, they could maximize average pressure for steam used. Material model knowledge, like propositional knowledge, depends on its context for a rendering of its meaning. As the context changes, so does the meaning. This fact points to a kind of hermeneutics of material representation. But, in respect to measurement, this "material hermeneutics" differs from textual hermeneutics because these material representations simultaneously are the field of action for working knowledge.

What is crucial, however, is that the indicator captured an instance of material agency—working knowledge—irrespective of how its users understood the model space within which the indicator's performance was displayed. This allowed the indicator to survive the theoretical ignorance of its birth and promote both the material and the theoretical advances now associated with it.

This also is one way material knowledge can promote scientific progress. The instrument presents a phenomenon, working knowledge, even when there is no way, or no way that is not contentious, to describe in words what the instrument is doing. That it is doing something, encapsulating working knowledge, is enough for it to lead to better material creations (better steam engines in the case at hand) and better theories (thermodynamics in the case at hand). This is why developing new instruments is a central epistemic goal for scientific progress. Theories come and go, but a new instrumentally created phenomenon endures. It can lead to better instruments—better material knowledge—and to better theories—better propositional knowledge.

2.4. Arguments

The stories I present here aim at philosophical argument. Lest the argument be lost in the historical details, I recapitulate. To begin, there are arguments from analogy. Material models, such as Watson and Crick's model of DNA, function analogously to theory. They can provide explanations and predictions. They can be confirmed or refuted by empirical evidence. Material model knowledge is epistemologically analogous to theoretical knowledge.

Working knowledge is different, and the arguments from analogy that support it are correspondingly different. We say someone knows how to ride a bicycle when he or she can consistently and successfully accomplish the task. A phenomenon, such as that exhibited by Faraday's motor, shares these features of consistency and success with what usually is called "know-how." To overanthropomorphize the situation, one might say Faraday's motor "knows how to make rotations." I prefer to say that the motor bears knowledge of a kind of material agency, and I call such knowledge "working knowledge."

This analogy runs deeper. Frequently enough we are unable to say, for example, how to ride a bicycle. The knowledge is tacit. Such is the case with Faraday's motor, and from two different points of view. The motor itself articulates nothing in words. But also it was difficult for Faraday to articulate how the phenomenon comes about. Yet, as in the case of bicycle riding, it is clear that the instrument works. The instrument bears this tacit working knowledge in its regular controlled action.

My stories also present a different collection of arguments that turn on what I call cognitive autonomy. Davenport learned something from his examination of Henry's electromagnet. He could take what he learned and use it to develop another, potentially commercially useful device. He did this while ignorant of theory and unable to express in words both what Henry's electromagnet taught him and what he was doing with the knowledge. We can cognitively interact with the material in a different and autonomous way from how we interact with the theoretical. James Watson's play with cardboard models of DNA bases demonstrates the same phenomenon. Cognitive content is not exhausted by theory. Epistemic content should not be exhausted by theory for the same reason.

My stories are drawn from history, but I intend modern morals from them. Contemporary instruments and devices are concatenations and combinations of model knowledge and working knowledge. Levels of complexity add a further dimension, but the underlying structure of the material knowledge remains the same. These historical examples provide simple but real cases that exemplify epistemologically significant kinds of thing knowledge.

At the end of the day the fundamental argument for the epistemological place of instruments is my articulation of how instruments do epistemological work. While each of the arguments I present aim to persuade readers that instruments should be understood as epistemologically on a par with theory, it is the overall picture that must seal the deal.

3. Making It Knowledge

3.1. The Problem

Thus far I have briefly sketched three epistemologically different kinds of instruments: instruments that provide material representations—model knowledge —instruments that present a phenomenon—working knowledge—and measuring instruments that integrate both model and working knowledge. I claim that these material products bear our knowledge. But this claim raises several pressing questions. The concept of knowledge is tied to the concepts of truth, justification, and detachment. These concepts are well suited to our discussion of propositions but not to our discussion of instruments. Here I sketch alternative concepts that can serve the same epistemological functions as truth, justification, and detachment.

3.2. Material Truth, Instrumental Function

Philosophers are accustomed to think of truth in terms of propositions or sentences. They ignore turns of phrase that point us in the right direction. Consider the expression "a true wheel." Amongst the more philosophically common senses of the word *true,* the dictionary also includes "9. Accurately shaped or fitted: *a true wheel.* 10. Accurately placed, delivered, or thrown" (*American Heritage College Dictionary* 1993). But a true wheel is not true simply because it properly conforms to a particular form; a true wheel spins properly, dependably, regularly. A wheel that is "out-of-true" wobbles and is not dependable. Ultimately it will fail. This sense of "truth" picks out those contrived constellations of materials that I call "working knowledge." A public, regular, reliable phenomenon over which we have material mastery bears working knowledge of the world and "runs true" in this material sense of truth.

The need for the wheel to spin properly immediately intertwines material truth with normative notions of function. Knowledge, expressed in propositions, provides fodder for further theoretical reflection. These resources—sentences with content—are manipulated linguistically, logically, and mathematically. Theoreticians are "concept-smiths," if you will, connecting, juxtaposing, generalizing, and deriving new propositional material from given propositional material. In the material world, functions are manipulated. In a spectrograph, photographic film is used to record spectral lines. An analyst, then, determines elemental concentrations from the intensities of the lines on photographic film. In a direct reading spectrometer, photo-multiplier tubes replace photographic film; condenser electronics replaces the analyst. These are functional substitutions. One material truth is substituted for another that serves the same function. Photo-multiplier tubes, instead of photographic film, perform the function of intensity recording. Condenser electronics performs the function of determining elemental concentrations. "Instrumenticians" are "function-smiths," developing, replacing, expanding, and connecting new instrumental functions from given functions.

Humanly contrived control over material agency—working knowledge—coupled with the instrumentician's purposive interest in the phenomena he or she seeks to develop and deploy, provides the best first analysis of a function. An instrumentician may have a detailed theoretical understanding of the phenomenon—the working knowledge—he or she manipulates. In a direct reading spectrometer condensers are used to collect the charge produced by the photo-multiplier tubes. A well-understood theory relating capacitance, resistance, and discharge time allowed the initial designer of the direct reading spectrometer, Jason Saunderson, to build electronics that would use discharge time to measure the amount of stored charge (Baird 2000a). But an instrumentician may not have a detailed theoretical understanding of the working knowledge he or she manipulates. Saunderson did not know why photo-multiplier tube output was as sensitive to precisely where light struck the tube's cathode. For whatever reason, it was. With a quartz plate, he "fuzzed" the light over the whole cathode and, through such material averaging, achieved dependable regular output from the tubes (Baird 2000a).

Instruments have purposes, and the various parts of instruments have purposes that contribute to the larger purpose of the instrument. Thus, to analyze the concept of an instrumental function, "instrumental purpose" must be coupled to the concept of working knowledge. This allows the substitution of one phenomenon for another radically different in nature but that can fulfill the same function. The electrical phenomenon of photo-multiplier tubes is not the same as the chemical phenomenon of photo emulsions. Human discriminatory skills differ from condenser electronics. Yet, both photo-multiplier tubes and photo emulsions can be used to sense light; both human visual sense and condenser electronics can be used to discriminate relative amounts of light.

3.3. Justification

There are a number of subtle and important issues that must be resolved for a fully satisfactory analysis of an instrumental function. I do not attempt to settle these difficulties here.[8] Instead, I employ a cruder analysis sufficient for my epistemological purposes. The most important aspect of an instrumental function, for my purposes, is that it encapsulates "working knowledge." Furthermore, instrumental functions serve purposes in the instrumental contexts in which they are used. Exactly how these purposes are understood, both in an individual case and as a general analytical question, is not important here. What is important is that the eliciting, stabilizing, routinizing, even black boxing, of a phenomenon is hard work. This is the work of justifying a putative instrumental material truth. Peter Galison (1987, 1997) has documented this in great and fascinating detail. Work by Hacking, Buchwald, Gooding, Latour, Pickering, and others speak to similar points.[9]

Justification of material truths—model knowledge, working knowledge, and accurate measurement—is a matter of developing and presenting material, theoretical, and experimental evidence that connects the behavior of a new material claim to truth with other material and linguistic claims to truth. In some cases, a phenomenon is sufficiently compelling on its own. Such was the case with Faraday's motor. But, more generally, it is important to connect the phenomena an instrument deploys with other instrumental, experimental, and/or theoretical knowledge. This situates new working knowledge in a field of material and theoretical knowledge. Such connecting work provides depth and justification to new knowledge. Thus, in a report on the first commercial use of a direct reading spectrometer for steel analysis, table 3.1 was included. The table shows first that, for the five elements measured, the range of concentration readings provided by the spectrometer centers on the concentrations found by

TABLE 3.1.

Elements	Chemical Analysis	Extremes, Spectrometer	Perfect Standard Deviation	
			Spectrograph	Spectrometer
Manganese	0.55	0.54–0.56	1.82	1.35
Silicon	0.28	0.27–0.29	1.97	2.46
Chrome	0.45	0.44–0.47	1.92	2.06
Nickel	1.69	1.68–1.71	1.85	0.79
Molybdenum	0.215	0.21–0.22	2.66	1.68

Source: Vance 1949, 30.

wet chemical analysis. Second, the table shows that for manganese, nickel, and molybdenum, the precision of the spectrometer, as understood in terms of percent standard deviation, is better than for the spectrograph. With silicon and chrome it is the reverse. The table, thus, connects the behavior of the new instrument with other techniques (wet chemistry) and instruments (a spectrograph).

Work such as that reported in table 3.1 justifies the subsequent use of the new instrument. Analysts can use the instrument with the degree of confidence justified by the data in the table. Another way of thinking about this is that such work justifies the transition to a different material form of knowledge. It assures the appropriate kind of stability through change. The instrument is calibrated, relative to other material and conceptual knowledge, for its range of appropriate—trustworthy—uses.

This is the work of creating instrumental functions or material truths. An instrument maker has to produce, refine, and stabilize a phenomenon—working knowledge—that serves some instrumental purpose. These instrumental functions, then, can be manipulated, conjoined, combined, adapted, and modified for the overall purpose of the instrument in question. The behavior of the resulting material device, then, is connected to established apparatus, theories, and experiments. The result is growth in material knowledge.

The fact that it is hard to establish an instrumental function materially has a corollary. Where truth serves as one regulative ideal for theory construction, the regularity and dependability of a phenomenon serves for instrument construction. This is where the material sense of "true"—as opposed to a wobbly wheel, out of true—points us in the right direction. Material truth—working knowledge combined with instrumental function—serves as a regulative ideal for material knowledge, just as theoretical truth serves as a regulative ideal for theoretical knowledge.

3.4. Detachment

If I want some information about plutonium I can look it up in an encyclopedia:

> Plutonium. Actinide radioactive metal, group 3 of the periodic table. Atomic number 94. Symbol Pu. This element does not occur in nature except in minute quantities as a result of the thermal neutron capture and subsequent beta decay of ^{238}U; all isotopes are radioactive; atomic weight tables list the atomic weight as [242]; the mass number of the second most stable isotope ($t_{1/2}$ = 3.8 × 10^5 years). The most stable isotope is ^{244}Pu ($t_{1/2}$ = 7.6 × 10^7 years). Electronic configuration. (Considine 1983, 2262)

I do not have to read about Glenn Seaborg's discovery of the element in 1940 through the deuteron bombardment of uranium in the University of California, Berkeley's sixty-inch cyclotron. Nor do I have to read about all of the various ways that the information above has been ascertained and justified. This information has been detached from its context of discovery. This is one key feature of—at least some—propositional knowledge. The concept of "black boxing" serves a similar function with respect to material knowledge.

In 1955 Jason Saunderson introduced a new instrument, "Spectromet," that could be used right on the foundry floor by personnel with no training in chemical analysis or spectroscopy (Baird 2000b). Spectromet was push-button simple to operate. A foundry worker could take a sample of molten steel in a steel pour, perform some simple preparatory steps on it, and insert it in the instrument. Moments later the instrument would provide information on the percentage concentration of the various important elements in the alloy. The information was available quickly enough to modify the steel in preparation. One no longer would have to perform post hoc analyses and scrap bad metal.

Spectromet is an example of black boxed technology. In *Pandora's Hope,* Bruno Latour defines "blackboxing" as follows: "An expression from the sociology of science that refers to the way scientific and technical work is made invisible by its own success. When a machine runs efficiently, when a matter of fact is settled, one need focus only on its inputs and outputs and not on its internal complexity. Thus, paradoxically, the more science and technology succeed, the more opaque and obscure they become" (Latour 1999, 304). In one sense, this is exactly right. The internal complexity of Spectromet is invisible. Indeed, most of its users could not understand it if it were "cut open and made visible." Once built and installed, Spectromet could provide useful information for people with no understanding of its operation.

But I understand this feature of science differently from Latour. Spectromet encapsulates knowledge that can be taken to a new setting—a new foundry— and used. The knowledge used in this context is tacit in the instrument in the sense that those using Spectromet may have no personal ("subjective") understanding of spectrochemistry. Recall Bucciarelli's telephones. But, with this instrument and its encapsulated knowledge, they can deploy spectrochemistry for their own purposes. This spectrochemical knowledge has become detached and now serves other technical and scientific purposes in a manner similar to how theoretical knowledge can detach from its context of discovery.

One of the central features of the recent explosion of "off-the-shelf" scientific instrumentation is just such encapsulation of knowledge into black boxes that then detach from their context of creation to serve other technical and scientific needs. In 1946 Ralph Müller foresaw and advocated just this development: "Lord Kelvin once said that, 'the human mind is never performing its highest function when it is doing the work of a calculating machine.' The same may be said of the [chemical] analyst and his chores. At present, under the compulsion of industry's pace, we are in a stage of extensive mechanization. That process cannot be stayed, however much the classical analyst bemoans the intrusion of the physicist and engineer upon our sacred domain. It is to be hoped, rather, that it will afford the analyst more time and better tools to investigate those obscure and neglected phenomena which, when developed, will be the analytical chemistry of tomorrow" (Müller 1946, 30A). The mechanization of science into detached black boxes has allowed science and technology to investigate other areas. The portentous and widely reported mapping of the human genome would not be possible without automatic methods of analysis, without knowledge encapsulated in detached black boxes. Medical technology has reached the point where few doctors understand the biophysics that produces the images with which they diagnose disease. While this situation does create the possibility (and, indeed, actuality) of misdiagnoses because of a failure to understand just how the images get made, there can be no question that our diagnostic capabilities are vastly improved with these new black boxes (Cohen and Baird 1999).

There is a tendency to associate the concept of black boxing with infallibility or perhaps inevitability. Much recent work in the sociology of science can be understood as opening up science's black boxes to show contents that are neither infallible nor inevitable. Things—in my case, quite literally "things"—could have been different. Mistakes are made. Ill-advised trade-offs are taken. Certain options may have been taken for whatever reason. Thus, it must be emphasized that my talk of detached knowledge and black boxing does not imply that science and engineering is infallible, not subject to (material or conceptual) revision. It does not imply that things could not have been different. An instrumental black box can be found wanting. Its results may not square with accepted theory; it may be less reliable than desired. It may produce data that are less precise than desired or that are simply not useful. It may cost too much. New devices or new theories are made to respond to such perceived failings. Such is the way we make and remake our knowledge of the world.

There is one last feature of detachment that emerges from the brief story about Spectromet. Knowledge, either in material or conceptual form, does not simply detach and float free, going just anywhere. The original direct reading spectrometers encapsulated spectrochemical knowledge. In the right circumstances, they would have operated fine. But they needed the right circumstances —operators who knew how to align the optics and an environment that would not disturb the aligned optics. These turned out not to be the circumstances into which the spectrometers found themselves, and they did not work well. Spectromet solved these problems, but Spectromet itself cannot be expected to work under any conditions. Spectromet can determine the concentrations of only preselected elements. Operators have to prepare samples for the instrument. The electronics have to be checked periodically. And so forth. The reliability of detached black boxes is an open-ended process of adjusting or refining the instrument and its context of use, its niche. Adjustment happens at both ends. Both the instrument and its context of use are brought into mutual consilience. Latour (1987, 249) tells just such a story as the "extension of Pasteur's laboratory to the farm." It is also a matter of moving beyond the laboratory, and it is an open-ended and fallible process of coadaptation.

4. A Neo-Popperian Picture of Objective Thing Knowledge

4.1. Popper's Objective Knowledge

Thus far I have argued that we need to understand instruments themselves as knowledge bearers, on a par with theory. I have articulated three different kinds of knowledge born by instruments, and I have offered several thing-centered substitute concepts for the key epistemological concepts of truth, justification, and detachment. Even so, I anticipate resistance to thing knowledge. Any materialist epistemology immediately faces the problem that virtually all epistemology has been rooted in beliefs or propositions—ideas, generally speaking. Whatever they are, instruments are not simply ideas or propositions. I close then with a more general epistemological picture that speaks of objective knowledge, borne by, among other things, things—scientific instruments.

Given the need for an objective epistemology that material instruments pose, Karl Popper's account of "objective knowledge" or "epistemology without a knowing subject" is the obvious place to look (Popper 1972, chap. 3). There are immediate problems, however, as Popper (1972, 118) restricts his epistemology to the "world of language, of conjectures, theories, and arguments." Thus, at best, what I offer here is a neo-Popperian account of objective knowledge.

Popper's ontology includes three distinct, largely autonomous but interacting "worlds" (Popper 1972, vii and chap. 3). The first is the material world of stones and stars—"the first world," or "world 1." Next is the world of human (or possibly animal) consciousness, of beliefs and desires—"the second world," or "world 2." Finally, Popper proposes a "third world," or "world 3," of objective knowledge. Popper's third world consists of the contents of the propositions that make up the flow of scientific discourse. Each world emerges from, and is largely autonomous from, its predecessor world. Conscious states may require material instantiation, but they are not explicable in purely material terms. Objective knowledge may depend on human consciousness, for conscious humans (typically) produce knowledge, but objective knowledge is not explicable purely in mental terms.

Popper's third world can sound dubiously metaphysical, but the kinds of objects he populates it with bring it down to earth. These include "theories published in journals and books and stored in libraries; discussions of such theories; difficulties or problems pointed out in connection with such theories; and so on" (Popper 1972, 73). It is not the physical marks on journal paper that Popper points to, but the assertions these physical marks express.

There are two crucial differences between the linguistic products with which Popper populates world 3 and the material products with which I am concerned. The first is truth: "With the descriptive function of human language, the regulative idea of *truth* emerges, that is, of a description which fits the facts" (Popper 1972, 120). We should seek true theories, and Popper (1972, chap. 2, sects. 8–10) offers an elaborate theory of verisimilitude as a central component to his method of "systematic rational criticism." But Popper's theory of verisimilitude has been beset by empirical and conceptual problems and has not been persuasive. Still, accuracy remains a regulative ideal of science. We want our theories to be accurate, true to what can be learned of the objects they claim to describe. Things need not describe; but we do want them to perform well, to bear our working knowledge, which can serve our instrumental functions. This problem was the subject of section 3.2.

The second difference is material: things are, while propositions are not. It makes some sense to think of language in immaterial terms. The proposition expressed by the sentence "There is no highest prime number" surely is a candidate for an immaterial object. It is quite natural to think of propositions, ontologically, as something akin to Plato's forms. The material products of science and technology with which I am concerned most certainly are material, and the "idea of a thing" cannot be identified with the thing itself. In Popper's terms, the material creations would seem to occupy world 1.

4.2. Subjective and Objective Again

Popper strongly criticizes those whom he calls "belief philosophers." These are philosophers who "studied knowledge . . . in a subjective sense—in the sense of the ordinary usage of the words 'I know.'" (Popper 1972, 108). Such a focus, says Popper, leads to irrelevancies. Our focus should be different: *knowledge or thought in an objective sense,* consisting of problems, theories, and arguments as such. Knowledge in this objective sense is totally independent of anybody's claim to know; it is also independent of anybody's belief, or disposition to assent; or to assert, or to act" (Popper 1972, 108–109). Imre Lakatos's rational reconstruction of scientific research programs presents an extreme vision—some empirically and historically minded scholars might say reductio ad absurdum—of Popper's proposal for objective epistemology (Lakatos 1970; see also Hacking 1983a, chap. 8).

I prefer a less extreme vision for objective epistemology. Popper focuses on problems, theories, and arguments, the stuff that might be found preserved in libraries. In most cases the sentences that comprise these problems, theories, and arguments are connected with beliefs held singly or jointly by the author(s) of the sentences. People with their subjective beliefs almost always are involved in one way or another with objective knowledge. Popper's example of tables of logarithms produced by machine and never used by humans is exceptional and likely is related to beliefs in a second-order analysis. In many cases, the sentences preserved in libraries are one way to understand the beliefs of the actors involved. Thus, my brief reconstruction of James Watt's beliefs about the value of his Indicator Diagram (Baird 1989) is based on the written historical record. But it goes beyond the specific sentences in the record, and Watt's beliefs are a useful historical category on which to pin the reconstruction.

There is a similar relationship between the things we make and a complex of human capacities that include skills, know-how, the ability to visualize, and, indeed, beliefs, the nexus that often is referred to as "tacit knowledge." David Gooding's (1990) discussion of the work Faraday did that led up to the making of his electromagnetic motor provides a nice insight into exactly this relationship. Gooding's reconstruction of Faraday's work, using the written record to direct reenacting this work, provides valuable insight into the motor Faraday ultimately made and its relationship to Faraday's skills, know-how, etc. In parallel with my reconstruction of Watt's beliefs about the Indicator Diagram, there is both a more objective epistemological object—Faraday's motor and contemporary reconstructions of it—and a more subjective epistemological object—Faraday's skills, etc. Gaining insight into either helps understand the other.

With these concessions to critics of Popper in mind, I nonetheless agree with the thrust of Popper's push for a focus on objective epistemological objects. There are several reasons for this. Objective epistemological objects, sentences and things, are public. In principle they are open to examination by anyone. For this reason they can provide insight into the more private domains of beliefs and skills. This surely is one of the reasons why work in artificial intelligence promises insight into natural intelligence. Artificial intelligence is public, open to scrutiny and manipulation in a way that natural intelligence is not. Harry Collins and Martin Kusch's (1998) theory of action presents a theory of public behaviors as a way to understand skill and know-how and our relationships with machines.

In a similar vein, historical reconstruction must depend on evidence that can be examined. For the most part these are texts, although artifacts have increasingly become important. Klaus Staubermann's (1998) work presents an interesting dialectic between objective and subjective epistemological objects. He re-creates Karl Friedrich Zöllner's nineteenth century astro-photometer, starting with the public record—both written and artifact. He makes a public object and uses it to rework Zöllner's experiments. The result is insight into Zöllner's skills, both in making and using the instrument. But Staubermann's insight into this subjective epistemological object then reflects back and provides a deeper understanding of the public materials, written and artifact, that were the basis of early astrophysics.[10]

With both historical reconstruction and the contemporary construction of theory, objective epistemological objects play an essential role. While we do not have to join Popper in abandoning the subjective epistemological objects, the fact that these objective objects can be shared is of fundamental importance.

Objective epistemological objects also are important because they are what can qualify as scientific knowledge. At best, individual beliefs might be called "candidate claims" to scientific knowledge.[11] Individual skills might lead to reliable instruments, but in themselves they do not qualify as scientific knowledge. It is the community that determines what is scientific knowledge and communities have to act on public—objective—objects.

Related to this point is the fact that scientific knowledge transcends the subjective beliefs and skills of any individual. This is true in the simple sense that there is more "known" than any single person could subjectively know. But it also is true in the more complicated sense that the tools we have for making beliefs public—speaking, writing, engaging in dialog—allow us to articulate beliefs in a way that is not possible purely subjectively. My beliefs about Watt's indicator, for instance, develop and crystallize as I write about Watt's indicator. Writing provides us the ability to build more content into our beliefs, creating objective epistemological objects in the process.

A similar point can be made about the tools we have developed to work materials. They dramatically exceed the level of our skills and know-how. As I write, cranes tower over the Boston skyline as Bostonians remake their city. They call it "the big dig." It is a vast project to build a tunnel under the city to remove traffic from city streets; elevated expressways will be a thing of the past.[12] Huge projects such as this, of course, involve a tremendous amount of politics. They involve selling visions of a future Boston to skeptics and those with large purses. They also involve machinery that vastly extends our skills in making things. If we want to understand the growth of our abilities to make things, we have to understand the development of our tools for doing so. The big dig is a visually and financially remarkable project. Our abilities to make tools capable of working at finer and finer degrees of precision, reliably mounting (for example) untold thousands of transistors into smaller and smaller integrated circuits is having and will have a much broader impact. Thing knowledge stands on the shoulders of giants, giant machines. Only at its peril can an epistemology of science and technology ignore these objective epistemological objects.

4.3. The Metaphysics of Thing Knowledge

I close with the ontological problem thing knowledge poses. Popper's world 3 is not a domain of things. For Popper, our material creations are world 1 applications of world 3 theories: "It cannot seriously be denied that the third world of mathematical and scientific theories exerts an immense influence upon the first world. It does so, for instance, through the intervention of technologists who effect changes in the first world by applying certain consequences of these theories" (Popper 1972, 155). Viewing the fruits of technology as applications or instantiations of theoretical knowledge will not stand historical scrutiny. Faraday's motor was not an instantiation of his theory of electromagnetism. But the alternative is perplexing. It would seem that including such world 1 objects in Popper's third world of objective knowledge produces ontological confusion. One would like to know what difference there is, if any, between my "third world–first world" objects and regular first-world objects.

The first thing to realize is that theories themselves require material expression. Popper speaks at length of libraries being the repositories of world 3. Anyone who has moved even a small collection of books knows that they are distressingly material! Popper (1972, 168) also writes of "paper and pencil operations" in the solution of problems (for example, the product of 777 and 111).

Paper and pencil operations are operations in the material world. Of course Popper might say that these operations are linguistic items with world 3 meaning, what he calls "third-world structural units." They are "capable of being grasped (or deciphered, or understood, or 'known')" (Popper 1972, 116). With instruments, tools, and other material products of human ingenuity, it is not quite the same. They do not have meaning in the same sense as propositions do. Yet it is possible to grasp "a meaning" of a material object. Davenport did so. Maurice Wilkins and Rosalind Franklin did so when they saw Watson and Crick's model of DNA.

The question remains. What distinguishes a part of world 1, which also is a part of world 3, from that which is not? What are the differences between a riverbed, a spiderweb, and the things I call thing knowledge? First of all, I insist on these differences. Whatever the analysis of them, we can distinguish objects of human manufacture—both linguistic and material—from other natural products of life. Both can be distinguished from the products of purely physical forces. A moonscape differs from a landscape, which in turn differs from a painting of a landscape. What an art historian writes about a landscape is different yet again.[13]

Second, I recall one of Ian Hacking's observations about phenomena: naturally occurring phenomena are rare. While various aspects of the heavens provide examples, and this may explain why astronomy is the oldest science, generally speaking, "in nature there is just complexity ... [while] ... most of the phenomena of modern physics are manufactured" (Hacking 1983a, 226, 228). The "expressivity" of instruments—how they are part of both world 3 and world 1—is a consequence of their comprising phenomena, working knowledge.

It is significant that both Hacking and Popper waffle on biological phenomena. Popper (1972, 145) sees many biological products as akin to theories: "The tentative solutions which animals and plants incorporate into their anatomy and their behavior are biological analogs of theories." Yet he does not include them in his world 3. Hacking (1983a, 227) writes, "Each species of plant and animal has its habits; I suppose each of those is a phenomenon. Perhaps natural history is as full of phenomena as the skies of night." Yet earlier, he writes: "It will be protested that the world is full of manifest phenomena. All sorts of pastoral remarks will be recalled. Yet these are chiefly mentioned by city-dwelling philosophers who have never reaped corn nor milked a goat in their lives. (Many of my reflections on the world's lack of phenomena derive from the early morning milkstand conversations with our goat, Medea. Years of daily study have failed to reveal any true generalization about Medea, except maybe, 'She's ornery often')" (Hacking 1983a, 227).

Hacking's and Popper's ambivalence about the putative phenomena of natural history suggests a further distinction. Thing knowledge, existing in a more refined constructed space, exhibits greater simplicity—although perhaps less robustness—than do the adaptive living creations of natural history. Perhaps more important, our material creations, through our various acts of calibration connecting them with each other and with theory, have a greater depth of justification than the animal phenomena that Popper likens to theory. Spiderwebs are well adapted to catch spider food in many circumstances. But they have no established connection with other possible and actual approaches to catching spider food. We can and do connect the action of direct reading spectrometers with other spectrographs, with wet chemical techniques, and with spectrochemical theory.

In the end, then, we have a material realm of thing knowledge, fallible and dynamic like Popper's world 3. A realm with objects we can think about and use to change our physical surroundings—a realm that interacts with worlds 2 and 1. This is a realm where we encapsulate different kinds of knowledge—theoretical, skillful, tacit, and material—into statements, performances, and material bearers of knowledge. It is a material realm that simultaneously is an epistemological realm.

NOTES

This essay presents in condensed form the primary thesis of my book, *Thing Knowledge: A Philosophy of Scientific Instruments,* forthcoming from University of California Press. Both this essay and my book draw on material from several other publications of mine. This is a work of long gestation, over which time many people have been very generous and helpful with critical and constructive commentary. At risk of neglecting to thank certain individuals explicitly—for which my apologies—I would like to thank my colleagues at MIT's Dibner Institute for the History of Science and Technology, particularly Ken Caneva, Yves Gingras, and Babak Ashrafi. I also would like to thank the participants in Hans Radder's June 2000 conference, "Towards a More Developed Philosophy of Scientific Experimentation," especially Hans Radder and Henk van den Belt, who provided a very insightful commentary on my presentation at the conference.

1. My brief remarks here can be supplemented by many fine histories of the discovery of DNA (Olby 1974; Judson 1979; Watson 1981; Crick 1988).

2. On the semantic view of theories see Suppes (1961, 1962, 1967); van Fraassen (1980); Giere (1988); and Hughes (1997).

3. According to the semantic view of theories, a theory identifies, or in some versions is coextensive with, a class of models. The models I focus on are singular; they are not classes of models. Which is to say they cannot be identified themselves as theories but, at best, as single elements of a class of models—that is, a theory. It is not clear, however, whether we must restrict knowledge claims to theories and not extend them to individual models. The semantic view takes no position on this point. Here I extend knowledge claims to the individual model.

4. My discussion of "working knowledge" bears strong similarities to Hans Radder's discussion of "material realization." See Radder 1996, particularly chap. 2.

5. Information on Faraday's electromagnetic motor is from Faraday (1821a; 1821b; 1822a; 1822b; 1822c; 1971), Gooding (1990), Baird (1995).

6. More detail on the indicator is in Baird (1989). Primary sources on the invention and development of the indicator, including its role in the development of thermodynamics, are Robison (1797); H. (1822). Good secondary sources on the indicator are Cardwell (1971) and Hills (1989).

7. Pressure is force per unit area, $P = F/A$, and hence force is pressure times area, $F = P \times A$. Work is the integral sum of force over a given distance, $W = \int F dx$. But $F = P \times A$, so $W = \int P \times A dx$. The area of a piston times the distance it moves is its change in volume, so $A dx = dV$. Consequently, the work done in moving a piston is the integral sum of the pressure through a given volume change, $W = \int P \, dV$. This is the area under the pressure-volume curve traced out by the indicator.

8. However, I would point to work by Peter Kroes on this topic as very useful and encouraging; see Kroes (1996, 1998, 2001).

9. See Hacking (1983b), Latour (1987), Gooding (1990), Buchwald (1994), and Pickering (1995).

10. Otto Sibum's work presents a similar dialectic (Sibum 1994, 1995).

11. On this point see Gooding (1990), Pitt (1999).

12. Thomas Hughes has written about the big dig in *Rescuing Prometheus* (1998).

13. There are mixed cases, such as, for example, a managed forest reserve. But it seems to me that we will be better equipped to deal with such cases once we consider the less difficult cases.

Peter Kroes

4 / Physics, Experiments, and the Concept of Nature

1. Preliminary Remarks about the Natural and the Artificial

My main concern will be the question of what kind of conception of nature underlies modern experimental physics. My strategy will be to compare the natural objects and phenomena studied in physical experiments with the artificial infrastructure (technological artifacts, artificial conditions) of those experiments in order to uncover differences between the domain of the natural and of the artificial. We will start with a few preliminary remarks about the distinction between the natural and the artificial. Then, the traditional view on the dividing line between the natural and the artificial in the context of physical experiments will be discussed. Thereafter, I will present a critical analysis of this traditional view. First, it will be pointed out that the modern conception of nature is based on a thoroughly teleological methodology because of the use of experiments. Secondly, Hacking's criticism of the traditional view, according to which phenomena are discovered, will be examined. Against the traditional view, Hacking maintains that phenomena are created in experiments. This raises the question whether, and if so, in what sense, those phenomena can still be considered to be natural phenomena. I will argue that Hacking's claim, when properly construed, in no way undermines the traditional view on the distinction between the natural and the artificial in experiments. I conclude with the observation that the notion of nature involved in modern physics stays rather obscure.

The difference between the natural and the artificial appears to be intuitively rather unproblematic. Starting from paradigmatic examples of natural objects or processes (for example, birds, trees, clouds, and tides) and of artificial objects and processes (for example, chairs, cars, television sets, and radio transmission), it appears to be self-evident that artificial things are human made and natural things are not. But it is a difficult task to turn this intuition into an explicit, coherent conceptualization of the difference between the natural and the artificial (see, for instance, Fehér 1993). Depending on whether or not human beings are considered to be part of nature, the artificial becomes either a subcategory of the natural or a category alongside the category of the natural. It is not clear whether the distinction is exhaustive—that is, whether any arbitrary object/process can be classified as either natural or artificial. Neither is it clear whether it is always possible to draw a sharp dividing line: Are genetically modified organisms natural objects or artifacts? Moreover, the notion of "human-made" is notoriously vague. A human footprint on a beach is human made, but is it an artifact? Is it necessary for an artifact to be the result of *intentional* human behavior? On closer inspection, the distinction appears to be rather complicated.

For our purposes it will be sufficient to have a clear point of reference within the domain of the artificial as a touchstone for clarifying the idea of the natural. Technological artifacts, such as cars, bicycles, television sets, etc., serve as such a reference point. They are the result of intentional human action: they are designed, made, and used in order to perform a practical function. This category of objects may be characterized as objects with a function whose physical structure is human made and based on a human design. A design is taken to be a scheme or plan that shows how a particular function may be realized. For what follows it is important to keep in mind: technological artifacts have a dual nature (Kroes 1994; Kroes and Meijers 2000). On the one hand, they are physical objects, which obey the so-called laws of nature; that is, their behavior can be explained causally in a nonteleological way. On the other hand, they are the physical embodiment of a design, which has a teleological character; the whole construction is intended to perform a practical function. This function is an integral part of a technological artifact; without taking its function into consideration, an object cannot be understood properly as a technological artifact. Think away the function of a technological artifact, and what is left is no longer a technological artifact but simply an artifact—that is, a human-made object with certain physical properties but with no functional properties.

A special category of objects consists of "technological pseudoartifacts." These are objects with a practical function; that is the reason for calling them "technological." But as physical objects they are not based on a human design and are not human made. Examples are a shell used as a cup for drinking water or a rock used as a hammer. Just as technological artifacts, these kind of objects have a dual nature. On the one hand, they are physical objects and as such are not real artifacts, since their physical structure is not the result of intentional human action; as physical objects, none of their properties are human made. Without their function, these objects would simply be natural objects (in the sense of not being human made). On the other hand, these objects are tools and thus objects with a function (in a specific context of action), which they perform on the basis of their physical properties. They are tools with a function by virtue of intentional human action and as such are artifacts. In view of this specific dual nature, these objects are called "pseudoartifacts": insofar as they are tools, they are artifacts; but insofar as they are physical objects, they are not.

The above remarks far from settle the issue about how to distinguish between the natural and the artificial in a general way.[1] In line with the Aristotelian characterization of nature, the basic idea underlying these remarks is that the distinction between the natural and the artificial is primarily a *genetic* one.[2] Roughly, the domain of the natural is taken to be the domain in which objects behave according to their own principles of change (motion) without any interference by human beings. Natural objects are not human made, and natural processes are processes without any human interference. When objects or processes are the (intended) result of intentional human action, they are artifacts: because they are human-made, they do not carry within themselves their principle of change. Now let us zoom in on experiments to see how these ideas on the natural and the artificial match the distinction between the natural and the artificial in experiments.[3]

2. The Traditional View on the Natural and the Artificial in Experiments

In experimental physics, nature is studied with the help of technology; nature is observed through "technological spectacles." Various kinds of technological artifacts are used, not only for generating the systems to be studied but also for making measurements. According to the traditional view of the role of experiments in physics, this technological (artificial) infrastructure is simply a means to generate new data. Once the new data are produced, technology has played its part and the real scientific work, theorizing and making models of nature, may begin. Technology may contribute to the generation of new data in three different ways:

1. It may help to overcome imperfections and limitations in human perception by providing measuring equipment—that is, it may extend and refine our sensory apparatus (Ackermann 1985, 127).

2. It may provide equipment for studying the behavior of systems under very special human-made (that is, artificial) conditions that do not occur spontaneously in nature.

3. It may provide equipment for preparing or producing a system to be studied (to be called "object system").

According to the traditional view, the technological/artificial means and processes by which data about nature are obtained leave no traces in our conception of nature (Lelas 1993, 423–424; Tiles 1992, 99). This does not mean that the use of technology in experiments is considered to be unproblematic. For several reasons, the technological equipment does not simply deliver reliable facts, nor does it tell us what nature looks like.

In the first place, it is quite common among experimentalists to speak of "artifacts" of the measurement equipment or of the experimental setup. They thereby refer to results that are generated by the artificial environment or artificial means of observation of the natural phenomenon under study (Franklin 1986, 3). Data are called "artifacts," as opposed to genuine data, when they carry no information, or unreliable information, about the object of study. Many measuring instruments produce artifacts; for instance, early telescopes produced colored fringes due to chromatic aberration. It is, of course, of prime importance in experiments to discriminate between artifacts and genuine results. The results of an experiment are always the outcome of the object system interacting with an artificial environment, and therefore it is always necessary to filter out the component in the results that tells us something about the object system. Although in practice this may be extremely difficult, and may involve long and intricate chains of reasoning and algorithmic processing, the traditional view maintains that it is always possible in experimental results to eliminate all effects that are due to the use of technology. There exists, in other words, an epistemology of experiment—that is, "a set of strategies that are used to provide rational belief in an experimental result. These strategies distinguish between a valid observation or measurement and an artifact created by the apparatus" (Franklin 1986, 192).

In the second place, the results of experiments may be untrustworthy due to malfunctioning or to incorrect handling of equipment. The performance of experiments usually requires a lot of (technological) skills on the part of the experimentalist. These skills are required not only for the use of "off-the-shelf" equipment, but also for fitting equipment to the particular needs of the case.

In the third place, like science, technology is dominated by theory; therefore, the evaluation and interpretation of the results of experiments may involve the use of much theory. For these reasons, the creation of experimental evidence (facts, data) in experimental practice is not a straightforward affair. But, according to the traditional view, when properly performed and interpreted, experiments will deliver reliable facts with the help of technology. These facts constitute the evidence for developing and evaluating theories about the natural world. It may be true that in experiments physicists look at nature through technological spectacles, but in principle these spectacles are completely transparent: they in no way influence the resulting picture of nature. As Lelas (1993, 432) remarks, in experiments, "anything artificial can be extracted, and its traces erased so that the natural shines out in its full splendor to the glassy essence of scientific apparatuses."

This traditional view is based on a strict separation of the natural and the artificial in physical experiments. The system to be studied, the conditions under which it is studied, and the measurement equipment may all be artificial (human made), but the *behavior* of that artificial system in its artificial environment observed through artificial measurement devices is nevertheless considered to be a natural phenomenon. The reason is that that behavior is taken to be the outcome of the laws governing that system. The experimental phenomena themselves are not taken to be artificial. They become manifest only under very special artificial conditions and when observed with special devices. With the aid of technology, the scientist gains access to parts of nature that would otherwise remain hidden due to limitations in human sense organs or due to contingent conditions prevailing in our universe.

On the one hand, this conception of the natural (nature) in the context of experiments is close to the Aristotelian one; on the other hand, it is far removed from it. From the Aristotelian conception it inherits the idea that the natural is somehow related to intrinsic principles of motion or change. In experiments, the behavior of the object system is the result of (artificial) initial and boundary conditions and the intrinsic dynamics of the system involved. Once the initial and boundary conditions have been set in place, the system is left to itself, and from then on it is up to these intrinsic dynamics—that is, the laws governing this kind of system—to determine its behavior. This behavior is considered to be a natural phenomenon, because these laws are the intrinsic principles of motion (change) of the system. They are the system's natural laws: being the kind of system it is, it is governed by its proper laws (a gravitational system by the laws of Newton and the law of gravitation, an electromagnetic system by the laws of Maxwell, etc.). Another important similarity with the Aristotelian view on nature and science is that in this conception of experiments the scientist remains essentially a passive spectator with regard to nature; he only prepares the stage (the initial and boundary conditions) for nature to show itself, but once this is done, he resumes again his role of a spectator (albeit with a whole battery of measuring and data-processing devices).[4]

At the same time, however, this view of nature and natural phenomena is far removed from the Aristotelian conception. According to the above view, any change of the object system is a natural phenomenon; it does not matter what kind of initial or boundary conditions are involved. It is, for instance, irrelevant whether these initial and boundary conditions are human made or occur spontaneously. For Aristotle this is not the case. Take the example of the motion of heavy bodies. Aristotle distinguishes between natural and violent motion of heavy bodies. Natural motion occurs when, in accordance with its intrinsic principle of motion, a heavy body moves toward its natural place—that is, it moves toward the center of the universe. When a heavy body is thrown in the direction opposite to the center of the universe, it performs an unnatural, violent motion. It is not a motion that is in accordance with its intrinsic principles of motion as a heavy body; it finds its origin outside the heavy body. Thus, only within special situations—that is, in special initial and boundary conditions—heavy bodies perform a natural motion. From the point of view of modern physical science, this distinction does not make any sense. Irrespective of the prevailing initial and boundary conditions, a heavy body always moves according to its intrinsic principles of motion and thus its motion is a natural phenomenon.

The deeper reason for this difference in view of what constitutes a natural motion (change) concerns the question whether or not the intrinsic principles of motion are goal oriented. For Aristotle the intrinsic principle of motion of an object, its nature, is directed toward the goal or telos of that object. For instance, the natural evolution of a beech tree—that is, its evolution according to its own intrinsic principles of motion—is to become a full-grown beech tree; its goal, as a beech tree, is not to be axed down and turned into a bed.[5] According to Aristotle's doctrine of the four causes, every natural thing has a *causa finalis;* it is the goal of a thing that makes it move. Modern physics has abandoned this idea. Within the mechanistic worldview emerging in the sixteenth and seventeenth centuries, every implicit or explicit reference to a goal or telos in describing nature was rejected. This view did not leave any room for attributing goals to natural objects. Within the ontology of the modern physical sciences, that is still the case. Thus, objects may have their intrinsic principles of motion, the laws governing their behavior, but these principles are not directed toward realizing an intrinsic goal of the object (or even any extrinsic goal whatsoever). The laws of nature are not teleological laws.

The traditional view of the natural and the artificial in experiments raises an interesting problem concerning the role of functions (more generally, teleological concepts) in the ontology and the methodology of the physical sciences: on the one hand, functions play no role in the description of the natural world, whereas, on the other hand, they are indispensable in analyzing experiments.

3. Functional and Structural Descriptions in Experiments

In the following, we will refer to descriptions of the world in which reference to functions is illegitimate (makes no sense) as "structural" descriptions of the world. "Functional" descriptions are descriptions in which function attribution to objects or processes occurs and does make sense.

There can be no doubt that the modern description of the physical world is a structural one (particularly at the most fundamental level of elementary particles) and that almost all empirical data upon which this structural description is based is gathered in experiments. It is a remarkable fact, however, that these experiments themselves (the experimental setup and the operation of equipment) are described and analyzed by physicists in terms of functions. Every experiment has a goal (to measure x or to detect y, or to show phenomenon z, etc.) and it is in relation to this goal that every part of an experimental setup is attributed a function, as well as actions performed during the experiment. For describing and understanding the experiment, reference to functions is unavoidable. In contrast, the description of the outcome of an experiment (the observations, data, measurements) is free of any reference to functions at all. The same is true for the way the outcomes of experiments are explained in terms of physical processes occurring in the object system. Thus, whereas experiments are described in a functional way, the description of the results of experiments and of physical reality as constructed on the basis of those results is of a structural kind. This means that the structural description of physical reality rests implicitly on a functional description of at least part of the world (Kroes 1991).

By itself, this situation need not be necessarily problematic. It is important to distinguish carefully between the properties of a process and the properties of the outcome of that process. A chemical process may produce a product that has the property of being inflammable, but it would be a category mistake to attribute this property to the chemical process itself. More or less the same may apply to experiments and the outcome of experiments. The latter may be described in a structural way, the former in a functional way.

But more seems to be at stake in the case of experiments. A characteristic feature of a functional description is that it allows normative statements with regard to the performance of functions.[6] A function may be performed well or badly; part of the equipment or measuring devices may malfunction or an action may be performed badly. Here I touch upon a far-reaching difference between the attribution of functional and of structural (physical) properties to objects. With regard to structural properties, it does not make sense to make normative statements. From a methodological point of view, reference to functions and normative statements regarding functions are of crucial importance in experiments. Any assessment of the reliability and validity of the outcome of an experiment requires an analysis and evaluation of the functioning of the experimental setup and of the actions performed. In other words, from the point of view of the methodology of experiments, a functional description of part of the world (the technological infrastructure of the experiment) is inevitable for generating reliable observations (data) for a structural description of the physical world. From an ontological point of view, physical science may have banned functions from the physical world; from a methodological point of view, reference to functions is unavoidable because of the experimental nature of physical science.[7] The question we are facing is whether there is a real conflict (or incoherence) between the methodological reliance on functional discourse and the idea of a structural ontology of the physical world.[8]

Any attempt to answer this question will have to deal with the problem of how to reconcile two different conceptualizations of the world with their corresponding modes of description: the conception of part of the world as consisting of material (physical/chemical) systems that interact through causal relations; those systems are described in a structural language; and the conception of part of the world as consisting of human beings who intentionally represent the world, who act on the basis of reasons, and who attribute functions to objects in the world; this intentional representation of the world makes use of a functional language.

The existence of these two different conceptualizations with their own mode of description would pose no problem as long as they would pertain to nonoverlapping parts of the world. That, however, is not the case. Take an object like a mercury thermometer. That object may be analyzed and described from a purely physical (structural) point of view. A structural description of this object specifies its "static" properties, such as the geometrical form of the glass, its mass, its center of gravity, the chemical composition of the glass and of the mercury, the vapor pressure of mercury, the electrical conductivity of mercury and of glass, the moment of inertia of the glass, etc. But it may also describe its dynamic properties—that is, the behavior of the glass and the mercury and its other parts under varying physical conditions; for instance, its behavior under variations of temperature or within a time-varying electromagnetic field, etc.

This dynamic behavior in turn can be explained in a purely physical way on the basis of physical laws and theories that are themselves described in a purely structural language. This structural description not only does not refer to any functional properties of the object or parts of it, but, moreover, there is nothing in this structural description by itself that would indicate that this object has a specific function (that is, that it is a mercury thermometer).

Let us change our perspective and describe the "same" object from an intentional (functional) point of view. The functional description looks very different, because of the simple fact that the object now has or is attributed a function. It represents the object as a means to a certain end, namely to measure a particular physical quantity called "temperature." It focuses on the design of the object. It says that the height of the mercury column is a measure of the temperature of the object measured; that the dimensions of the glass device, containing the mercury, is so designed that there is much more mercury in the glass bulb at the bottom than in the glass tube. This design feature is explained with reference to its function: this is to ensure that the height of the mercury column is a reliable indicator for temperature. The functional description also contains the manual of operation. It says under what conditions and how the thermometer is to be used in order to ensure that a temperature measurement with this thermometer is valid (it specifies, for example, the period of time the thermometer has to be brought into contact with the object of which the temperature is measured).

The first thing to be noted is that the functional description also makes use of structural concepts; it makes reference to structural properties of the object, not only in describing but also in explaining the design of the object. The functional description may be considered to be some kind of extension of the structural description.[9] Secondly, there is a striking difference in the kind of things to be explained from a structural and a functional point of view. For instance, from a functional point of view, the dimensioning of the glass device containing the mercury, in particular the volume of the glass bulb versus the volume of the glass column, is something to be explained. This design feature of the glass device can only be explained by reference to its function. But from a structural point of view, this design feature is just some physical property of the object, namely its geometrical form. Physics has no resources for explaining this geometrical form. From a structural point of view, the geometric configuration of the object is a completely contingent feature of the object.

Thus, we have two different kinds of descriptions of the "same" object, namely a structural and a functional one. The structural description represents the object as a physical system, whereas the functional description represents the object as a technological artifact. The two different descriptions of the object called "mercury thermometer" correspond to two different roles this object may play in experiments. On the one hand, the thermometer may be the object of study in physical experiments; then the structural description will be the outcome of analyzing this object. On the other hand, the object may be used as a measuring device in physical experiments; the result of analyzing this object as a measuring device will be the functional description. The rather strange situation is that, although physics aims at a purely structural description of the world, it cannot do without the functional description as long as it allows measuring devices such as mercury thermometers in experiments. Both the structural description of the object called "mercury thermometer" and its functional description are inevitable for doing physics, but only the structural description is compatible with the physicist's ontological conception of the world.

The foregoing analysis applies not only to a mercury thermometer but also to any kind of measuring device used in physics. The same can be said of the equipment used in experiments; a detailed understanding of the physical operation of a piece of equipment is necessary in order to be able to evaluate whether or not it is functioning adequately in an experiment. From a methodological point of view, the use of experimental equipment and measuring devices requires a functional and structural description of objects; whereas from an ontological point of view, only the structural one makes sense. In other words, from a methodological point of view, the description of certain physical systems as technological artifacts is unavoidable. Within experimental physics, the description of nature appears to presuppose somehow the description of technological artifacts and, vice versa, the use of technological artifacts in experimental physics requires structural descriptions of these artifacts.

We appear to be trapped in a vicious circle. One way out of it is to accept functions as part of the furniture of the universe and to recognize functions as entities sui generis relative to physical entities. Then, the functional description becomes an autonomous mode of describing the world relative to the structural mode; more precisely, it becomes an autonomous extension of the structural mode of description.[10] Within this approach, a structural description of the world (which is the aim of physics) is essentially an incomplete description and rests from a methodological point of view upon a functional description of the world. The tension between methodology and ontology disappears, but at a considerable price, namely a bifurcated picture of the world using two modes of description, of which it is not clear how they hang together. Another option is to deny that there is a fundamental difference between the functional and structural mode of describing the world. That is a defensible position if it can be argued, for instance, that the functional mode of description is not an autonomous extension of the structural mode but can be reduced to the latter. Then, the meaning of functional statements can be analyzed in terms of structural concepts. It is highly questionable whether this option is a viable one.[11]

With the birth of modern physical science, teleological (functional) elements may have disappeared from its ontological stage, but because of its experimental character they have entered the methodological stage. Within experimental physics, the description of the natural world (the structural mode of description) and the description of the artificial world (the functional mode of description) appear to be deeply rooted in each other.

4. Experiments and the Creation of Phenomena

The traditional view on the role of the natural and the artificial in experiments assumes that natural phenomena are *discovered* in experiments.[12] In this view, the physicist is essentially a passive observer in experiments: once the stage is set he just observes (discovers) what is going to happen. This point of view has been strongly criticized by Ian Hacking. In his book *Representing and Intervening* he claims that in experiments phenomena are created by the scientist (Hacking 1983a). He rejects the idea that experimental scientists discover phenomena in the world. "To experiment," in his own words, "is to create, produce, refine and stabilize phenomena," a phenomenon being "something public, regular, possibly law-like, but perhaps exceptional" (Hacking 1983a, 222, 230). Discussing the example of the Hall effect, he states that this effect was not found by Hall simply because it did not exist before Hall succeeded in producing this effect in the laboratory. It was literally created by him, because this effect does not exist without the appropriate experimental setup. If science should have taken another historical path, the Hall effect might never have been created. Note that according to Hacking the creation of phenomena does not involve the creation of the physical objects involved in these phenomena (Hacking 1992, 37): "I did not think of electrons being created, but did think of the photoelectric effect being created, in a pure state."

The idea that phenomena are created does not imply, according to Hacking, some sort of subjectivism or relativism in the sense that all is possible. The experimentalist cannot create phenomena at will. In his interaction with the world, he is subjected to all kinds of constraints: relativism is barred by the fact that the world seldom does what the experimentalist wants (Hacking 1989a).

Hacking combines the idea of the creation of phenomena in experiments with a "hard-headed" scientific realism about unobservable or theoretical entities. In his opinion, it is not because we can perform successful experiments on them that such entities are real. They become real as soon as they can be *manipulated* to produce new phenomena: "Experimental work provides the strongest evidence for scientific realism. This is not because we test hypotheses about entities. It is because entities that in principle cannot be 'observed' are regularly manipulated to produce a new phenomena and to investigate other aspects of nature. They are tools, instruments not for thinking but for doing" (Hacking 1983a, 262). According to this line of thought, the best evidence for the existence of electrons is the common television set using cathode ray tubes (CRT). In a CRT, electrons are produced by an electron gun, accelerated and deflected so as to hit appropriate spots on the TV screen, where they cause a phenomenon known as electroluminescence: the electrons hit a substance that thereby starts to produce light of a certain wavelength. Here, indeed, electrons have become an element of engineering, and from an engineering point of view they are as real as the directly observable deflection coils in a CRT.

The most striking feature of Hacking's view is that it suggests a rather strong parallel between experimental science and technology. The classic distinction between the two is related to the idea that the engineer creates (invents), whereas the scientist discovers.[13] But according to Hacking, not only the engineer creates things (technical artifacts); the experimental scientist also creates things (phenomena). And not only technical artifacts are tools for doing; scientific entities are also tools for doing. Furthermore, neither the engineer nor the scientist can create technical artifacts, respectively phenomena, at will; they are restricted in this by all kinds of constraints. But does this parallel imply that experimental phenomena are created in just the same sense as technical artifacts are created? Are experimental phenomena therefore also artificial phenomena? A similar question arises with regard to Hacking's claim that scientific entities are "tools." This claim seems hardly compatible with the idea that scientific entities are natural entities. Tools are entities with a function, and functions have no place within the ontology of the natural world. If indeed scientific entities are tools, does this imply that they are of the same kind as technical artifacts? The problem with Hacking's view is that it becomes hard to maintain a distinction between the natural and the nonnatural.

Alongside the traditional role of scientists as spectators *in* this world, Hacking puts the role of experimental scientists as creators *of* this world. How do these two roles of spectator and creator fit together? On the one hand, scientists consider themselves to be spectators in this world in the sense that they aim at describing and analyzing as accurately as possible a physical world that they take to be independent of their own presence and actions. In no way is the physical world a world of their own making. On the other hand, in their experiments, scientists are manipulating and intervening in the physical world; in studying that world, they are not behaving as spectators at all but are constantly intervening in it. On the basis of the outcomes of their experiments they construct descriptions (representations) of the physical world. How does the fact that they intervene in the world affect their descriptions of that world? Is the resulting description a description of a world that is indeed independent of the actions of scientists or of a world (partly) of their own making? The answer according to the traditional view on experiments is that scientists remain spectators and that experimental phenomena are not created but discovered. But what if phenomena are indeed created, as Hacking claims?

Let us be very clear about what is involved here. There are various reasons why it could be maintained that our *representations* of the physical world contain elements of our own making in the sense that these elements do not refer to aspects of physical reality itself. For instance, due to the empirical nature of the physical sciences, the only way for scientists to study the physical world is on the basis of how that world manifests itself to them through their senses. Consequently, anthropomorphic elements, due to the specific nature of their sense organs, may creep into their descriptions of the physical world. Scientists are not able to observe the world from a "God's-eye point of view." Their intention may be to avoid anthropomorphic elements, but they will never be sure that they succeed.[14] From this perspective, the description of the physical world by scientists may contain anthropomorphic elements and thus be partly of their own making. But that is a consequence, not of the experimental nature of modern physical science, but of the simple fact that the observations, on which descriptions of physical reality are based, are always observations made by humans. It is a problem that arises on the part of the scientist as a spectator. Apart from the presence of anthropomorphic elements, it has been argued by Poincaré, Duhem, and Reichenbach, for instance, that conventions play an important role in our representations of the physical world. These conventions are human constructions and thus in this respect our representations of the physical world may contain human-made elements. If Hacking were claiming that our representations of the physical world are partly of our own making in the sense that they contain elements that find their origin in the knowing subject and not in the physical world itself, then his claim would not be very exciting. That is an epistemological claim that has been discussed extensively within philosophy (philosophy of science).

As I interpret Hacking, when he claims that phenomena are created in experiments, he is not making an epistemological but an ontological claim. Physical phenomena are literally created in experiments, and consequently our representations of those phenomena are representations of things of our own making, just as a representation of a house is a representation of a human-made object.

He rejects the idea that experimental phenomena are given phenomena waiting to be discovered. *Given* here means preexisting, unaffected by actions of scientists. Scientists relate to these given phenomena only as observers or spectators. They are in no way their makers or creators. In that case, phenomena may be studied as they exist "by and from themselves" without any human intervention. It seems that for Hacking experimental phenomena are created more or less in the same sense as a technological artifact or a piece of art is created. It is the scientist's activity that brings experimental phenomena into existence. Just as technological artifacts and pieces of art are human made, so also experimental phenomena are human made. Given Hacking's interpretation of experiments, the question arises whether the object of study in an experiment is still a natural object or not. Can a scientist *create* a natural phenomenon? In what sense can it be said that scientists are studying nature or natural phenomena in experiments?

Before we enter into these issues, it is important to note that Hacking's position is also of interest with regard to the problem about the role of functions in modern physical science. For Hacking the crucial criterion for the existence of (unobservable) physical entities is that they can be manipulated to produce other phenomena. It is not the testing of theories in experiments that leads to the final verdict about the reality or nonreality of entities, but engineering. Whether or not a hypothetical entity can become an effective tool in the hands of engineers and experimentalists determines whether or not that entity is real. In other words, as long as the hypothetical entity is itself part of the object of study in experiments, and thus is described in a structural way, the reality of that entity remains problematic. As soon as that same hypothetical entity becomes part of the experimental setup, in which it becomes a tool to achieve a certain end and in which it, being a tool, is described in a functional way, it becomes real. This means that our knowledge about the reality of the constituents of the physical world, which are themselves described in a structural way, rests upon functional descriptions of the same constituents. It is engineering that provides knowledge about the reality of physical entities, not theorizing or methods such as inference to the best explanation. Theorizing and inference to the best explanation squarely stay within the domain of the structural. So, Hacking's position reinforces the problem discussed above: although physics aims at a purely structural description of the world, functional descriptions play an indispensable role in it.

With regard to the issue of the conception of nature in modern physical science, Hacking makes two interesting claims:

1. Experimental phenomena are created; this claim undermines the traditional view on the natural and artificial in experiments: scientists create not only the technological infrastructure of an experiment but also the experimental phenomena themselves.

2. Scientific entities are tools for doing, not for thinking; this claim undermines the idea that functions play no role in the description of physical reality.

Elsewhere I have argued that Hacking's view about the creation of phenomena may be interpreted in a way that comes very close to the traditional view on the natural and the artificial in experiments (Kroes 1994). All depends on the interpretation of the expression "creating phenomena." It may be interpreted in a weak or a strong sense. In the weak sense, creating phenomena means creating the *occurrence* of phenomena—that is, creating the proper initial and boundary conditions for a phenomenon to take place. In other words, creating a phenomenon means creating an instantiation of a phenomenon. This leads to an interpretation of Hacking's position that is very close to the traditional one. In the strong sense, creating phenomena means not only that the occurrence of a phenomenon is triggered by creating the appropriate conditions, but also that all or some of the properties of the phenomenon itself are created by the experimentalist. Although Hacking stresses that in his opinion phenomena are literally created, there are strong indications that this expression should be taken in the weak sense. Otherwise, he would be driven into "some sort of ultimate idealism in which *we* make the phenomena" (Hacking 1983a, 220). If the expression "creating phenomena" is interpreted in the strong sense, then the notion of nature (natural phenomenon) becomes problematic because those phenomena become artifacts with or without a technical function.

I take the opportunity to clarify here in more detail the distinction between creating in the weak and the strong sense by comparing the creation of an experimental phenomenon with the creation of a technical artifact. Consider again the object called "mercury thermometer." Taken as an object system in an experiment for studying a particular physical phenomenon, namely the expansion of mercury as a function of temperature, the geometric configuration of this physical object functions as a boundary condition (a specific container for the mercury). Clearly, the initial and boundary conditions involved in this experiment (the geometric form of the object, the varying temperature conditions, etc.) are created by the scientist; as a matter of fact, those specific conditions do not occur spontaneously in our world. So, it could be maintained that the occurrence of the phenomenon of the expansion of mercury as a function of temperature in this physical object is created by the scientist. But obviously this does not mean that the phenomenon of the expansion of mercury as a function of temperature is itself a creation of the scientist. It only means that this phenomenon is created in the sense that the scientist has realized the proper boundary and initial conditions for this phenomenon to occur. All this is in line with the traditional interpretation of experiments.

Now, let us compare this with the creation of a technical artifact. Again we take the object called "mercury thermometer," but this time it is considered to be a measuring device—that is, a technical artifact with a function and based on a design. First note that whether the object under consideration is taken to be a technical artifact or an object system in an experiment, under the same conditions exactly the same physical phenomena take place in it. And just as the scientist creates these physical phenomena by realizing the appropriate conditions in experiments, so does the instrument maker by making and using a measuring device. Thus, the physical phenomena on which the operation of the mercury thermometer is based are created, in the weak sense, by both the instrument maker and the scientist.

The creation of these physical phenomena, however, should not be confused with the creation of the technical artifact. Much more is involved in creating a technical artifact than realizing a specific set of boundary conditions that triggers the occurrence of certain physical phenomena. The function of the mercury thermometer, to measure temperature, is an essential aspect of that object as a technical artifact. But the functional properties of the artifact (as a whole and of its parts) and its design are not created by realizing the appropriate physical phenomena in a physical object. A creative or inventive act is necessary to exploit a physical phenomenon (object) and to turn it into a tool for realizing a goal.[15] The design underlying the mercury thermometer and its function are created by human beings in the sense that they are invented and attributed to the physical object by human beings. This type of creative activity is called creation in the strong sense. It is different from creation in the weak sense in that some of the properties of the thing created are attributed by human beings.

With regard to creation in the strong sense, two situations must be distinguished, depending on the kind of properties attributed. When natural (physical) properties are involved, the claim that phenomena are created in the strong sense may come close to some form of idealism (or relativism) in case properties are attributed to phenomena, which violate the laws of physics. Take the creation of the phenomenon of free fall in a vacuum in a laboratory. The strong sense of creation would imply that some of the properties of the phenomenon of free fall in a vacuum are attributed freely by humans. So, it would be possible, for example, to create a phenomenon of free fall in which not all objects fall in the same way. In case the properties involved are functional (that is, nonnatural) properties, the situation is quite different. This is what happens when technical artifacts are created: functional properties are attributed to human-made objects and phenomena. By attributing a technical function to a physical object (phenomenon), that object (phenomenon) with its function is ipso facto taken out of the domain of the physical world and turned into a technical artifact. Note that, just as in the case of the attribution of physical properties, there is no absolute freedom in the attribution of functional properties. In technical artifacts, there is a strong coupling between function and physical structure.

There can be no doubt that Hacking does not intend his claim, that phenomena are created, to be construed in the strong sense. As Hacking remarks, phenomena cannot be created at will. So, it seems that we have to interpret the notion of creation in the weak sense. But then Hacking's claim loses much of its novelty, since, as argued above, the claim that phenomena are created in the weak sense is compatible with the traditional "discovery view" of experiments.

Let us now briefly turn to Hacking's second claim, that scientific entities are tools. For Hacking, not theory but experimental work provides the best evidence for the reality of (unobservable) scientific entities. Electrons are real because they are tools. Given the distinction between the structural and functional mode of describing the world, this claim sounds rather strange: scientific objects are characterized as tools—that is, as objects with a function—whereas the notion of function has no place within the physical worldview. In a nutshell, Hacking's line of argument is the following: scientific entities are real when they can be used *as* tools for doing (other experiments). From this it does not follow, however, that they *are* tools.[16]

Hacking's argument for the reality of scientific entities may be rephrased with the help of the notion of a technical pseudoartifact in the following way. Scientific entities like electrons are real when they can be turned into (effective) technical pseudoartifacts in some context of action. Like a shell, an electron is not human made; it is a natural object that by virtue of its physical properties (or "causal powers," in Hacking's words) can be turned into an object with a technical function in a particular context of action. The difference between the shell and the electron is that the reality of the shell as a natural object in no way depends on its use as a technical pseudoartifact. According to Hacking, the crucial criterion for assessing the reality of electrons is their being effective pseudoartifacts in some context of action. But electrons as technical pseudoartifacts have a dual nature: on the one hand, they are physical objects with physical properties; on the other hand, they are objects with a particular function (in some context of action). The fact that the reality of electrons as scientific entities is assessed on the basis of the criterion of being tools (that is, being technical pseudoartifacts) in some context of action does not imply that electrons, as scientific entities, are tools (technical pseudoartifacts) as such. Electrons are artifacts only insofar as they perform a certain function in a particular context of action, and they can perform that function only on the basis of their physical properties. In this regard, there seems to be no fundamental difference between electrons and the object called "mercury thermometer." When we keep in mind the dual nature of technical pseudoartifacts, the use of scientific entities as tools in some context of action poses no threat for the traditional distinction between the natural and the artificial in experiments.

5. Conclusion

In the context of the modern physical sciences, the distinction between natural and nonnatural phenomena (objects) loses its meaning: the behavior of any kind of system (whether human made or not) under whatever conditions (human made or not) is a natural phenomenon. Similarly, all objects, human made or not, fall in the same category: a helium atom, a buckyball, a steam engine—from a physical point of view, they are all of the same kind. Within prescientific thinking the helium atom may be considered a natural object and the steam engine an artificial object. Within the ontology of the physical universe, however, the distinction between natural and artificial objects/processes has no relevance, because there is no room in physical science for a functional description of the physical world. Functional descriptions of part of the world are nevertheless necessary within the context of experimental physics. From a methodological point of view, the description of certain physical systems as technological artifacts appears to be inevitable.

It must be emphasized in closing that the notion of nature involved in modern experimental physics stays rather obscure. To conclude that the behavior of any kind of system (whether human made or not) under whatever conditions (human made or not) is a natural phenomenon is nothing short of concluding that the notion of "nature" has lost all meaning within physics: if every behavior is natural, because it is in conformity with the laws of physics and violation of these laws is by definition excluded, nothing distinctive is claimed by saying that a particular behavior is natural. The natural-artificial distinction loses its meaning. Nevertheless, experimental physics cannot do without the distinction between the natural and the artificial, in spite of the fact that physics is unable to provide a firm foundation for this distinction.

NOTES

I thank Larry Bucciarelli and the members of the Department of Philosophy of the University of Technology Delft for their valuable comments on earlier versions of this chapter.
    1. For a detailed discussion of this issue, see e.g. Dipert (1993).
    2. Aristotle distinguishes between things that exist by nature and those that exist by other causes. Things that exist by nature (i.e., natural things) are characterized by the fact that they

carry within themselves a principle of change. Nature, according to Aristotle, is "a source or cause of being moved and of being at rest" within a thing, which it has by virtue of being that thing (*Physica,* bk. 2, 192^b). In other words, a natural thing is a thing that, by being that thing, carries within itself its own principle of motion or change (or of being at rest). This principle of motion is called the nature of that thing.

3. The following section is partly based on Kroes (1994).

4. For Aristotle, science is essentially a contemplative, theoretical activity; the scientist is a passive observer who tries to grasp necessary and eternal truths about the world.

5. Aristotle, *Physica*, bk. 2, 193^a: "if you planted a bed and the rotting wood acquired the power of sending up a shoot, it would not be a bed that would come up, but *wood.*"

6. For an analysis of the notion of function and its normative aspects, see, among others, Wright (1973), Cummins (1975), Bigelow and Pargetter (1987), Preston (1998).

7. Maybe the same conclusion applies to empirical science in general, even if it is not experimental; assessing the reliability of unaided observation may require analysis and evaluation of the functioning of sense organs.

8. It seems that a real conflict occurs in the context of physicalist-reductionist positions because of a complete mismatch between ontology and methodology: from an ontological point of view, functions would be superfluous for describing the *totality* of the world, whereas, from a methodological point of view, they would be inevitable.

9. Note that, generally speaking, purely functional descriptions of objects are possible.

10. This approach implies a rejection of the physicalist-reductionist position.

11. Normative statements with regard to functional properties of objects constitute a great obstacle for such a reduction; for more details, see Kroes (2001). But apart from that, there are other problems. See, for instance, Mumford (1998); he discusses in detail the distinction between categorical and dispositional properties, which is very close to our distinction between structural and functional properties. In his chapter 8 he analyzes the arguments for and against reduction of dispositional concepts to categorical ones (and vice versa), and reaches the conclusion that "there is insufficient warrant for any form of reductionism" (Mumford 1998, 190).

12. Parts of this section are adapted from Kroes (1994).

13. Clearly, technical artifacts may be discovered in the same sense as physical phenomena: an archaeologist may discover a hitherto unknown ancient technical artifact. That artifact is considered to exist independently of the archaeologist just as physical phenomena are considered to exist independently of the physicist. But the discovery that an object is an artifact implies that that object *as an artifact* was invented (made) by a human being, not discovered.

14. An interesting analysis of the role of anthropomorphic elements in physical theories is contained in Planck (1909).

15. Note that this is also true for technical pseudoartifacts; a creative act is necessary to turn a natural object like a shell into a technical pseudoartifact (a tool for drinking).

16. For more details, see Kroes (1994).

Jim Woodward

5 / Experimentation, Causal Inference,

and Instrumental Realism

Although there is a large philosophical literature on experimentation and an even larger literature on causation and causal inference, there has been remarkably little contact between the two. This is so despite the fact that it is a common view in many areas of science that experiments are a particularly reliable way of finding out about causal relationships and of distinguishing causal from merely correlational relationships. Researchers in the biomedical, behavioral, and social sciences in particular often tend to think of experimentation as the "gold standard" for establishing causal relationships, and they tend to draw invidious contrasts between causal claims that are supported by experiments and those that are based only on passive observation. Nonetheless, most philosophical discussions of experimentation have little to say about the difference between using experiments to test claims that have a specifically causal content and those that do not have a specifically causal content. Similarly, most philosophical treatments of causation leave the connection between the content of causal claims and the role of experimentation completely opaque. Many philosophers (for example, Von Wright 1971; Menzies and Price 1993) who have focused on the connection between causal claims and the results of experimental manipulation (as in so-called agency or manipulability accounts of causation) have advanced highly anthropomorphic and subjective accounts of causation that do not seem to capture the content of the sorts of causal claims that figure in science.

1. A Manipulability Theory of Causation: Basic Ideas

The basic idea of a manipulability theory of causation is that causes are potential means for manipulating or controlling their effects: very roughly, if *C* causes *E,* then if we could wiggle *C* in the right way and in the right circumstances, *E* would wiggle too, and if wiggling *C* in the right way and in the right circumstances is associated with wiggles in *E,* then *C* causes *E.* By and large, the assessment of such theories in the philosophical literature has been quite negative: it is widely claimed that they are naively anthropomorphic and unilluminatingly circular. Few of the philosophical critics of manipulability theories appear to be aware of the existence of a large literature on experimental design and on problems of causal inference in the biomedical, behavioral, and social sciences that emphasizes the close connection between causation and manipulation. Claims to the effect that causal relationships are just those relationships that are potentially exploitable for purposes of manipulation and control are ubiquitous in this literature; here are a few representative quotations. In their seminal text *Quasi-Experimentation* (1979), Cook and Campbell write: "The paradigmatic assertion in causal relationships is that manipulation of a cause will result in the manipulation of an effect. Causation implies that by varying one factor I can make another vary" (36). A similar point of view is expressed by the statistician David Freedman (1997, 116) in the following passage: "Descriptive statistics tell you about the correlations that happen to hold in the data: causal models claim to tell you what will happen to *Y* if you change *X.*" Very similar ideas can be found among economists. For example, Kevin Hoover writes in his survey of neoclassical economics (1988, 173) that the following "definition of cause is widely acknowledged. . . . *A* causes *B* if control of *A* renders *B* controllable."

Are claims of this sort simply naive or confused, or is there perhaps more to be said in favor of a broadly manipulationist account of causation than the philosophical critics allow? In what follows, I will argue for the latter position.

The versions of the manipulability theory developed by philosophers and by the statisticians and the theorists of experimental design quoted above have very different goals. Philosophical defenders of manipulability theories have typically attempted to turn the connection between causation and manipulability into a reductive analysis: their strategy has been to take as primitive the notion of manipulation (or some related notion like agency or bringing about an outcome as a result of a free action) to argue that this notion is not itself causal (or at least does not presuppose all of the features of causality we are trying to analyze), and to then attempt to use this notion to construct a noncircular reductive definition of what it is for a relationship to be causal. Philosophical critics have generally assessed such approaches in terms of this aspiration and have found the claim of a successful reduction unconvincing. By contrast, statisticians and other nonphilosophers who have explored the link between causation and manipulation generally have not had reductionist aspirations—instead their interest has been in unpacking what causal claims mean and how they figure in inference by tracing how they link up with the notion of manipulation, but without suggesting that the notion of manipulation is itself a causally innocent notion. While I fully agree with the philosophical critics that the reductionist version of the manipulability theory is unsuccessful, I will try to show how, if we are willing to give up the goal of reduction, we may develop a manipulability conception that is genuinely illuminating and not unacceptably anthropomorphic or subjectivist. I believe that it is something very like this conception that the statisticians and social scientists quoted above have had in mind.

2. Causal Claims and Interventions

As a point of departure, let us consider the following highly schematic causal inference problem. A researcher observes that two variables X and Y are correlated. She is able to rule out the possibility that this correlation is merely a coincidence and also the possibility that Y causes X, but she is uncertain whether X causes Y or whether instead the correlation between X and Y is due to the operation of some common cause or set of such causes, Z.[1] In a stock philosophical example, X and Y might be, respectively, variables representing the reading of a barometer, B, and the occurrence or nonoccurrence of a storm, S, and the researcher's concern may be that this correlation is entirely due to the common cause of atmospheric pressure, A. Or X might represent whether or not a subject is treated with some drug, Y recovery, and the researcher's concern may be that an observed correlation between X and Y is due to the fact that physicians give the drug to those patients who are healthier or more likely to recover. Another example is provided by the apparently robust correlation in the U.S. population between attendance at a private rather than public school and higher level of scholastic achievement (see, for example, Coleman and Hoffer 1987). Does this correlation show that attendance at private schools causes an increase in scholastic achievement, as many researchers have claimed, or is the correlation instead entirely due to operation of some third set of factors that operate as common causes of both? For example, is the correlation perhaps due entirely to the fact that parents who send their children to private schools tend to be wealthier and more concerned with their children's education than parents who send their children to public schools, and do these variables affect both school choice and scholastic achievement?

One obvious thing a researcher faced with this problem might do is to perform an experiment. In the case of the barometer/storm correlation, for example, one might imagine an experiment in which the position of the barometer dial, B, is manipulated in a way that is appropriately independent of such possible common causes as the atmospheric pressure, and it is then observed whether the position of the dial and the probability of occurrence of a storm, S, remain correlated under this operation. Just what the phrase "manipulate in a way that is appropriately independent" means is an issue that will occupy us below. For the present, let us note that intuitively this sort of independent manipulation might be accomplished by, for example, consulting the output of some random device that is causally and statistically independent of the position of the value of the atmospheric pressure and, depending just on the output of this device, physically fixing the position of the barometer dial in such a way that it is no longer free to move in response to changes in atmospheric pressure. Similarly, one way in which we might try to test the claim that private school attendance causes a boost in scholastic achievement is by conducting an experiment in which students are assigned randomly to one of two groups, in such a way that their attendance at either private or public school is determined entirely by this random assignment and scholastic achievement in the two groups is then measured.

How shall we conceptualize what is going on in these experiments? Why are they good ways of determining whether the causal claims under investigation are correct? Let me begin by observing that it is extremely natural to think of the content of these causal claims as closely connected to claims about what would happen to the putative effect variable Y if experimental manipulations of the putative cause variable X of the sort described above were actually carried out. For example, if changes in the position of the barometer dial were to be carried out in the way described above and if we were to find that the probability of whether a storm occurs changes depending on the position of the dial, this would be excellent evidence that B does indeed cause S. Conversely, if under all such manipulations of B there were no corresponding changes in S, this would be very strong prima facie evidence that B does not in fact cause S. Similarly for the other experimental manipulations described above. The version of the manipulability theory that I want to consider explains this connection by thinking of it as built into the content of the causal claims under test. Thus, in the sort of case under discussion, to say that B causes S means nothing more and nothing less than that if an appropriately designed experimental manipulation of B were to be carried out, S (or the probability of S) would change in value. If the claim that B causes S did not at least imply this conditional, it is hard to see why the experiment described above would be an appropriate way of testing the claim. If the claim that B causes S had some additional content over and above what it implies about the outcome of this experimental manipulation, again it would be hard to see why we should take (as we in fact do) the presence of changes in *S* under manipulation of *B* as sufficient to establish that *B* causes *S*.

What do we mean by an appropriately designed experimental manipulation of *B*? It is clear that some ways of changing *B* are inappropriate for the purposes of determining whether *B* causes *S*. For example, if it is indeed true that atmospheric pressure, *A*, causes both *B* and *S* and we manipulate *B* by manipulating *A*, this will lead to a corresponding change in *S* even if *B* does not cause *S*. Similarly, it would be inappropriate to manipulate *B* via some process that directly changes *S*, via a causal route that is independent of and does not go through *B*. If we are to formulate even a prima facie plausible version of a manipulability theory, we need to characterize the notion of an "ideal experimental manipulation" in such a way that it excludes possibilities of this sort. I will follow an important current in the recent literature on causation in referring to the notion of an ideal manipulation we are looking for as an "intervention." There have been a number of attempts to characterize the notion of an intervention in this literature recently, including Cartwright and Jones (1991); Spirtes, Glymour, and Scheines (1993); and Pearl (2000). I have offered my own suggestions in Woodward (1997) and Woodward (forthcoming a). For purposes of this discussion, I believe that the following informal characterization will do:

> (IN) An intervention *I* on *X* with respect to *Y* (for the purposes of determining whether *X* causes *Y*) is an exogenous causal process that completely determines the value of *X* in such a way that if any change occurs in the value of *Y* it occurs only in virtue of *Y's* relationship to *X* and not in any other way. This means, among other things, that *I* is not correlated with any other variable that also causes *Y* and does not lie on the causal route (if such a route exists) from *I* to *X* to *Y* and that *I* does not cause *Y* via a route that does not go through *X*.

The basic idea is illustrated by the experimental manipulation of the barometer dial described above: what the intervention does is replace the previously existing situation in which the barometer reading is determined endogenously by the atmospheric pressure with a new situation, having a different causal structure, in which the barometer reading is determined exogenously entirely by the randomization procedure, the procedure itself having been carefully designed in such a way that any change in *S* that occurs must result from the change in *B* and cannot be produced in some other way.

Before proceeding, several comments and clarifications regarding interventions and their use to characterize causal relationships will be helpful. First, the notion of "cause" under discussion is the notion of variable *X* such that, given other causal factors or background circumstances that need not be explicitly described or represented, changes in the value of X will result in changes in the value of some other variable Y. This corresponds to what is sometimes called a "partial" rather than a "total" (type) cause. Thus, in the barometer example, it is obviously true that other factors besides the atmospheric pressure A are causally relevant to the position of the barometer dial; manipulation of A can establish that A is a cause of B, but A is certainly not the only cause of B.

Second, note that the connection that I have formulated relates causal claims to claims about what will happen under *some*, rather than *all*, interventions: if X causes Y, then there is some way of intervening on X (some possible intervention on X) that in some appropriate background circumstances will change Y, and if there is such an intervention, then X causes Y. I do not claim (and it is plainly not true) that if X causes Y, then *all* interventions on X will change Y. For one thing, even if X causes Y, it will usually be the case that there are some changes in the value of X that do not change the value of Y because the former changes are outside the range in which the causal relationship between X and Y holds—outside what I have elsewhere (Woodward forthcoming a) called the "range of invariance" of the relationship. For example, although, within a certain range, changes in the extension X of a spring will cause changes in the restoring force F it exerts, in accordance with the familiar Hooke's law relationship $F = -kX$, this will not be true if we extend the spring too much. Under a sufficiently large extension, the spring will break and this will sever any causal relationship between X and F. The formulation in terms of "some" interventions given above correctly concludes that the relationship between X and F is causal, as long as some manipulations of X will change F.[2]

A related issue concerns a choice that we face concerning the characterization of the notion of an intervention.[3] Consider again a manipulation of X that breaks the spring in the above example. One possible view that is in fact taken by several writers (Cartwright and Jones 1991; Pearl 2000) is that such a manipulation should not count as a bona fide intervention—in general, manipulations of X that alter or destroy the mechanism by which X affects Y should not count as "admissible ways of changing X for the purposes of determining whether X causes Y." These writers suggest that we should build into the notion of an intervention the requirement that an intervention on X should not change or destroy the causal mechanism or relationship (if any) connecting X to Y. Call this the mechanism preserving requirement. Obviously (IN), as formulated above, does not impose such a requirement.

The motivation for the mechanism preserving requirement may seem obvious: if our manipulation of X destroys the causal mechanism that connects X to Y, so that Y does not change under manipulation of X, then we may be misled into thinking that there is no causal relationship between X and Y when in fact such a relationship exists.[4] Note, though, that we will make this mistaken inference only if we formulate the connection between causation and manipulation as a claim about what will happen under "all" rather than "some" interventions. If we formulate the connection in terms of "some" interventions, we will not be justified in concluding that X does not cause Y just because there is some intervention on X (a manipulation that destroys the mechanism connecting X to Y) that does not change Y.

One apparent consequence of adopting the mechanism preserving requirement is that in order to determine whether we have carried out an intervention on X, we must have some basis for determining whether our manipulation of X has disrupted the mechanism, if any, connecting X to Y. This in turn seems to require that we already have a certain amount of information about the causal relationship, if any, between X and Y. This introduces a worry about "circularity." To put the matter unsympathetically, on this conception of an intervention, it looks as though we already must know about the character of the causal relationship, if any, between X and Y before we can learn about that relationship by intervening on X. When expressed in this bald way, I think that the "circularity" worry is overstated—for one thing, one can sometimes recognize that a contemplated manipulation of X is likely to disrupt any causal relationship between X and Y, should any exist, without knowing whether there in fact is such a relationship,[5] so that the circularity just described is not always or automatically vicious. Nonetheless, I think that it is also true that we sometimes find out about whether there is a causal relationship between X and Y and about its characteristics by means of relatively "black box" experiments—by manipulating X in circumstances in which we have little, if any, prior information about this causal relationship. In part for this reason, in part because it is often unclear how to determine whether an intervention on a putative cause has disrupted the mechanism connecting it to its effects, and in part because, in view of my discussion in the preceding paragraph, there is no overriding reason for adopting the mechanism preserving requirement, I prefer *not* to build this requirement into the characterization of an intervention and instead stick with just the characterization (IN). If we adopt a notion of intervention like (IN) that does not impose the mechanism preserving requirement, then we will say, for example, that extensions that break the spring can be genuine interventions—it is just that the generalization connecting X to F is not invariant or does not continue to hold under such interventions, although it does hold under others. However, I also think that virtually all of my claims about the connection between causation and intervention in this chapter can also be expressed in a framework in which the preferred notion of intervention conforms to the mechanism preserving requirement, as long as that requirement is understood in a way that allows that one can sometimes determine whether a manipulation on X is mechanism preserving without already knowing whether X causes Y. As we shall see, different notions of intervention are associated with different ways of describing or conceptualizing some of the inductive risks associated with experimentation, but perhaps are interchangeable in other respects.

A third point to notice is that the characterization (IN) is framed entirely in terms of notions like cause and correlation and makes no reference to human beings or what they can or cannot do. Some experimental manipulations carried out by human beings will qualify as interventions, but if so, this will be because of their causal and correlational characteristics and not because they involve human action. Other manipulations carried out by human beings will not qualify as interventions because they lack the features described under (IN). Moreover, processes that do not involve human action or design will qualify as interventions as long as they have the right causal/correlational characteristics. Indeed, the important and philosophically neglected category of "natural experiments" typically involves the occurrence of processes in nature that have the characteristics of an intervention but do not involve human action or at least are not brought about by deliberate human design. As we will see in more detail below, by avoiding reference to human action in the characterization of an intervention, we can avoid the anthropomorphism and subjectivism that has infected the philosophical versions of manipulability accounts of causation. This is not an ad hoc move whose sole motivation is to avoid these difficulties. We need the causal/correlational characterization of interventions (rather than a characterization in terms of some notion like "free human action") if we are to make sense of basic principles of experimental design and if we are to capture the idea that natural experiments as well as those that involve deliberate human contrivance can tell us about causal relationships.

Because the characterization (IN) makes use of causal information, we cannot use it in service of the project, characteristic of traditional manipulability theories of causation, of appealing to notions like agency and manipulation to give a reductive analysis in terms of noncausal concepts of what it is for X to cause Y. Nonetheless, if we adopt the characterization (IN), the claim that X causes Y if and only if the value of Y will change under some intervention (in appropriate circumstances) on X is not viciously circular because the causal information required for the characterization of an intervention on X does not make reference to the presence or absence of a causal connection between X and Y. Instead, the required causal information is information about *other* causal relationships—for example, about the relationship between I and other causes of Y besides X. In other words, to reliably infer in the above examples on the basis of the kind of experimental intervention under discussion, that X causes Y, one must have considerable causal background knowledge, but one does not already have to know the very thing that one is trying to establish—whether X causes Y.[6]

Fourth, it is important to understand that the connection between causation and manipulation being proposed is regulative and hypothetical. The form of the connection is roughly this: X causes Y if and only if there is some possible intervention on X such that if it were to occur, Y would change. It is *not* part of the view I am proposing that X causes Y only if an intervention on X actually occurs. Moreover, while I will not try to explain here what "possible" means in "there is some possible intervention," I take it to be clear that the relevant notion of possibility has nothing to do with presently available technology or what human beings can or cannot do (see section 6). It also must mean something broader than "is physically possible." Thus causal claims concerning the effects of the moon's orbit on the tides are perfectly legitimate even though human beings cannot at present manipulate the position of the moon and may never be able to do so. Similarly, causal claims concerning the past are meaningful even though there is an obvious sense in which, under the assumption of determinism, if the relevant interventions did not occur in the past, they could not have done so (since their occurrence would be inconsistent with laws of nature and the actually occurring initial conditions).[7]

For similar reasons, it is also not part of the view that I am recommending that causal claims can only be established on the basis of experiments. Instead, the connection between causation and manipulation is proposed as an interpretation of what causal claims mean (or what must be the case for them to be true). The idea is that meaningful causal claims must have an interpretation in terms of what would happen in hypothetical experiments and that if a causal claim is true, what is predicted to happen in the relevant hypothetical experiment must in fact be what would happen. Thus, when one engages in causal inference on the basis of nonexperimental data or on the basis of data produced by a nonideal experimental manipulation that does not satisfy the conditions for an intervention, one should think of oneself as trying to establish on the basis of such data and other assumptions what would happen in an ideal experiment in which an intervention does occur. For example, when researchers attempt to establish (as they do) on the basis of purely observational evidence whether attendance at private schools causes an increase in scholastic achievement, we should think of them as trying to predict what would happen in a hypothetical experiment of the sort described above, in which students are randomly assigned to schools. It is only if the data in the observational study and other plausible assumptions provide a basis for such a prediction that we should accept the researchers' causal conclusions.

The advantages of interpreting causal claims as claims about the outcomes of hypothetical experiments are several fold. First, as already suggested, it provides a natural explanation of the connection between experimentation and causal inference when we can do the relevant experiments. Many alternative philosophical accounts of causation fail to do this.[8] Second, particularly within the biomedical, behavioral, and social sciences, it is often quite unclear what researchers mean when they claim the existence of various causal relationships (see section 5). In such contexts, a manipulability account of causation helps to force researchers to clarify and make determinate what causal claims mean and serves to distinguish causal from merely correlational, descriptive, or classificatory claims. The fact that manipulability accounts can be heuristically useful in this way is, I believe, one of the main reasons why the researchers quoted above are attracted to them. For very similar reasons, manipulability accounts are preferable to philosophical treatments that take causation to be an undefined primitive or an "amiable jumble" (in Brian Skyrms's [1984] words) or a "cluster concept" of different criteria that are weighted differently (and can come apart in various ways) in different contexts. One obvious objection to such cluster theories, at least from the point of view of practical methodology, is that they permit or even encourage researchers to be unclear or noncommittal about what the causal claims they are making mean. Finally, while it is not part of the theory I am proposing that states or properties that cannot as a practical matter be manipulated are defective candidates for causes, I think that it is true (see section 5 below) that causal claims to which we cannot even in principle give an interpretation in terms of hypothetical experiments (because, for example, the notion of manipulation of the putative cause seems conceptually ill-defined or because we are given no information about what would follow from such a manipulation) will often lack a clear meaning.

I said above that X causes Z if and only if there is a possible intervention on X that, *in the appropriate circumstances,* would change Z. The italicized phrase must be understood to include, among other things, possible additional interventions on other variables. Consider the following example: X causes Z via two routes—a direct route and an indirect route that passes through Y, as in figure 5.1. (I adopt here the familiar device of representing causal relationships by means of directed graphs—an arrow from X directed into Y means that X is a direct cause of Y. See Spirtes, Glymour, and Scheines [1993] and Pearl [2000] for additional discussion.)

Suppose that, as it happens, the direct causal influence of X on Z is exactly canceled by the indirect causal influence mediated by Y, so that the net influence of X on Z is null.[9] (Cases of this sort may seem artificial, but in fact, a well-

FIG. 5.1

known dispute over the interpretation of a social experiment designed to measure the effect of cash payments on recidivism turned on the existence of just this sort of structure).[10] In cases of this sort, there are compelling theoretical reasons for regarding X as a genuine cause of Z (in some sense of "cause"), even though there are no interventions on X alone that will change Z. As I have explained elsewhere (Woodward 2001; forthcoming a) it is possible to capture the claim that X causes Z in structures like figure 5.1 by talking about what would happen to Z under combinations of interventions, one of which is an intervention on X. In particular, a structure like figure 5.1, with exact cancelation present, will be a correct representation of the causal facts only if there is a combination of possible interventions, one of which involves changing the value of X and the other of which involves freezing the value of Y in a way that makes its value independent of the value of X, which will change Z. Other complex causal systems, such as those involving backup mechanisms that come into play only when some primary mechanism is disrupted, can also be treated in terms of combinations of interventions.[11] Thus more complex structures also carry implications for what will happen in hypothetical experiments, but the experiments in question will be more complicated than those described above.

3. Experimentation and Causal Inference

With these considerations as background, let us return to the question of just how experimental interventions figure in causal inference. Following Pearl (1995) and Spirtes, Glymour, and Scheines (1993, 2000), let us restrict our attention to interventions that may be represented graphically as follows:[12] the intervention on a variable V breaks all arrows previously directed into V, replacing these with a single arrow directed from the intervention variable I into V. This corresponds to the idea that the value of V is now determined entirely by I while preserving all other arrows in the system of interest, including those, if any, that are directed out of V. The latter arrows should be, in the terminology of Woodward (forthcoming a), invariant under the intervention. Thus, assuming that prior to the intervention the ABS system has the structure represented in figure 5.2, an intervention on B should replace this structure with the structure in figure 5.3.

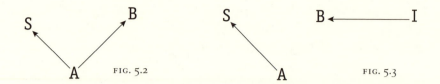

FIG. 5.2   FIG. 5.3

The intervention breaks the arrow directed into B from A but should not break the arrow from A into S (it should not disrupt the causal relationship between A and S), for if it did, the value of S might change even for a fixed value of A under an intervention on B, and this would make it look as though the intervention on B had changed S and, hence, that there is a causal connection between B and S even if no such connection existed. By contrast, an intervention on the atmospheric pressure variable A should preserve the arrows directed out of A into B and S, signifying that we would expect the values of B and S to change under this intervention.

The basic logic of this sort of experiment is exhibited in the following biological example, which is taken from a chapter ("Testing Adaptation by Manipulation") in Rose and Lauder (1996). Thornhill (1992) experimentally manipulated both forewing asymmetry W and pheromone release R (by covering the dispersal gland with glue) in male scorpion flies in order to determine the factors affecting female mate choice F. Relatively low levels of asymmetry are correlated with high levels of pheromone, and females prefer males with both of these characteristics. Thornhill found that when pheromone release was experimentally blocked, females no longer preferred more symmetric males. When pheromone was not blocked and wing asymmetry was experimentally manipulated, females did not prefer more symmetric males. Because of the correlation between W and R, passive observation cannot tell us which of these variables causally influence female choice. By contrast, the experiments seem to reveal that R and not W is causally influential. Commenting in a more general way on the significance of experiments of this sort, Rose and Lauder write:

> Manipulations of the phenotype are particularly useful in the study of signal-receiver relationships because a single character of an individual can be altered in order to determine whether that one character alone is a target of sexual selection and thus affects fitness. One important experimental advantage of manipulation of a male signal is that phenotypic and genetic correlations between the trait of interest and other male traits can be broken. In many cases, phenotypic manipulations have allowed hypotheses to be tested that would otherwise be difficult to evaluate. For example, when male traits covary, and it is not possible to identify all traits that might be important to females, statistical association between female responses to males and male characteristics can lead to misleading inferences about which traits are targets of female choice. (Rose and Lauder 1996, 163–164)

We can represent Thornhill's experiment graphically as follows. In the original system, the level of pheromone release R and wing asymmetry W are correlated—most likely because both are influenced by a cause or set of causes G. These might be either a common suite of genes or some common environmental influence. Another possibility—presumably less likely—is that W directly influences R. These two possibilities are represented in figures 5.4 and 5.5.

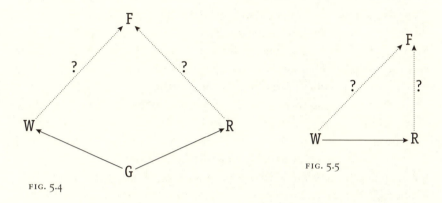

FIG. 5.4

FIG. 5.5

Both R and W are in turn correlated with female preference F, and the question is whether either R or W causally influences F. (We represent this by means of a dotted arrow punctuated by a question mark from W to F and from R to F). If figure 5.4 is the correct structure, it is reasonable to assume that the experimental intervention on R will break the causal connection between G and R, since the level of R is now set by the experimenter's glue. Similarly, if the correct structure is figure 5.5, the intervention on R will break the connection between W and R. In both cases, it is assumed—again with considerable plausibility— that the intervention on R does not also affect and is not correlated with W. Under these conditions, if—as in fact is the case—F changes under manipulation of R, we are justified in concluding that this change in F could only be due to the change in R—hence that R causally influences F. Notice that, as is suggested above, to establish this conclusion it is sufficient to show that some intervention on R alters F, not that all such interventions do.

Let us turn now to the possibility that forewing asymmetry W influences F. If figure 5.4 is the correct structure, we expect that the intervention that manipulates W will break the causal connection between G and W and also will not be correlated with R so that if F does change under this intervention, we may conclude that the change is caused by W. If figure 5.5 is the correct structure, we need to be concerned in principle about the possibility that the intervention on W also changes R (which we know from the above experiment also influences

F), but we can see that this possibility is not realized, since the result of the experiment is that F does not change under the intervention on W. Note that on the account of the connection between causation and behavior under interventions that I have advocated, the failure of F to change under any single intervention on W is not in itself decisive evidence that W does not cause F, since it is compatible with the possibility that there is some other manipulation of W that would influence F. To establish the conclusion that W does not causally influence F, one would need to examine a range of different manipulations of W, showing that none of them are associated with changes in F and/or provide a reason to think that the failure of F to change under the manipulations of W actually carried out would also hold for other interventions on W—at least for those interventions that fall within some ecologically natural range.[13] Alternatively, if one adopts the mechanism preserving requirement on interventions, one needs to consider the possibility that although F does not change under the interventions on W that the experimenter carries out, W nonetheless affects F, but (contrary to what the experimenter apparently supposed) the chosen way of experimentally manipulating W somehow interferes with the mechanism or means by which W affects F. Whichever conception of intervention is adopted, it seems that the appropriate response to this worry will be largely the same—explore other manipulations of W to see if they are associated with changes in F and consider whether there are any reasons to suppose that if the particular interventions on W that have actually been carried out do not change F, then no other interventions will. (Information about the means or mechanism by which W might influence F, assuming that it influences F at all, will be relevant to this latter issue.)

Of course whether any actual experimental manipulation has all of the features of an intervention is an empirical question. It is entirely possible that the actual manipulation of some variable V with respect to some putative effect E is not "surgical" in the way required by the characterization of an intervention and that this manipulation affects other variables besides V (and those that are causally downstream from it) that in turn effect E or that the intervention affects E via some route that does not go through V. To the extent that this is so and the experimenter is unaware of these additional effects or unable to correct for them, causal inferences drawn from the experiment may be misleading. This would happen in the example described above if the manipulation that the experimenter believed just changed R somehow also altered W and in fact W but not R causally influences F. Like every other experiment, Thornhill's rests on causal background assumptions that may well prove to be mistaken.

For this reason, among others, it is not plausible to claim, as some methodologists do, that experiments are infallible or automatically reliable ways of finding out about causal relationships or that they are always superior to causal inference based on passive observation supplemented by appropriate background assumptions. Rather, what can be said on behalf of the use of experiments in causal inference is this: they can sometimes provide us with evidence about causal relationships that is not available when we are confined to a passive observation and have no opportunity for manipulation. They do this by physically altering or rearranging the natural world in such a way that it provides evidence about causal relationships that is not provided by nature in its unaltered or unmanipulated form. Moreover, the background assumptions whose truth is required for reliable causal inference on the basis of experiments (that is, assumptions about the causal characteristics of the experimental manipulation) are often different from the background assumptions required for reliable causal inference from nonexperimental data. It is not uncommon for assumptions of the former sort to be more secure than assumptions of the latter sort, and when this is the case experimentation will be a superior way of finding out about causal relationships.

In his classic discussion of experimentation in *Representing and Intervening,* Ian Hacking (1983a) emphasizes the role of physical manipulation of nature in the creation of new phenomena—phenomena that may not have previously existed. The idea that I have been trying to emphasize is closely related in one respect but different in another. My suggestion, which I take to be illustrated by the above examples and which is nicely captured by the arrow-breaking representation of the effects of interventions, is that often causal factors as they happen to occur in nature will be arranged in such a way that purely observational evidence consisting of correlations can be misleading about causal relationships —the presence of correlations can suggest causal relationships that are not in fact present, as when the correlation between wing asymmetry and female mate preference wrongly suggests that the former influences the latter. It is also possible for the absence of correlations to suggest that causal relationships are absent when in fact they are present, as in the example described in figure 5.1. In both sorts of cases, an important role for experimentation is to break or disrupt certain causal relationships that occur naturally and in this way change correlational relationships so that other causal relationships (or their absence) can reveal themselves in nonmisleading correlational evidence. It is just this idea that we find expressed in the passage about the advantage afforded by phenotypic manipulation in the study of sexual selection quoted above.

In one respect this feature of experimentation involves, as Hacking claims, the creation of something new, or at least the destruction of something old, for patterns of correlational relationships are changed with the experimental manipulation, and some previously existing causal relationships are disrupted. However, as I see it, the way in which we should conceive of the aim or underlying point of the experiment is that all this is in the service of discovering causal relationships that are assumed to be there all along—relationships that are revealed but not created by the experiment. That is, what the experiment aims at is not the systematic creation of an entirely new set of causal relationships but rather the disruption of some previously existing relationships and the preservation of others in such a way that the latter become more readily detectable. Again, the arrow-breaking representation that tells us to think of the intervention as breaking some arrows while preserving others makes this idea explicit. Of course, as emphasized above, it is an empirical question whether any particular experimental system bears the relationship to an unmanipulated system described by the arrow-breaking idea. But it is hard to see how experimentation could ever be a useful way of finding out about causes if it were always the case that all of the causal relationships in the manipulated system were created or altered by the intervention or if the manipulated system bore no systematic relationship to the unmanipulated system. Using experiments to learn about causes requires that some relationships remain stable or invariant across the manipulated and unmanipulated systems.

To put the same point in a slightly different way: when we engage in the sorts of experiments described above, we assume we can learn about causal relationships in naturally occurring systems by creating or finding systems that are "artificial" or have been "disrupted" in various ways and observing what happens in them. We assume that while such artificial systems will lack certain causal relationships that hold in natural or unmanipulated systems and may in this respect be far from exact replicas of those systems, they nonetheless will still contain other causal relationships that are identical or very similar to those at work in naturally occurring systems and whose operation is revealed in the experimental setting. This assumption, which at least some commentators see as one of the key methodological changes that characterizes the kind of modern experimental science that begins with Galileo, contrasts with the idea, sometimes ascribed to Aristotle, that artificial systems created by manipulation have little to teach us about naturally occurring processes.[14] Thus the "Galilean" idea that I have been defending assumes that not all ways of intervening to manipulate some variable X automatically disrupt the causal relationship between X and its effects—for if it did, experimental investigation of the causal powers of X by manipulating it would make no sense. Similarly, it assumes that intervening on X does not automatically create a new causal relationship between X and Y where none previously existed.[15] More generally, the experimental methodology under discussion goes along with a picture of nature according to which it (or at least those portions of nature that are susceptible to the kind of experimental investigation under study) is composed of separate mechanisms or separate sets of causal relationships that can be locally interfered with in a way that does not automatically disrupt other mechanisms or causal relationships.[16] To the extent that systems lack this kind of modularity or decomposability (at least at the level at which we are modeling and investigating them), there will be limits on the extent to which we can learn about causal relationships in them experimentally.

4. Two Inference Problems

I can make this account of the role of experimentation clearer by distinguishing two kinds of problems that are often conflated in discussions of testing and inductive evidence. Consider the following causal structures.

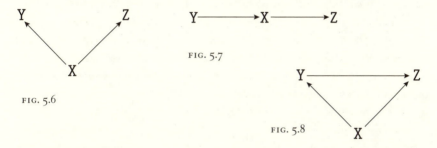

FIG. 5.6

FIG. 5.7

FIG. 5.8

All of these systems are capable of generating the same pattern of correlations among X, Y, and Z. Indeed, even if we make the standard assumption linking causal structure to conditional independence relationships—that the effects of a common cause are independent conditional on the common cause and so on[17]—we cannot determine on the basis of purely correlational information which of figures 5.6 or 5.7 is correct, although we can distinguish both from figure 5.8. In both figures 5.6 and 5.7, Y and Z will be correlated and become independent conditional on X. Suppose now that a researcher observes the correlations among X, Y, and Z but does not intervene in whatever system is generating these correlations. One problem faced by the researcher is this: what reason, if any, is there to suppose that these correlations will continue to hold in the future? This is the standard problem of induction, as it is understood by many philosophers: why should we expect that (and in what respect will it be true that) the future will be like the past? Which of the correlations we now observe are *projectable* (in the sense of Goodman 1955) into the future? This is a special kind of "prediction problem"—a problem of prediction given the assumption that there are no interventions or other sorts of changes in causal structure and that whatever causal structure generates the correlations that we see remains stable.

It is important to understand that this prediction problem is fundamentally different from the problem of ascertaining which of the causal structures underlying the observed patterns is correct or is responsible for generating that pattern of correlations. To have good grounds for believing that the observed pattern of correlation between X, Y, and Z will continue into the future, I need not know what causal structure is generating those correlations—I need only have good reason to believe that whatever structure is responsible will persist undisturbed into the future and that the correlations I observe are representative in some appropriate way. As an illustration of this sort of prediction problem, consider predicting on the basis of a small random sample drawn from the U.S. population a week before a presidential election what the outcome of that election will be. Such predictions can be surprisingly accurate even though they are not accompanied or supported by any causal theory of why people vote as they do. Or consider inferences from relatively local cosmological observations conjoined with some sort of Copernican principle to the effect that there is nothing special about our position as observers to conclusions about uniformities (for example, about the mass/energy distribution at a sufficiently large scale) in the universe as a whole—uniformities that are not themselves laws. Or consider projecting the regular pattern exemplified by alteration of the seasons into the future in the absence of any understanding of the causes of this alteration.

As these examples illustrate, even if I know exactly which correlations will persist into the future, or even if I know that the correlations I observe locally will hold very widely in other spatial regions, this information need not (and on plausible assumptions will not—see below) tell me which causal structure is correct (that is, which structure generates those correlations). This is because, on the assumption that the manipulationist account of causation developed above is correct, the problem of determining the correct causal structure is the quite different problem of predicting what would happen (what different correlations we would observe) if whatever structure is generating the observed correlations were to be disturbed or changed in certain ways—for example, by an intervention. Thus, on a manipulationist account of causation, what figure 5.6 claims is that there is an intervention on X that will change both Y and Z but that an intervention on Y will not change Z or X. By contrast, figure 5.7 claims that an intervention on X will change Z but not Y and that an intervention on Y will change both X and Z. Although figure 5.6 and 5.7 agree about the statistical dependence and independence relations that we will observe, assuming that no interventions occur and both systems continue to operate undisturbed, they disagree about what would happen if various interventions were to occur. It is only if we make the very strong assumption that the right sorts of interventions or other sorts of changes in figures 5.6 and 5.7 will occur (or if we make other assumptions about the way that figures 5.6 and 5.7 happen to be embedded in larger patterns of correlation)[18] that we can be assured that correlational evidence that will discriminate between figure 5.6 and figure 5.7 will occur. Of course this may happen—for example if there is an experimenter around with the desire to intervene and the right technology for doing so, or if nature should just happen to produce the right sort of natural experiment—but nothing guarantees that it must happen. To the extent that the requisite interventions or other sorts of changes that would produce the correlational relationships that would distinguish different causal structures remain merely possible and not actual, the problem of inferring causal structure cannot be identified with the problem of determining which correlational relationships are projectable. More generally, it is simply a mistake, although an extremely pervasive one in philosophy of science, to identify the problem of determining which generalizations describe causal relationships with the problem of determining which generalizations are projectable. One of the virtues of a manipulability account of causation is that it allows us to see this clearly.

The contrast between figures 5.6 and 5.7 illustrates two of the central ideas of this paper. First, we can cash out the content of different sets of claims about causal relationships in terms of the different claims they make about what would happen under different hypothetical interventions. Second, experimentation can sometimes allow an investigator to discriminate among different causal structures in circumstances in which evidence from passive observation does not.

5. Causal Claims as Predictions About the Outcomes of Hypothetical Experiments

A manipulability theory requires that causal claims should be interpretable as claims about the outcomes of hypothetical experiments. This in turn suggests a way of drawing the contrast between causal (or, if you like, explanatory) and merely descriptive or classificatory theories. It also suggests that to the extent that the hypothetical experiment associated with a causal claim is left unclear—either because the relevant manipulation is not well defined or because what is being claimed about what would happen under this manipulation is left unspecified—the causal claim itself will be unclear and ambiguous. While we should not be skeptical of causal claims merely because the associated manipulations are not technologically possible, we should be skeptical of such claims when we cannot specify even in principle what would happen under some associated hypothetical manipulation.

The idea that one mark of the difference between descriptive and causal theories is that the latter but not the former provide information that is potentially relevant to manipulation is a commonplace within some areas of social science (recall the quotations at the beginning of this chapter) and also in molecular biology. A typical statement of the biological version of this idea can be found in Robert Weinberg's (1985) discussion of recent developments within molecular biology. He tells us that "[b]iology has traditionally been a descriptive science," but that because of recent advances, particularly in instrumentation and experimental technique, it is now appropriate to think of molecular biology as providing "explanations" and identifying "causal mechanisms." What does this contrast between description and the identification of causal mechanisms consist in? Weinberg explicitly links the ability of molecular biology to identify causal mechanisms with the fact that it provides information of a sort that could in principle be used for purposes of manipulation and control. According to Weinberg, biology is now an explanatory science because we have discovered theories and experimental techniques that provide information about how to intervene in and manipulate biological systems. Earlier biological theories—for example, traditional systems of classification of plants and animals—fail to provide such information and for this reason are merely descriptive. As Weinberg puts it, molecular biologists correctly think that "the invisible submicroscopic agents they study can *explain,* at one essential level, the complexity of life," because by manipulating those agents it is now "possible to *change* critical elements of the biological blueprint at will" (Weinberg 1985, 48; my emphasis). These remarks exemplify the view of causation (and explanation) characteristic of a manipulability theory.[19]

Some theories in the biological or social sciences are descriptive in the sense that they do not even purport to make causal claims. In many other cases, however, theories may seem to make (something like) causal claims, but it is unclear how to associate well-defined hypothetical experiments with those claims. This is true, for example, of claims that attribute causal efficacy to being a member of a particular biological species or racial or ethnic group. Thus, on a manipulability theory the claim that being a raven "causes" some particular animal to be black lacks a clear meaning, given the plausible assumption that there is no well-defined operation of intervening to change a particular raven into a member of some other species. It would be more perspicuous to replace this claim with a claim that instead makes reference to features of the raven (genes, regulatory networks, biochemical pathways) for which there is a well-defined notion of manipulation and which are such that altering these will alter pigmentation.

In other cases (for example, being female causes one to be discriminated against in hiring and/or salary), the problem is not so much that under all interpretations of the putative cause ("being female") we lack any clear idea of what it would be like to manipulate it, but rather that there are several rather different things that might be meant by manipulation of "being female" (which is to say that there are several quite different variables we might have in mind when we talk about being female as a cause), and the consequences for salary discrimination of manipulating each of these may be quite different. Distinguishing among these different hypothetical experiments can help clarify the meaning of the original causal claim. For example, one natural suggestion is that the intended causal variable in the above example is not really gender but rather something like the beliefs and attitudes of prospective employers and that we should interpret the original causal claim as the claim that if one were to intervene to change an employer's belief that a candidate is male (female) to a belief that the candidate is female (male), in a way that is independent of what else the employer believes about the applicant's qualifications, credentials, background, and so on, the result would be to decrease (increase) the probability that the candidate is hired and decrease (increase) the salary that the candidate is offered. Yet another interpretation is that the intended causal variable is something like a set of employment practices and an associated legal, institutional, and cultural framework that may be either discriminatory or nondiscriminatory, and that manipulating these would change salaries and who gets hired. Needless to say, these two hypothetical experiments are quite different from one another and both are in turn different from a hypothetical experiment that involves a literal manipulation of gender, which itself might take several different forms (sex-change operations, genetic manipulation after conception that replaces the X with a Y chromosome, etc.).

Although I lack the space for detailed discussion, I believe that many other theories in the behavioral social sciences are similarly unclear in the causal claims they make because we are given no sense of what would be involved in or what would happen if the alleged causal factors to which they appeal were manipulated. This is true, for example, of many psychological theories that purport to explain certain behavior by reference to systems of internal representations but have nothing informative to say about the conditions under which those representations would change or how behavior would change if they were to do so. It is also true of many theories of development, both at the individual and societal level, that postulate "normal" stages of development but have nothing informative to say, even in principle, about the factors involved in "abnormal" development.[20] Such theories lack the feature to which Weinberg draws attention—they do not identify (at least with any precision) factors such that if we were able to change them we could change other outcomes. Thus, on a manipulability theory, they fail to perspicuously identify causal relationships.

6. Agency Theories of Causation

In previous sections I have assumed that claims about what would happen if various hypothetical experiments or interventions were to occur (and hence causal claims) can possess truth values independently of whether those experiments are in fact carried out. This assumption raises the more general issue of the relationship between a manipulability account of causation and "realism" about causal relationships. This issue has several different strands. One has to do with the question just alluded to: whether on a manipulationist account of causation we should think of causal claims as true or false in circumstances in which the relevant manipulations are not and perhaps could not be carried out. More generally, there is the question of whether the truth of causal claims, conceived along manipulationist lines, is independent of the activities and thoughts of human beings.

Many philosophers have advocated accounts of causation that are antirealist or subjectivist in the sense that they think of causal relationships as built out of two distinct components—a component that is objectively "out there" in nature (this is usually taken to have to do with the existence of certain regularities or correlations, perhaps supplemented with facts about spatiotemporal relationships) and a component that is in some way made up or added or projected onto nature by us and which has to do with facts about our psychology or mental organization (for example, our expectations that certain regularities and not others will persist, or our inclination to organize our experience in certain ways and not others). The core idea of such accounts is that the difference between those regularities that reflect causal relationships and those that do not is not an intrinsic, objective difference but rather has to do with a difference in *us*—in our beliefs or attitudes regarding these two classes of regularities.

Philosophical defenders of manipulability theories have generally thought that such theories lead directly to this sort of subjectivism. For example, Menzies and Price (1993) take their version of a manipulability theory to imply that causation is a "projection" onto the world of our experience of human agency and that causation is thus a "secondary quality" like color—they argue that claims about causal relationships cannot be understood without reference to the standpoint human beings adopt when they act as agents, just as "red" must be understood by reference to a characteristic experience that human perceivers have. Interestingly, critics have also regarded manipulability theories as committed to subjectivism and have taken this to be a powerful reason to reject such theories.

In contrast to both groups, I think that the most plausible version of the idea that causal relationships are relationships that are potentially exploitable for purposes of manipulation leads directly to a conception of causation that is *not* subjectivist in the sense described above but rather realist. Indeed, on a manipulability theory, the sort of subjectivism about causation just described looks not just wrong but crazy.

Consider again the examples discussed in section 2. We want to know whether the observed correlation between X and Y is due to the fact that X causes Y or is instead entirely due to the operation of some set of common causes for X and Y. On the account I have advocated, we should interpret the claim that X causes Y as a claim about the outcome of a hypothetical experiment. I take it that there is an obvious intuitive sense in which it is facts about how the world is and not facts about our expectations or projective activities that determine whether claims of this sort are true or false. It is not the experimenter's beliefs or expectations that determine whether manipulating the barometer dial will affect the probability of the storm or whether manipulating the level of pheromone release will alter mate preference. Indeed, it seems very hard to make sense of the activity of conducting experiments to assess the correctness of causal claims if the truth of those claims is somehow the product of the experimenter's beliefs or expectations or if claims about the outcomes of experiments lack truth values if those experiments are not performed. If the "objective" core of the content of the claim that X causes Y is just the claim that X and Y are correlated and all else is the product of some agent's projective activities, what sense can we make of experiments designed to distinguish the claim that X causes Y from the claim that they are correlated because of the operation of some common cause or because Y causes X? Are such experiments simply roundabout ways of finding out about the experimenter's (or the scientific community's) projective activities? Within a subjectivist framework, what is the rationale for the features of experimental design described above in which we take care to ensure that the experimental manipulation of X is not correlated with other causes of Y and so on? Of course it is true that my views and expectations will influence my *beliefs* about what the outcome of this experiment will be, but (special circumstances aside) this is not to say that they will influence the outcome of the experiment itself, were it to be carried out.[21]

We may drive this point home by focusing again on the structure of the notions of intervention and invariance. Consider a case, not involving psychological causation in an ordinary, unproblematic sense, of an agent who wishes to change Y but cannot change it "directly," who can change X directly and who wonders whether she can change Y by changing X. While there is (let us suppose) a sense in which it may be up to some agent who is interested in whether manipulating X will change Y (and hence dependent on her beliefs and attitudes) whether she chooses to bring about X, it is a presupposition of her deliberation that it is not also up to her whether or not it is true that if X occurs, Y will occur—whether X is or is not a means to Y. Instead, it is a presupposition of her deliberation that if it is possible to change Y by intervening on X, then there must be an independently existing, invariant relationship between X and Y that the agent makes use of when she changes X and, in doing so, changes Y—a relationship that would exist and have whatever characteristics it has even if the agent were unable to manipulate X or chose not to manipulate X or did not exist. In other words, it is built into the whole notion of a manipulation that the agent's activities, manipulative or otherwise, do not somehow create or influence or constitute whether there is a relationship between X and Y that allows us to manipulate Y by manipulating X. The idea of manipulating Y by intervening on X requires that there be something—the X-Y relationship—that is independent of the agent and her mind and not created or constituted by these. For example, when I deliberate about whether to dash across the street in heavy traffic and reflect that if I were to be hit by the speeding bus bearing down on me, I would be seriously injured, I assume that the truth value of this last claim is independent of my choices, deliberative activities, and state of mind. Thus, surprising as it may seem, and contrary to what many philosophers have supposed, a commitment to some version of realism about causation (in the sense that causal relationships exist independently of our minds and practical limitations on our ability to manipulate) seems to be built into any plausible version of a manipulability theory.

On this way of thinking about the philosophical significance of manipulation, rather than (as traditional versions of manipulability theories claim) somehow serving as a reductive basis out of which causal relationships are created or constructed, our practical interests as agents serves, so to speak, to pick out the kind of (independently existing) relationship between X and Y that we are interested in when we worry about whether that relation is causal. What the relationships that we label as causal have in common is that they support potential manipulations in the way described above.

This version of a manipulability theory agrees with philosophers like Von Wright, Price, and Menzies that our notion of causality developed in response to the need to distinguish those situations in which X and Y are correlated and manipulation of X is a way of changing Y from those situations in which X and Y are correlated and manipulating X is not a way of changing Y. However, unlike these authors, I think that it is built into the notion of a relationship that will support manipulations in this way that such relationships would continue to be present even if we do not or cannot manipulate X, or if our beliefs and attitudes were different, or even if we did not exist at all. If it is asked why this sort of agent independence is built into our notion of causation, my response is that any other view of the matter would involve a bizarre and magical way of thinking, according to which our ability to manipulate X or our practical interest in manipulating X or our beliefs about the results of manipulating X somehow makes it the case that a means-end connection comes into existence between X and Y, where this connection or capacity would not exist if we did not have the ability or interest or beliefs in question.

The superiority of this objectivist version of a manipulability theory can be further brought out by a more detailed look at the subjectivist version advocated by Price and Menzies. These writers claim that "an event A is a cause of a distinct event B just in case bringing about the occurrence of A would be an effective means by which a free agent could bring about the occurrence of B" (Menzies and Price 1993, 187). They also claim that "from an early age, we all have direct experience of acting as agents" and that it is this "commonplace experience that licenses what amounts to an ostensive definition of the notion of 'bringing about.' In other words, these cases provide direct nonlinguistic acquaintance with the concept of bringing about an event; acquaintance which does not depend on prior acquisition of any causal notion." They suggest that in this way "agency theory thus escapes the threat of circularity" (Menzies and Price 1993, 194–195). The fundamental problem with this suggestion is that to capture the effects of A we require the supposition that A is realized by an intervention rather than the supposition that "A is realized as a free act" and the former notion cannot be characterized noncausally. Suppose that A is realized via a free act (in the sense of "free" associated either with soft determinism or with libertarianism) and remains correlated with B when produced in this way, but that A is also correlated with C, another cause of B. Then, it need not be true that A causes B. Suppose then that we respond to this difficulty by adding to our characterization of a free act the idea that A must not itself be correlated with any other cause of B. (Passages in Price 1991 suggest such an additional proviso, although the condition in question seems to have nothing to do with the usual understanding of free action.) Even with this proviso, it need not be the case that A causes B if A remains correlated with B when A is produced by an act that is free in this sense, since it still remains possible that the free act that produces A also directly causes B. (Think of a case in which the experimenter's act of administering a drug has a direct effect on recovery via a placebo effect.) Similarly, an action that causes A but is unfree (because it is caused or caused in the wrong sort of way) can nonetheless qualify as an intervention as long as it has the right causal characteristics. Indeed, as we have already noted, causal processes that produce changes in A but involve no human actions at all can qualify as interventions.

Menzies and Price's underlying idea that A causes B if the association between A and B persists when A is given the right sort of independent causal history is correct, but the relevant notion of an independent causal history is a rather complex notion, which is given by the notion of an intervention rather than the notion of a free action. We require the notion of an intervention and not just the notion of a free action if we are to capture the role of experimental manipulation in causal inference.

7. Instrumental Realism

I turn now, by way of conclusion, to some more general issues concerning the connection between experimentation and "realism" and to a brief description of "instrumental realism."[22] The view that I have been presenting takes causal claims to be claims about the outcomes of hypothetical experiments. Theories that make causal claims thus should be understood as codifying a set of relationships that are potentially relevant to manipulation and control. The question I now want to raise is this: suppose that a theory correctly or approximately correctly describes a set of such manipulability relationships—it tells us what will happen in some set of actual or hypothetical experiments. What, if anything, follows from this about the correctness of the claims the theory makes about which entities exist? That is, can we argue from the claim that a theory correctly tells us what will happen in some range of experiments to the conclusion that some of its ontological claims are correct? Or, even if there is no reasoning of this sort that constitutes an argument from independently acceptable premises, is it perhaps true that the correctness of a theory's claims about manipulability relationships requires or presupposes that its associated claims about entities must be interpreted realistically?

Claims about the connection between experimental or manipulative success and realism about entities have been the subject of a great deal of discussion, again much of it prompted by Hacking (1983a). I do not want to enter into the details of this debate. Instead, I want to explore, or at least suggest as a possibility, a different connection between manipulative success and a variety of realism that has to do not with entities but rather with the claims that theories make about causal relationships. The basic idea is this: whatever one's ultimate assessment of the experimental "argument" for realism about entities, it seems fairly uncontroversial that theories are often manipulatively successful even though, from the perspective of later theories, they are fundamentally mistaken about just what the entities are that are being manipulated. Very roughly, theory T_1 may tell us to conceptualize or interpret an experiment or successful manipulation in terms of producing changes in entities of sort E_1, which in turn produces changes in entities of sort E_2. A subsequent (or perhaps merely different) theory T_2 may then reinterpret this experiment as involving changes in entities of a quite different sort E_3 and E_4. Nonetheless, there will be a recognizable sense in which what T_1 and T_2 predict about the measured outcome of this experiment will be the same. In this sense, some portion of what we can describe as the dependency or manipulability relationships recognized by T_1 and T_2 will remain at least roughly the same across changes in fundamental ontology.

Consider, for example, eighteenth-century particle theories of light. These made a number of essentially correct claims about dependency relations that are embodied in Snell's law of refraction, the law of the equality of the angles of incidence and reflection for reflected light, and so on. According to such theories, when we physically manipulate lenses and mirrors or alter boundaries between materials through which light passes, we alter forces exerted by these media and in this way alter the trajectories of the light particles on which these forces are incident. When particle theories were displaced by wave theories in the nineteenth century, beliefs about the ontology or composition of light changed in a fundamental way, but the above claims about dependency relations were retained and supplemented by new claims about dependency relations involving interference and diffraction phenomena. Over the course of the nineteenth century, there were further radical shifts in scientists' conceptions of the nature of light—from longitudinal to transverse waves and from conceptions that treat light as vibrations in a mechanical ether to the late-nineteenth-century conception of light as a series of displacements in an electromagnetic field that need not be supported by any mechanical substratum. There are corresponding changes in ideas about what the fundamental entities are that are being manipulated in optical experiments. In the twentieth century, there are further revolutionary changes in physicists' conception of light—it comes to be seen as made up of photons and as a quantum-mechanical phenomenon. With each such change in theory, new dependency relations are introduced or represented, but dependency relations in earlier theories are in very large measure retained, although they may be revealed as only approximately correct or as having a more circumscribed domain of application than formerly supposed.

As a second example, consider the relationship between general relativity (GR) and an equivalent theory that postulates a flat space-time background and a separate gravitational potential. Actual or hypothetical experiments that alter the distribution of mass-energy within some region of space-time and in this way alter the trajectory of a free particle will be described within the framework of GR as manipulations of the affine structure of space-time. By contrast, such manipulations will be viewed within the flat space-time theory, not as manipulations of the structure of space-time (this is assumed to be fixed and unmanipulable) but rather as manipulations of the gravitational field alone. Many philosophers who are (entity) realists about space-time structure think that there must be a fact of the matter about which of these descriptions is correct. By contrast, the instrumental realist will think that if (as I am assuming) both theories make exactly the same predictions about what will happen in all possible hypothetical experiments, they should be viewed as equivalent representations of the same set of dependency relationships. Again, what matters is that both theories get these relationships right, not which entities they postulate.

If examples of this sort are at all representative, a characteristic pattern in the development of scientific theories is considerable stability and cumulativity in information about what the outcome of various actual and possible manipulations would be, but considerable instability in claims about fundamental ontology. This in turn suggests a view of theories according to which what is worth taking most literally and realistically about them are not their claims about what exists but rather their claims about relational structures and patterns and particularly their claims about how changing one quantity, property, or feature will change some other quantity. On this view, the causal or explanatory adequacy of a theory will have more to do with getting such relational features right than with getting the fundamental ontology right. It is no doubt true that one cannot talk of changing this or that property or feature without having some conception of what is being changed, and to this extent there will be some limits on how misguided a theory can be about what exists and still get some causal or dependency relations right. Nonetheless, as the above examples illustrate, the stability of procedures of detection and measurement across changes in theory and the fact that claims about the value taken by some quantity within one theoretical framework will often be translatable into corresponding or analogous claims within another framework will mean that we can often compare claims about dependency relationships across theories with different ontologies and determine when they are roughly the same.

This attitude toward theories combines elements of both realism and instrumentalism as traditionally conceived. It takes from instrumentalism the view that there are important aspects of what makes a theory good (success in capturing manipulability relationships) that are independent of whether it is right about the fundamental entities in its domain and also a certain skepticism about the ontological claims of even very manipulatively successful theories. More generally, it is "instrumentalist" in the sense of seeing theories as instruments for controlling or manipulating nature, and in its belief that theories that fully agree about manipulability relationships but disagree about ontology do not really represent different ways the world might be. In other respects, however, this point of view contains elements that look realist. To begin with, and in contrast to traditional instrumentalism, there is nothing about the view that requires a sharp distinction between what can or cannot be "observed," in the sense in which that not-very-clear word has come to be used in the philosophical litera-ture. Relatedly, it does not require a fundamental difference in epistemic atti-tude toward observables and unobservables. Instrumental realism is consistent with the view, which I have defended elsewhere (Bogen and Woodward 1992), that while the contrast between what can be detected or measured (at present) and what cannot is often of considerable importance in science, the contrast be-tween what can and cannot be perceived is usually not epistemically significant. It also fits with the view that while measurements of the value of a quantity are sometimes epistemically more secure than determination of its value via "infer-ence" from other information, there is certainly no automatic rule that this is always so; it all depends on the details of the case. The kind of instrumental realism I am describing focuses only on whether a theory gets manipulability relationships right and does not privilege observable over unobservable quan-tities.

An even more fundamental difference between instrumental realism and traditional instrumentalism is this: instrumentalists have generally been skeptical of notions like "cause" and "explanation." They often advocate replacing talk of causes with talk of correlations, or at least they think that what is objective and scientifically legitimate in the notion of causation is captured by the notion of correlation. Relatedly, they tend to think that the task of science is to describe rather than to explain and that scientific theories should be judged just by how well they capture or represent what actually happens (usually just in the realm of observables). (For example, this is the position adopted in Van Fraassen 1980.) On this view of the matter, scientific theories should not be in the business of making claims about what would happen under various counterfactual possi-bilities that will (or may) never be realized. In general, the modal or counter-factual implications of scientific theories should not be taken literally or, if taken literally, should not be believed as true. By contrast, as I have attempted to ex-plain, instrumental realism distinguishes causal from merely correlational claims and thinks that causal claims have an important role to play in science, although it adopts different criteria for the correctness of such claims than for traditional realism. In particular, traditional realists tend to see explanation and the suc-cessful identification of causal relationships as closely tied to getting the ontol-ogy right, while instrumental realism denies this connection. According to the instrumental realist, any scientific theory that makes causal claims (or that of-fers explanations) is inevitably in the business of making claims about what would happen if various counterfactual possibilities were to be realized, and we often have good reason to believe that such claims are true. Indeed, we cannot make sense of experimentation without taking counterfactuals seriously. From this point of view, both traditional instrumentalism and traditional realism place the investigator in the role of a passive observer of nature, able to record the passing scene but not able to intervene in it. They disagree only about how much of this scene may be reliably recorded. When we take intervention and experiment seriously, we are led to a very different understanding of science.

NOTES

Thanks to Hans Radder and the participants in the Amsterdam Conference on the Philosophy of Scientific Experimentation for a number of helpful comments on this paper.

1. My concern throughout this paper will be with so-called type rather than token causal claims.

2. Another case illustrating the need for the existential formulation is a case in which the functional relationship between X and Y maps a number of different values of X into the same value of Y, as when any change in the angular displacement of a dial below ninety degrees leads to a light's being off and any displacement above ninety degrees leads to its being on. Again there is a causal relationship between displacement and the state of the light even though it is not true that all changes in the displacement will change the state of the light. See Woodward and Hitchcock (forthcoming) for additional discussion.

3. I thank Hans Radder for raising an issue in correspondence that led to the inclusion of these remarks. Radder's concern was that the procedure described above, in which the position of the barometer dial is fixed independently of the atmospheric pressure (in an earlier draft I suggested that this might be accomplished by driving a nail through the dial), in effect destroys the barometer (or at least its usefulness as an instrument). How can such a destructive experiment be an appropriate way of learning about nature? In answering this question, we need to distinguish between different kinds of destructive interference. One thing that fixing the position of the barometer dial does is break any causal connection between the position of the dial and its previous causes. For reasons explained above, I think that this sort of destructive interference is not just permissible but often methodologically desirable in an experiment. As the example of the gluing of the pheromone dispersal gland in scorpion flies described below illustrates, this sort of interference helps to assure us that any change we see in a putative effect variable is really due to manipulation of the cause variable rather than being the result of some common cause that influences both. Another sort of destructive interference involves altering the mechanism by which a putative cause may affect its effect. For reasons described in the text, I think that we should also accept as legitimate experiments that do or may disrupt such a mechanism.

4. This motivation is explicit in Cartwright and Jones (1991).

5. An example: You see a wire running from a switch to a light and wonder whether flipping the switch causes the light to go on and off. You may not know whether this causal claim is true—that is what you want to find out—but it is a very plausible guess that *if* the position of the switch causally affects the light, it does so via the wire. Thus an experimental manipulation of the switch that involves severing the wire will not be illuminating for the purposes of determining whether the position of the switch affects the light.

6. Both Rainer Lange (this volume, chap. 6) and Hans Radder (in correspondence) have objected that while it may be acceptable to provide conditions for *testing* the claim that X causes Y that make reference to other sorts of causal information in the manner described above, it is objectionably circular to characterize what it *means* to say that X causes Y in this

way. Instead, for this second enterprise a reduction of "cause" to noncausal primitives is required. This issue is addressed in more detail in Woodward (forthcoming a). Here I will confine myself to two brief remarks. First, every attempt of which I am aware to provide such a reduction has failed, and yet we seem to understand and grasp the meaning of causal claims perfectly well. This suggests that a reductive analysis is not necessary for providing an account of the meaning of causal claims. Second, I do not think that there is any defensible motivation for demanding such a reduction—this demand is rooted in assumptions about meaning and concept-acquisition that are independently objectionable.

7. For additional discussion of what "possible intervention" means in this context, see Woodward (forthcoming a).

8. Consider, for example, the suggestion that what is distinctive about causal processes is that they are continuous in space and time or that they involve the transmission of energy and momentum in accord with a conservation law. Suppose that we have a group of subjects who ingest an ineffective drug but recover anyway, because they are healthier than those in a control group. The ingestion of the drug will involve the transmission of energy via a spatio-temporally continuous process to the subjects who recover. Within the causality as transmission framework, what explains why the design of this experiment is defective and that the drug does not cause recovery?

9. As a more specific illustration, suppose that the causal relationships in the diagram are represented by the equations $Y = aX$ and $Z = bX + cY$. If the coefficients happen to take values such that $b = -ac$, then there will be cancellation along the two routes connecting X to Z and changes in X will have no overall or total influence on Z.

10. See Rossi, Berk, and Lenihan (1980) and Zeisel (1982). In an experiment involving newly released felons, the incidence of recidivism among those receiving cash payments was the same as those in a control group who did not receive such payments. The sociologists who conducted the study argued that payments did reduce recidivism but that unemployment, which was a condition for receiving the payments, increased recidivism and that the two effects exactly cancelled. For discussion, see Glymour et al. (1987) and Cartwright (1989).

11. The stock philosophical example is the backup assassin who will shoot if and only if the primary assassin fails to shoot. Systems exhibiting this sort of redundancy are quite common in biology and pose special problems of experimental design, since finding actual experimental manipulations that will act as (or mimic) combinations of interventions is often not easy. See Keller (2000) for additional discussion.

12. I have glossed over some complications here in order to keep the exposition smooth. In view of our earlier discussion, we cannot assume that all interventions on X will preserve arrows directed out of X. Instead, what is true is that some interventions on X must do this, if the arrows correctly represent causal relationships between X and its effects. In what follows I restrict attention to just these interventions.

13. We need to distinguish between the question of whether R and W causally affect F within experiments of the sort carried out by Thornhill and the question of whether R and W affect F in the wild—that is, in the naturally occurring environment of scorpion flies. The former is an issue about "internal" validity and the latter an issue about "external" validity—see Cook and Campbell (1979) and Woodward (forthcoming a) for additional discussion. My interest in this paper is largely in questions of internal validity, but obviously, if, like Thornhill, we are interested in using the experiment to answer questions about what happens in the wild, we will want to focus on forewing manipulations that correspond to the range of variation that occurs in natural environments.

14. See, for example, Torretti (1999, 3). I take no stand on whether this is an accurate representation of Aristotle's views.

15. This is a nontrivial assumption in many biological contexts.

16. This is called a "modularity" assumption in Woodward (1999, forthcoming b).

17. This is the so-called Causal Markov Condition. See Spirtes, Glymour, and Scheines (1993, 2000) and Hausman and Woodward (1999).

18. For a discussion of the conditions under which causal structures can be identified on the basis of correlational information, see Spirtes, Glymour, and Scheines (1993, 2000).

19. The central role of practical concerns with manipulation and their connection with causal and explanatory claims in molecular biology is also emphasized in Keller (2000).

20. For additional discussion, see Woodward (1980).

21. Of course beliefs sometimes do exert a perfectly ordinary causal influence of a metaphysically uninteresting sort on the outcome of an experiment, as when a patient's beliefs about the likelihood of recovery causally influence whether she recovers. Needless to say, this observation does not support metaphysical antirealism about causation.

22. After presenting an earlier draft of this paper at the Amsterdam conference in 2000, I learned from Davis Baird that he had used the phrase "instrumental realism" in the title of a much earlier paper (Baird 1988). The view that Baird describes by this phrase and defends in his paper is similar in a number of respects to the position outlined above, but readers should consult Baird's (very worthwhile) paper for a more detailed comparison.

I also emphasize that "instrumental realism," as I understand it, describes one attitude or interpretive stance (among many) that one may take toward scientific theories. Whether that attitude (or any other) is justifiable in the case of any particular theory will of course depend on the details of the particular theory and the evidence for it. In particular, I do not claim that stronger forms of entity realism are never justifiable but merely that manipulative success does not in itself justify such realism.

Rainer Lange

6 / Technology as Basis and Object
of Experimental Practices

It is widely acknowledged that the relation between science and technology must be an important topic of any philosophical account of experimental science (Radder, this volume, chaps. 1, 8). Experimental scientists, especially in the natural sciences, construct and use instruments in order to produce, control, and register the phenomena that their publications refer to. With this action and production aspect of experimentation in mind, it is only natural to think of what experimental scientists do in terms of a special kind of learned craftsmanship, thus emphasizing the parallels between science and technology. In the extreme, one might view experimentation as just a branch of technology that aims at the production of those technical effects we call experimental phenomena, just as chemical technology, for example, aims at the production of certain dyes or drugs. Identifying science with experimentation, it might then even seem incorrect to ask for the relation between science and technology because that would imply that science is something external to technology. Instead one would ask, What special kind of technology is it that we call science? In line with this, some social scientists have argued that there is no special kind of "scientific" rationality that distinguishes science as a kind of productive endeavor from extrascientific practices (e.g., Knorr-Cetina 1981). Such a reductionist approach seems implausible for several reasons, one of which is that it fails to take account of the role of theory in experimental science. Nonetheless, it has exerted considerable influence on parts of the sociological branch of science studies. This makes it appropriate to argue against any simplistic, reductionist version of the "science as technology" thesis from a philosophical point of view.

On the other hand, the consequences of the fact that there is indeed a lot in experimental science that roots in technology remain to be charted. A philosophical account of experimentation that does not succumb to a reductionist reading of the "science as technology" slogan but attempts to reap some insights from it would therefore need to characterize experimental science both by what it has in common with nonscientific technology and by what sets it apart.

1. Experimental Instructions as a Special Kind of Recipe

In order to clarify the relation between experimental and extrascientific technological practices, it is helpful to take a look at modes of action that could count as extrascientific analogues of scientific experiments. In doing so, I do not mean to suggest that whenever we talk of "experimenting" in an everyday context rudimentary scientific activities are performed. There is, however, a class of actions that bear a close resemblance to scientific experiments, and it is to these that I shall turn. These kinds of actions belong to a human being's repertory for problem solving or, more precisely, for solving technical problems. One main constituent of a technical problem is an aim or purpose—that is, a state of affairs the agent desires to be realized. But this is not enough to identify a technical problem. Equally important is the agent's present situation, which includes various kinds of constraints, such as other purposes that may not be violated in pursuing the one presently in question, limits set by time and physical possibilities, and available means. The solution to a technical problem is a course of action that allows the agent to realize his aims given the circumstances that define the particular problem. A practical problem, by contrast, is a problem that calls for a modification in the agent's overall set of aims and purposes.[1]

Many technical problems can be solved by drawing parallels to previous ones. The new problem is classified as one more instance of a known problem type and a particular bit of know-how is applied more or less straightforwardly. Much more interesting, however, are "new" problems representing either genuinely new problem types or instances of known ones that for some reason cannot be recognized as such. Several strategies are available in such a case. Most simply, one can proceed by trial and error, knocking here, turning that screw, thumping on the back of the television, etc. Being able to learn from experience, human beings extend their problem-solving capacities as a result of such processes. What is important in the present context is that they do so not only individually but also collectively, and for that purpose they make use of their most conspicuous invention, language.

When confronted with a technical problem they do not know a solution to, people do not in general resort to trial and error. Rather, they address others who are more competent, or at least expected to be so, and ask for instructions. Traditions of craftsmanship depend on the ability to produce schematic instructions applicable to all instances of a particular problem type and hand them down from masters of their trade to new generations of apprentices. These schematic instructions may be called recipes. They rely on the capacity to understand language in the prescriptive mode. Usually, recipes consist of constant and variable parts. For instance, a recipe for a meal might contain a variable designating the number of people who are to be fed and specify the amounts of ingredients as functions of that variable. There is a number of a priori norms for recipes—for example, the principle of methodical order,[2] which requires that if to reach one's goal it is necessary to perform several steps in a particular sequence, that sequence must not be violated in the recipe (neither ought a description of the outcome of that course of action imply that the sequence has been violated).

Now what does this have to do with experiments? If we take into account the embedding of recipes in complex social practices, it allows us to define a type of action that might be regarded as representing an extrascientific or Lebenswelt analogue of scientific experiments. Recipes have to be invented, tested, modified, and corroborated. An extrascientific or Lebenswelt experiment, I suggest, is an action that is performed in order to establish a new recipe. More precisely, it is a kind of mock performance of the course of action prescribed by the recipe designed to exhibit faults in the recipe that result from incompleteness, redundancy, bad definitions, or, most important, unfeasibility.

The difference between Lebenswelt and scientific experiments is, of course, that the latter are not overtly performed in order to establish recipes but in order to establish scientific laws. There is a position in the philosophy of science from which this difference can seem negligible. It identifies laws with recipes by means of semantic reduction, amalgamating experimental practices from both spheres in the process.[3] That position, however, which might be labeled naive operationalism, has long been known to be a failure. What I want to suggest is that this is due not only to its reductionist pretensions but also to its simplistic view of experiments.

If we take a closer look at what scientific experimenters do, it is easy to see parallels with extrascientific or Lebenswelt experiments. Typically, we can distinguish three phases in an experiment. First, we have a phase of preparation in which complex devices are built, arranged, and adjusted to the particular task;

in addition, a test object—for example, a test person or a laboratory animal—is prepared and brought into a specified relation to the apparatus. This whole process results in what might be called an experimental setup. The next phase begins with an intervention by the experimenter that counts as the starting of the experiment, after which it is left alone for a certain time span.[4] At the end of that process—an end that either is effected by a second intervention on the experimenter's part or is defined by some kind of predetermined event—the last phase of the experiment begins when some aspects of the changes that have occurred in the experimental arrangement are recorded as the results of the experiment, a process sometimes called data collection. A single instance of starting an experiment, letting it proceed along its expected course, and recording the results is called an experimental run. It is the atom of experimental practices, as it were. Obviously, a single experimental run is meaningless to the scientist.

For the moment, let me notice one important thing: the whole procedure I have just described, very schematically, can also be the subject of a special kind of prescription, a recipe for an experiment I shall call an "experimental instruction." Such an instruction is not identical with an experimental law, and it is not the goal of experimental science to produce new experimental instructions. They are the contents of laboratory handbooks, they are featured in textbooks, and rudimentary versions can be found in the "materials and methods" sections of published articles in the laboratory sciences. They clearly belong in the realm of technology, listing the devices to be used and giving concrete values for temperatures, intervals of time, etc. Although it is debatable whether this is true in practice, in theory, at least, an experimental instruction should be so detailed as to enable a student or lab technician to reproduce the experiment in question autonomously even if he or she does not fully grasp the theoretical background.

It is instructive to contrast the notion of an experimental instruction with Radder's notion of the description of the material realization of an experiment (Radder 1996, 12–14). He asks us to imagine an experimental scientist who delegates the actual carrying out of a particular experiment to someone who is a layperson in the respective field of science. For that purpose, he has to find a way of instructing his aide in simple, Lebenswelt terms that do not presuppose any knowledge of experimental science—an instruction, as Radder writes, in a "common language" that is presumably very close to ordinary language, although it may include some special terms experimenter and layperson agree upon. Obviously, this sort of instruction is rarely used in practice, just as the situation Radder asks us to imagine rarely occurs in actual experimental science. The purpose of the imaginative exercise is to illustrate the distinction between the theoretical description and the material realization of an experiment. By contrast, experimental instructions as I understand them are an integral part of actual experimental practices. They are not addressed to laypersons but are instructions by and for specialists who do make use of specialist vocabulary ("plasmid prep," "EcoRI," "ultracentrifuge," etc.) and require some sort of training on the addressee's part. What they do not require, however, is commitment to the theoretical suppositions that the original experimenter invests in his claim that his experiment is a successful test of a particular experimental law. They are not completely theory-free, but they are, as one might say, maximally noncommittal. They make it possible for scientists to separate two issues: whether the experiment has been correctly performed and whether it has been correctly interpreted.

There are a number of philosophical points to be made about experimental instructions. First, they are conspicuously stable under theory change. This is implied by the thesis that they are maximally noncommittal and is part of the empirical basis of the claim, captured in a well-known quote of Hacking (1983a): "experimentation has a life of its own." Secondly, just as with extrascientific recipes, it takes a lot of testing and other work to develop a new experimental setup and establish a reliable instruction for reproducing it. As has been said above, this is not the overall purpose of experimental practices as constituents of experimental science. Nonetheless, there are criteria that have to be met by experimental instructions and that can be applied relatively independently from the scientific success or failure that ultimately lies in the validation or falsification of scientific hypotheses. And in line with this, there are experiments in the sense of individual experimental runs that serve no other purpose than to establish such an instruction—that is, a special kind of recipe—for future use. For instance, a purification procedure for membrane proteins is supposed to produce the raw material for an analysis by electrophoresis in a polyacrylamide gel, which in turn is meant to produce clear, sharp, and therefore easy-to-interpret bands. An important factor in this process is the use of detergents in order to bring into solution proteins that are immersed in the lipid bilayer of the membrane. Which particular detergents are to be used in which concentrations and in which particular sequence during such a complex procedure, however, is a question that cannot be answered by deducing a recipe from general principles. Of course, experience provides rules of thumb; but whenever a new class of proteins or a new kind of membrane is under investigation, researchers have to adapt their procedure to the new task in a process of trial and error. The individual trials in this process take the shape of an experimental run, but they do not result in the corroboration or falsification of a scientific hypothesis. The experimenter does not even need to claim that the outcome of each of these experimental runs is reproducible. They are just steps in the process of learning how to cope with the new experimental process in a way that allows the experimenter to produce reproducible experimental results in the future and instruct others to do so.

Finally, acknowledging the role of experiments in the scientific process, we can reformulate one of the central questions for a philosophy of experiment as follows: How are experimental instructions related to the theories the scientists test? In order to address that question directly, a detailed account both of experimental instructions and of scientific theories is needed. Even without providing these, however, much can be learned about the importance of experimental practices and their relation to technology by looking at the relation between experiments, as prescribed in experimental instructions, and experimental laws. On the one hand, experimental laws, I suggest, are not identical with theoretical laws. Rather, they are empirical generalizations covering classes of phenomena, some instances of which can be produced in an experimental setup. They can be interpreted as causal laws. On the other hand, they are not identical with recipes or instructions for experiments either. So we can break down our question and ask, first, How are experimental instructions related to experimental laws, and, second, How are experimental laws related to scientific theory? In the rest of this chapter, only the first question will be addressed. This is not meant to imply that experimental practices can be fully understood without taking into account the role of theoretical science. However, focusing on the relation between experimental instructions and experimental laws will suffice to explicate the thesis that experimental practices are technological practices of a special kind, characterized by the phrase that technology is both their basis and their object.

In order to explicate the relation between experimental instructions and experimental laws, we will now take a look at the criteria of success that govern the establishing of an experimental instruction, which in turn provide the connection to extrascientific technology. The crucial link is an interventionist conception of causality.

2. Technology Self-Applied: The Role of the Principle of Causality

Historically, the interventionist conception of causality has been developed in an attempt to cope with certain problems that the empiricist so-called regularity conception seemed unable to deal with. Most important, it seemed that this conception, which has its roots in Hume's discussions in the *Treatise,* could not save the distinction between causal regularities on the one hand and epiphenomenal or accidental regularities on the other. Nor did it seem to capture the asymmetry between cause and effect. In order to overcome these perceived defects, Gasking invented and Von Wright further developed a position that explicated the meaning of "cause" and "effect" in terms of agency and related concepts (Gasking 1955; Von Wright 1971).

In essence, Von Wright's account of causality goes like this: If we want to find out about the causal processes in a specified system, we need to know of two states, α and a, of the system with the following properties: a is the initial state of a development of the system (a tree of paths branching from a single trunk) we want to investigate; α is a state necessary but not sufficient for a; and if the system has been put to rest at α, we can induce a transition from α to a. Then by repeatedly observing the development of the system subsequent to its being in state α, either with or without manipulating it to bring about a, we learn which states of the system are necessary consequences of a and which other external states β, γ, ... are sufficient for some of the states following a. But this is possible only if we have the ability to control the system's transition from α to a. Consequently, Von Wright maintains that "we cannot understand causation, nor the distinction between nomic connections and accidental uniformities of nature, without resorting to ideas about doing things and intentionally interfering with the course of nature" (Von Wright 1971, 65–66).

In order to give precise expression to this connection, Von Wright proposes the following distinction between ways of describing actions. We can describe an action by saying what we are doing, such as opening a window or throwing a ball; such descriptions entail state descriptions, and the states thus described are called results of the respective actions. But we can also describe an action by saying what we bring about with it, like letting fresh air into the room or smashing the window pane. Here it makes sense to ask, How did you do that?, implying that there is a way of describing the action in question that does not imply a description of the state in question, and the state brought about is called a consequence of the action.

Given this terminology, Von Wright proceeds to formulate his distinction between cause and effect (Von Wright 1971, 70): "p is a cause relative to q, and q an effect relative to p, if and only if by doing p we could bring about q or by suppressing p we could remove q or prevent it from happening. In the first case the cause-factor is a sufficient, in the second it is a necessary condition of the effect-factor. The factors can become 'relativized' to an environment of other factors. Then the cause is not 'by itself,' but only 'under the circumstances,' a sufficient or necessary condition of the effect." It is important to note that Von Wright anticipated the charge of regress that might render the distinction between doing and bringing about vacuous, and that he had an answer to it: for him there are basic actions of which it is incorrect to say that they have been brought about by some other action. Specifically, it is incorrect to say that we bring about the movement of our limbs by contracting certain muscles, for we do not know how to contract these muscles except by moving our limbs. By this claim, Von Wright means to argue against the supposition that his root distinction can only be saved by invoking causal notions, rendering his analysis circular.

In this respect, Von Wright's account stands in sharp contrast with that put forward by Jim Woodward. Quite correctly, Woodward points to the fact that an interventionist account of causality needs to explicate what an intervention is. He goes on to characterize an intervention as a special kind of causal process.[5] Interestingly, he maintains that this will not render his analysis viciously circular because the characterization he gives of an intervention I on X with respect to Y does not make reference to a causal relation between X and Y, and that is the only one that must not be referred to. But this will not do. After all, what we are after is the very concept of causality; we are not asking for a test that tells us whether or not a causal relation between X and Y obtains, but for an explication of what it means to say that X is causally related to Y. Indeed, this is what Woodward himself repeatedly says in his essay. And for that purpose, to take recourse to a concept of manipulation that in turn has to be explicated in terms of a concept of intervention is all right only as long as no reference is made to causality when explaining what the term "intervention" is supposed to mean.

While there is a circularity here, it might be argued that it is not a vicious circularity—indeed, something of this kind is exactly what is to be expected of an approach to analysis that is meant to be nonreductive. For the sake of clarity, it would perhaps be better to be frank about the holistic consequences, but that is not what I will discuss here. Rather, I would like to pursue the question of why Woodward considers it to be so important to take a nonreductive approach to the analysis of the concept of causality. His reasons for doing so are important because they are not general reasons against reductive analyses per se. What he criticizes a philosopher like Von Wright for, who in his view offers a reductive version of the manipulability account, is not that he is trying to perform a semantic reduction but that his account of causality is "highly anthropomorphic and subjective" and, in lieu of this, does "not seem to capture the content of the sorts of causal claims that figure in science." That criticism is unjustified.

First, let us turn to the presumed subjectivity of Von Wright's concept of causality. I take it that by criticizing that concept as "subjective" Woodward means that it implies a kind of relativist position to the effect that whether or not there is a causal relation between X and Y is a question that receives an answer only relative to an epistemic subject S. But in Von Wright's analysis, the central phrase says that there is a causal relation "if and only if by doing p we *could* bring about q," etc. Von Wright's is thus a modal account, making no reference to any actual subject S that attempts to bring about some particular event

$q.$[6] In other words, it is realistic in the sense that the truth-value of the causal claim under consideration is not dependent upon whether or not there has been, is, or will be someone who tries to find out about it. So, any complaint that an interventionist conception of causality like Von Wright's is subjective is unjustified, if by "subjective" we mean a position of the relativist type. But what about the complaint about "anthropomorphism"?

That complaint, I take it, is just a loose formulation of another one that really brings us to the heart of the matter. Although Woodward declares it a special advantage of his characterization of an intervention that it makes "no reference to human beings or what they can or cannot do," what he really seems to criticize in Von Wright's and similar positions is not the fact that they make reference to human beings, which is doubtful anyway. For the classical interventionist account of causality it is only essential that there is something any epistemic subject can do—that is, something that is in the power of an agent, be it human, animal, or from outer space. The important thing is not that we are talking about members of a particular biological species, but that we are talking about an agent and the special kind of "doing" that is typical of agents, namely, intentional agency. In Von Wright's view, it is not correct to translate "to do" into "to effect," making the position of an agent open to any kind of causally effective process or event. He is quite serious about the intentionalistic vocabulary. In other words, his position and others akin to his retain the connection between the concepts of manipulation and/or intervention, on the one hand, and agency, on the other. And they are right to do so.

Recall that in Von Wright's analysis, it is essential that the system under consideration be put to rest at a state α that is necessary but not sufficient for another state a, into which the system can then either be transformed or prevented from transforming itself by a suitable intervention of the investigator. The point of the exercise is that by varying the circumstances under which a is produced, we can test which of the events that occur after a has come into existence on a particular occasion can generally be brought about by doing a and which others are independent from a. And this is only intelligible as part of an interventionist explication of the concept of causality if the expression "can be brought about" is read as denoting something that someone can succeed in doing or fail to do. Success or failure, in turn, can only be predicated of intentional agency; it does not make sense to speak of a causal process as succeeding to reach its effect state. But on Woodward's analysis, the transition from α to a is characterized in exclusively causal terms. The question then is no longer what can be brought about by bringing the system into state a. Instead, what is tested is which other states follow regularly whenever the system goes from state α into state a. But this is just the typical question of the classical, Humean regularity account, so on Woodward's account the interventionist conception of causality collapses into a version of what it was supposed to provide an alternative to.

To provide an interventionist account of causality that is frank about its intentionalistic core is not only a good strategy to overcome the defects of a regularity theory of causation; it also helps to understand why we are interested in causes at all. It makes it natural to think of experimental science with its interest in causes as growing out of prescientific, technological practices and receiving both its sense of purpose and its core concepts out of that connection.

Historically, there can be no doubt that technology has preceded modern, experimental science. People have always wanted to solve their problems by technical means instead of modifying their ends. There is however also a methodical order here. Experience shows that recipes that have been good enough to solve a problem once or twice do not work all of the times, not under all circumstances. If a particular instance of applying a recipe results in failure, intervening factors or disturbances are sought for. Following the maxim "make sure that it is always possible to bring about q," people start to investigate the sources of such disturbances systematically and eliminate them in an attempt to improve their recipes and adapt them to variable circumstances. In other words, they start performing what has been characterized as extrascientific or Lebenswelt experiments in section 1. As long as this practice is driven and is simultaneously limited by the desire to solve Lebenswelt problems, it remains in the realm of technology. For technical purposes, it is sufficient to have a recipe that allows one to achieve one's purpose most of the time, without bothering about the whys and whens of a low percentage of failed attempts.[7] However, this is not the end of the matter. Generalizing on the concept of a disturbance, technology as the practice of improving means for specified ends can become recursive. That is to say, actual or anticipated disturbances that have no practical importance because of their rate of incidence can nevertheless become regarded as a challenge to the technologist in the pursuit of an abstract ideal of a perfect technology. We thus have a sense of improving on existing technology that transcends the immediate connection with Lebenswelt practices.

It is at that stage at which instruments and processes that are "good enough for practical purposes" are no longer perceived to be satisfactory that technological experiments give birth to experimental science. The crucial step is assuming the principle of causality. Roughly, it says "same causes, same effects," so, by contraposition, if effects differ, there must be a difference in causes. There is a well-known philosophical dispute over the status of this principle (Hartmann and Lange 2000, 88–90). One of the options is to regard it as a universally valid empirical proposition or even a natural law. This is implausible, however, since there is no conceivable situation in which the principle of causality could be proven invalid; any attempt at falsification would have to be regarded as a faulty experiment. Therefore, the principle cannot have any empirical content. Alternatively, the principle might have a transcendental status. The argument for this thesis would hinge on the assumption that its validity is a condition of the possibility of empirical science. Again, this is implausible. No more is needed for experimental science to be possible than that there exist some domains in which causal connections between types of events can be established. This falls short of the universal validity that is required of a transcendental principle. What I suggest instead is to regard the principle of causality as a methodological principle that is constitutive of experimental sciences. As a methodological principle, it has normative status. It tells experimenters what to do—namely, to look for instances of processes that have been started at what have been presumed to be identical conditions but that have nevertheless produced different results. If they find such sets of events, the norm says they ought to identify the factors that have to be controlled in order to (re-)produce identical results on future instances. This is fully compatible with the thesis that there can be events that violate the principle of causality—that is, in which the strategy just described does not work. The principle functions as a norm that delimits the domain of the experimental sciences, and while this domain is constantly and monotonously growing, there is no reason to suppose that one day "everything" will have been subjected to it. The situation in which each and every process complies to the principle of causality is implicit in the practice of experimental science as a regulative ideal, the principle itself can therefore be described as a regulative norm.

A consequence of the principle of causality is that processes and events have to be classified in ways that are nonintuitive. Situations that seem to be identical from a Lebenswelt point of view turn out to differ in causal properties, and others that seem to differ turn out to be equivalent in that respect. It is this pressure that the principle of causality exerts on our everyday classifications that triggers the development of an alternative way of describing processes and events —that is, the mode of description that is usually called scientific and features in experimental laws. By producing descriptions of processes in terms of causes and effects that are not correlated one to one with descriptions of experimental setups, starting events and results as recorded on the level of instructions, experimenters make room for tolerance; they learn to tolerate disturbances of their experimental processes in the sense that, instead of viewing them as something to be prevented or suppressed by any means, which is what the engineer should want to do, they learn to distinguish annoying disturbances from productive ones. While the former have to be dealt with at the level of instructions, the latter mark the point of departure from which new developments take off. The crucial move is to interpret failure of control as indicative of fundamental differences between experimental runs that cannot be captured in the language being used for experimental instructions.

This is where naive operationalism fails. The language of causality allows experimenters to say that although they have done "the same" in the technical sense of having adhered to a particular instruction, they have failed to reproduce the cause of the intended effect. In that sense, the point of the notion of a cause is to introduce a new equivalence relation between actions. But that equivalence has to be validated. For that purpose, new experimental procedures have to be developed that correlate new instructions with the underlying events that hypothetically explain the different outcomes of previous experimental runs. So the thesis that technology becomes recursive in experimental science means that "failed" experiments and causal hypotheses together engender new experiments in the permanent drive for improvement of our control of experimental processes. While the principle of causality can never be perfectly realized at any given moment, it triggers off the development of experimental science as we know it and enforces its permanent growth.

A correlate of this picture of experimental science is a thesis about the relation between science and its so-called applications. According to the traditional, linear model of innovation, the immediate products of science are natural laws—that is, validated universal propositions. Put simply, applied science is supposed to predict the possibility of new processes of technical interest by substituting realistic values for the variables in those laws and combining them into models. It then proceeds to realize in experimental gear what later is developed into a new, preferably marketable product or process. By contrast, on the picture sketched above, the main products of experimental science are not natural laws but the new processes themselves. The validation of laws is an aim that is internal to the practice of science and that is responsible for its continuous growth. In the process of what has been described above as technology becoming recursive, new processes are established and instruments that make it possible to control them developed, and it is to these processes and instruments that are primarily needed for the internal purposes of experimental science which one should turn one's attention if in search of the sources of technical innovation.[8] So while experimental science is an offspring of technology in the historical dimension, improved technology nowadays is a spin-off of the practice of experimental sciences rather than a direct application of its results. The role of applied science is not to apply the laws that fundamental science establishes, but to monitor the development of experimental science and identify those new experimental processes that might prove technically useful.

3. The Nonlocality of Experimental Practices

It has been mentioned in the first section that in technological contexts, recipes serve to transfer bits of know-how from one individual to another. Likewise, in experimental practices, know-how is transported via experimental instructions. But that claim, which has been elaborated in the first section, might raise one worry: have experimental practices not been assimilated too closely to technological ones once again? Can we really get anything recognizably scientific out of them merely by invoking the principle of causality? After all, while recipes are a means to transfer know-how from one person to another, there is no reason to suppose that this always works. Recipes or not, technical know-how might well remain (temporally and spatially) local. And if experimental practices fundamentally share their mode of self-reproduction with technological ones, how can they sustain transsubjective or nonlocal knowledge claims?[9] This worry is enhanced by the thesis from section 1, according to which experimental instructions are instructions by specialists addressed to specialists. If this is so, how do the practices of scientists differ at all from any kind of esoteric ones?

In order to address this question, I will return to a topic that had to be postponed above—namely, the embedding of single experimental runs in what are here called "experimental practices." As I use that term, these practices have much in common with Rheinberger's experimental systems. These are characterized as the "smallest integral working units of research," which, according to Rheinberger, are "inseparably and at one and the same time, local, individual, social, institutional, technical, instrumental, and, above all, epistemic units" (Rheinberger 1997, 28). Since experimental systems lack clear identity conditions, it is difficult to decide whether or not they coincide with what I call experimental practices. In particular, I would deny that experimental practices are local. Also, I would prefer to reserve the term "experimental system" for certain material configurations that are only one intermediate, although reasonably stable, product of experimental practices. Therefore, I will stick to my terminology.

Note first that, in contrast with Pickering (1995, 4), I use "practice" as a count noun. Thus, there is on my concept of practice a number (not necessarily definite) of experimental practices that together comprise, for example, the experimental science of biology. Each of these practices is a historical entity. It encompasses specific instruments, experimental objects, and techniques—what I suggest to describe as an experimental system (that is, the set of material objects and practical abilities necessary to produce instances of a particular experimental setup) —but in addition to these, also modes of cooperation and communication, of apprenticeship and specialist education, etc. Why it is necessary to include these among the constituents of an experimental practice will be the topic of this section.

In order to contribute an original set of results to an experimental science, it does not suffice to establish a new experimental system, demonstrate that it runs, and then produce data. To claim that a proposition such as that describing the results of an experiment is scientific implies that its validity is transsubjective. Transsubjective validity in turn requires that everyone can substantiate for himself what the original claimant said. With respect to an experimental claim this means that everyone must have the chance to perform the experiment in question, at least in principle. And since this is implicit in the claim that the experimental results are scientific, it is in the responsibility of the one who established the new experiment to make sure that it can be reproduced by everybody else, any time, any place. Of course, "everybody" here refers, not to every single individual in the world, but to those who are willing and able to become scientists themselves. While at any particular point in time this is only a subset of all human beings, it is nonetheless an open, indefinitely extensible set. It is this sort of extensibility that sets scientific practices apart from artistic or esoteric ones that can afford to view themselves as a kind of "closed shop."

The means by which scientists ensure that their experiments can indeed be reproduced by others are communication and, importantly, cooperation. Communication of instructions, as of recipes in general, serves exactly that goal. It does not suffice, however. Novel experiments usually require new competencies on the experimenters' part, including practical skills that cannot be acquired merely by reading an instruction. Instructions can only help in learning a new type of action by referring to other, more basic types that the addressee must be supposed to have mastered already. In many instances, these chains terminate in types of action that belong to the repertory of every human being who is able to cope with everyday life, so the instruction can indeed serve its purpose on its own. However, many more complex and specific skills cannot easily be described if one restricts oneself to such a Lebenswelt repertory.[10] The criteria of success for individual performances may be impossible to describe to someone who is not yet acquainted with the whole practice, or they may refer to a unit of practice that is too complex to be transparent to the learner, leaving him to guess whether success or failure are due to his or her skills or to incidental circumstances. These skills thus have to be learned by emulating the example of a competent practitioner. This is where direct, hands-on cooperation comes in. How important it is can be seen from the traditions of summer schools and of mutual laboratory visits, typically by junior scientists who transfer practical knowledge from one place to another and at the same time add to their standing as scientists. The labs of people who have invented a new method—that is, a new experiment type—are usually the preferred goals of such visits.

In addition to skills, there can also be material objects that need to be shared in order to make experiments reproducible in other places. A case in point is the practice of genetic mapping as established in the *Drosophila* system by the Morgan group.[11] An important step in establishing that system was the production of a standard organism, the so-called wild-type *Drosophila,* and a large number of mutants. Each of these stocks had to be bred in the laboratory and thereby literally be reproduced in order to conduct cross-breeding experiments which were the basis for genetic mappings. But since such mutant stocks are, logically, historical individuals (they consist of the descendants of a particular pair of flies), neither the exact reproduction of an experiment involving such a stock nor any variation upon it is possible for someone who has no access to *Drosophila* individuals from that particular stock. The *Drosophila* people therefore had good reason to invest much effort in establishing what Kohler calls the Drosophila Exchange Network and describes as having had such a great importance that its rules are aptly called part of the "moral economy" of their community.

If communication and cooperation ensure the synchronous extensibility of experimental practices in space, specialist education ensures their diachronous persistence. An important component of any course of studies in the experimental sciences are practical exercises. There are didactic reasons for this, but there are other reasons as well. Many of the skills that one needs to acquire in order to be a competent member of a community of experimental scientists bear only a distant relation to Lebenswelt practices and therefore cannot be learned from instructions by the untutored candidate. For instance, a student of biology or of chemistry has to learn how to pipette liquids and how to weigh substances within narrow limits of tolerance. For both these types of action, attention has to be paid to so many details of performance—for example, the speed with which the liquid is filled into the pipette and released from it again, the angle at which it is held when taking a reading of the volume, the way one deals with liquids of different viscosity—that purely verbal instruction is practically impossible. Importantly, the claim that experimental results are transsubjectively valid implies that it will remain valid through time and thus requires that the experimental practice in question can indeed be successfully perpetuated through time by just that kind of specialist apprenticeship.

Experimental practices are thus historical entities that make provision for their own indefinite extensibility both in space and in time. But this is not enough. As has been described above, experimental practices generate the need for instrumental control of hitherto unknown processes. In order to acquire that kind of know-how, new experiments have to be established. At first, we only have a couple of observations that hint at hitherto unknown phenomena. As such, these observations do not belong to experimental science but rather to natural history. Collecting observations that might prove useful in learning more about those phenomena is part of a heuristics for experimental innovation. With respect to the processes later produced in a new experimental setup, it might be called the "phase of exploration."

Exploration is followed by the phase in which new experiment types are established by developing new instruments, learning to prepare new kinds of experimental objects, or combining those already available into innovative setups. It aims at providing effective, nonredundant instructions for producing the controlled, laboratory equivalent of the phenomena observed in the exploration phase. Initially, it is usually not the case that repeated instances of following a particular instruction invariably produce the same outcome. Rather, experimenters have to have a lot of hands-on experience with a particular experimental setup until they learn which kinds of disturbances can occur and how they can be prevented. Systematic variation of actions and means are required in order to know which aspects are necessary and sufficient for an effective and nonredundant instruction. If, for example, the starting of an experiment is schematically identified with the production of a situation S_1 from a setup S_0, the effectiveness of the starting action must be secured by trying all sorts of ways in which S_1 can be produced. The instruction "to start the experiment, produce S_1," is only effective if the result S_2 can be reproduced no matter how exactly S_1 is achieved. Consider an experiment in toxicology. If the starting of the experiment is supposed to consist in oral intake of a certain amount of a substance X, it should be irrelevant whether intake occurs in pure form or mixed with food or other substances. As is well known, in reality this is far from irrelevant, which is why the exact form of administration including combination with other substances has to be recorded in the instruction.

Ineffectiveness is only one threat to new instructions. The other is redundancy. In order to secure nonredundancy, the process of preparing and starting the experiment has to be varied in ways that make the result differ from S_1 in crucial aspects that are determined with respect to the special interest guiding the particular experiment. For instance, in order to check whether or not administering a pharmaceutical is redundant, a control group is given a placebo treatment instead.

Once an experiment can be controlled this far, classes of experiments can be defined by differentiating among the conditions that have to be controlled and classifying them into by-conditions, which are held constant, on the one hand,

and initial conditions, which are systematically varied, on the other. Only now, by systematically investigating which change in the results can be brought about by which kind of variation in the initial conditions, are experimental laws being established. Those experimental runs that are performed for that purpose might be called "experiments proper,"[12] whereas the experimental runs, which were needed in order to learn which factors have to be controlled to reliably reproduce results at all—that is, the experimental runs belonging to the phase of establishing a new experiment type—might be called pretests instead. It is in this phase that the principle of causality is made operative.

It is constitutive of experimental practices that there are modes of communication and cooperation plus traditions of apprenticeship that enable people other than the original experimenters to check whether their claims are right or wrong. As they have been described above, however—the case of the junior scientist, the provision of practicals—it was implied that the teacher already knew that she could control the process under investigation. With new experiments, how can she be so confident? In other words, if a new experiment is established in some laboratory that fails to be reproduced by colleagues elsewhere, who is to say whether it is the latter who are incompetent to follow the instruction correctly or the people in the former who have succumbed to an illusion? The circularity that ensues from the mutual dependence of claims to truth and claims to competency has been described as the experimenters' regress by Harry Collins (1985). It is open to debate whether or not he is right to regard this kind of regress as an instance of Wittgenstein's rule following problem. The rule following considerations are absolutely general and do not solely apply to practices that have truth claims as their main product. For technical practices with more tangible goals, however, it seems much more easy for the individual to judge whether or not he or she is competent to reach those goals. There is not the same kind of circularity involved here as is referred to by the experimenters' regress. Be that as it may, Collins's diagnosis seems fundamentally correct. In his view, the experimenters' regress is a consequence of an "algorithmic" model of science, according to which the setup and starting procedure of an experiment can be completely determined by a suitable, "complete" instruction. This, however, is not the case; the successful reproduction of an experiment depends on the possession, by the experimenters, of certain nonverbal skills. Attribution of those skills is equivalent with conceding to the candidate membership in the community of competent experimenters. So any experimental results have to be relativized, according to Collins, to such communities of experimenters.

Does this mean that any experimental knowledge is only locally valid? It does not, because the claim that an experiment reproducibly leads to a certain result is not relativized to a community of researchers in the sense of a finite set of real individuals. Rather, it is relativized to an experimental practice. An experimental practice is incomplete if it does not also provide the means to extend itself in time and space, including those needed to teach any skills necessary to perform its experiments and thus gain new members for the community of practitioners. This is whereby it transcends the local community of those who presently are members of that practice and ensures the transsubjective validity of its results. For the process of establishing a new experiment, this means that it is only finished when the skills necessary to perform it can reliably be taught to others. In the terminology of the experimenters' regress, the experimenter can only be confident that he himself has mastered the new method if he knows how to tell somebody else how to do it.

NOTES

1. Obviously, this is a distinction for analytical purposes only. In reality, problems do not come preclassified but are, at least in principle, always open for being treated either way.

2. The principle of methodical order is a central theme of the Erlangen constructivism tradition of philosophy of science, which has its roots in protophysics (see Janich 1985 and 1997 for details). Protophysics deals with the question of how to establish measuring procedures that can be proved to lead to unique results (proportions, not absolute values) that can be reproduced without taking recourse to a prototype of the measuring instrument, thus making it possible to understand how the practice of scientific measuring could get started at all. The task that protophysics set itself was to put down a set of operational norms on the basis of which such a procedure could be implemented. It was argued that in designing such a procedure for a specific dimension, the norms that define it must not make use of values or proportions of measures belonging to that dimension. For example, the defining norms for a clock must not allude to equal intervals of time. The protagonists of protophysics argued that this rule is just a special case of the more general principle of methodical order, which has to be obeyed whenever a set of norms is meant to be operational.

3. While it is true that Hugo Dingler, whose operationalistic philosophy of experiment anticipated much of the contemporary debate, stipulated a necessary connection between experimental laws and prescriptions for experimental operations (cf. Dingler 1928), he did not equate laws with recipes. Rather, what he claimed was that laws are meaningless unless there are recipes that tell us how to "realize" (that is, to "make real," bring into being) phenomena of the type that are the subject of those laws.

4. At least, this is what experimenters say. In fact, they guard their experiment continually, preventing disturbances and thereby assuring that the processes under observation proceed "naturally."

5. "An intervention I on X with respect to Y (for the purposes of determining whether X causes Y) is an exogenous causal process that completely determines the value of X in such a way that if any change occurs in the value of Y it occurs only in virtue of Y's relationship to X and not in any other way" (Woodward, this volume, 91).

6. Note that it is not only for stylistic reasons that he uses the first-person plural here; the "we" of the quoted passage denotes any actual or possible epistemic subject.

7. While this essay is inspired by the view that the desire to prevent or remove disturbances relative to intended outcomes—the concept of *Störungsbeseitigungswissen* (cf. Janich 1996b; Lange 1996)—has to play a central role in the reconstruction of empirical sciences, it is here presented as a distinctive characteristic of *scientific* knowledge that the practices in which it is validated are in a certain way detached from immediate Lebenswelt or technical purposes.

8. Starting from empirical evidence rather than conceptual considerations, Price (1984) criticizes the linear model of innovation and suggests an alternative picture that is similar to the one sketched in this chapter. According to Price, fundamental science and the development of innovative products and processes that is usually regarded as the aim of applied science are separate practices and proceed according to their respective purposes. The systematic point of contact is where new instruments developed in one of these practices can be put to (usually nonintended) uses in the other. Price also adduces evidence for the thesis that this process really cuts both ways—that is, science advances by appropriating technology developed in an applied context just as much as technology profits from spin-offs of experimental practices.

9. The idea to characterize the issue in question in this section as the issue of the nonlocality of experimental science is, of course, Hans Radder's. His terminology fits my approach here very well, so I take the liberty to borrow it from him. On transsubjective validity as a hallmark of scientific knowledge that replaces the concept of universal validity, see Janich (1996b).

10. If cultural differences are taken into account, it may well turn out that the truly universal repertory is very restricted.

11. The history of the experimental practice of classical *Drosophila* genetics is described in R. Kohler's (1994) instructive *Lords of the Fly*. For a short summary, see Lange (1999, 185–204).

12. Or they might be called "parts of an experiment proper." While experimental runs are the atoms of experimental practices, they are not individually meaningful. Not only is their history—the establishing of the respective experiment type—important; quite often, even after the experiment type has been established, a single token experiment consists in a series of experimental runs under varied initial conditions, including, for example, controls.

Michael Heidelberger

7 / Theory-Ladenness and Scientific Instruments
in Experimentation

Since the late 1950s one of the most important and influential views of post-positivist philosophy of science has been the theory-ladenness of observation. It comes in at least two forms: either as a psychological law pertaining to human perception (whether scientific or not) or as conceptual insight concerning the nature and functioning of scientific language and its meaning. According to its psychological form, perceptions of scientists, as perceptions of humans generally, are guided by prior beliefs and expectations, and perception has a peculiar holist character. In its conceptual form it maintains that scientists' observations rest on the theories they accept and that the meaning of the observational terms involved depends upon the theoretical context in which they occur. Frequently, these two versions are combined with each other and give rise to a constructivist view of scientific knowledge (I shall use the term "constructivism" roughly in the same way as Golinski [1998, chap. 1]). According to this outlook, our experience is categorized and preconditioned by prior belief since the process of gaining knowledge through science always involves the use of concepts from some theory or other. This view can easily be strengthened to serve as the corner-stone of a constructivist and antiempiricist account of science: The categories in terms of which we carve up our experience are not read off from the external world but follow from prior theoretical commitments.

The implications of theory-ladenness for a view of scientific experimentation are straightforward: If observations are theory-laden and if experimentation involves observation, then experimentation has to be theory-laden, too. Since experiments, according to this view, make sense only in relation to some theoretical background, they cannot play a role that is theory-independent. That means that an experiment can make sense only on the basis of some prior theory.

In the first part of this chapter I shall discuss the view of theory-ladenness as it appeared in the work of its originators and draw a distinction between three different meanings of the term. In the second part, I shall develop a classification of instruments that reflects these different meanings and specifies the different roles instruments come to play in experiments. Before instruments can be employed in a theoretical framework, they have first to be employed causally. I shall illustrate my view of instruments, and thereby of experiment, by drawing on Kuhn's discussion of Wilhelm Roentgen's discovery of the X rays and by referring to the way Georg Simon Ohm developed the law governing the flow of electricity in conductors that bears his name.

1. Three Conceptions of Theory-Ladenness

In order to develop the causal view of experiment and to investigate how it fares in relation to the theory-ladenness of observation, we first have to get an overview of the different meanings of the latter notion. In discussing theory-ladenness, the average postpositivist is likely to refer to Norwood Russell Hanson's book *Patterns of Discovery*. Strangely enough, the reader will almost never bother to go beyond the first chapter of this book, which is entitled "Observation." This chapter provides ample material for quotations that can be used in defending theory-ladenness against stubborn positivists. "[S]eeing is a 'theory-laden' undertaking," we read. "Observation of x is shaped by prior knowledge of x. Another influence on observations rests in the language or notation used to express what we know, and without which there would be little we could recognize as knowledge" (Hanson 1958, 19).

It should be noticed, however, that in chapter 3, entitled "Causality," Hanson's view receives a peculiar twist that must not be overlooked. He tells us there that theory-laden talk in science is mainly causal talk, talk in which causes and effects and the connections between them are identified. Hanson further maintains that this way of talking is to be contrasted with the use of sense-datum language, which is devoid of any causal meaning. The only way science fulfills its major goal, explanation, is by invoking causality: "Notice the dissimilarity between 'theory-loaded' nouns and verbs, without which no causal account could be given, and those of a phenomenal variety, such as 'solaroid disc,' 'horizoid patch,' 'from left to right,' 'disappearing,' 'bitter.' In a pure sense-datum language causal connexions could not be expressed. All words would be on the same logical level: no one of them would have explanatory power sufficient to serve in a causal account of neighbour-events" (Hanson 1958, 59). This quotation shows two things. First, against all claims to the contrary, there *can* be perceptual accounts according to Hanson that are free from theory. It is another matter that he attributes to them no great use in science since their phenomenal nature prevents them from having any explanatory content. Second, theory-ladenness in science primarily means "causality-ladenness" for Hanson, being loaded with causal meaning: "The notions behind 'the cause x' and 'the effect y' are intelligible only against a pattern of theory, namely one which puts guarantees on inferences from x to y. Such guarantees distinguish truly causal sequences from mere coincidence" (Hanson 1958, 64).

Before we can use this insight for a workable account of the nature of experiment, let us consider the next "founding father" of theory-ladenness, Pierre Duhem, and his work *La Théorie physique—Son objet et sa structure*. Duhem discriminates between a fact and its theoretical interpretation, or, as he says, between a "concrete" and a "theoretical fact." He tells us that an experiment in physics involves two parts:

> In the first place, it consists in the observation of certain facts; in order to make this observation it suffices for you to be attentive and alert enough with your senses. It is not necessary to know physics; the director of the laboratory may be less skillful in this matter of observation than the assistant. In the second place, it consists in the interpretation of the observed facts; in order to make this interpretation it does not suffice to have an alert attention and practiced eye; it is necessary to know the accepted theories and to know how to apply them, in short, to be a physicist. (Duhem 1974, 145)

> An experiment in physics is the precise observation of phenomena accompanied by an *interpretation* of these phenomena; this interpretation substitutes for the concrete data really gathered by observation abstract and symbolic representations which correspond to them by virtue of the theories admitted by the observer. . . . The result of an experiment in physics is an abstract and symbolic judgment. (Duhem 1974, 147)

This quotation demonstrates that Duhem's conception of theory-ladenness is clearly different from Hanson's. For him, experiments in physics are done on a level of the scientific enterprise that is not explanatory. As becomes evident, it would not be sufficient for Duhem simply to place practical facts into a web of causal relations in order to make them theoretical: "The result of common experience is the perception of a relation between diverse concrete facts. Such a fact having been artificially produced some other fact has resulted from it. For instance, a frog has been decapitated, and the left leg has been pricked with a needle, the right leg has been set into motion and has tried to move away from the needle: there you have the result of an experiment in physiology. It is a recital of concrete and obvious facts, and in order to understand it, not a word of physiology need be known" (Duhem 1974, 147). This description is as causal as it could be, but its causal nature is obviously not enough for Duhem to make it a theoretical fact. For Duhem, theory-ladenness has therefore little to do with causality, but with inscribing phenomena in the terms of an abstract and symbolic structure: "The result of the operations in which an experimental physicist is engaged is by no means the perception of a group of concrete facts; it is the formulation of a judgment interrelating certain abstract and symbolic ideas which theories alone correlate with the facts really observed" (Duhem 1974, 147).

In order to understand what Duhem has in mind here we need to clarify his distinction between experimentation at an advanced level of theory and experimentation at the level of "common experience," which is not theoretical at all. A theory is advanced, according to Duhem, when it provides an interpretation of experimental laws by substituting abstract and symbolic representations for them. In less advanced sciences like physiology or certain branches of chemistry, "where mathematical theory has not yet introduced its symbolic representations," the experimenter can reason "directly on the facts by a method which is only common sense brought to greater attentiveness" (Duhem 1974, 180; remember the example of the frog above!). In order to specify the rules that are in operation in this common sense reasoning Duhem quotes at length from his countryman, the physiologist Claude Bernard. Duhem thus clearly admits the possibility of observations and experiments that are free from theory, although only at a less advanced level of science. Note, however, that being theory-free means something different for him than for Hanson.

To record our results so far: Hanson and Duhem have different conceptions of theory-ladenness. Whereas for Hanson any injection of causality into the mere registering of facts is bound to render them theoretical, it is, for Duhem, with the representation of (causal) relations in an abstract, noncausal structure that theory begins.

Let us have a look now at the third advocate of theory-ladenness, Thomas Kuhn. For Kuhn, theory-ladenness is first of all "paradigm-ladenness": The normal-scientific tradition in which one has been trained, and the experiences that this has brought about determine how the scientist sees his world:

> [S]omething like a paradigm is prerequisite to perception itself. What a man sees depends both upon what he looks at and also upon what his previous visual-conceptual experience has taught him to see. (Kuhn 1970, 113)

[P]aradigm changes do cause scientists to see the world of their research-engagement differently. (Kuhn 1970, 111)

Within the new paradigm, old terms, concepts and experiments fall into new relationships one with the other. (Kuhn 1970, 149)

The proponents of different theories are like the members of different language-culture communities. (Kuhn 1970, 205)

Here we have the fusion I referred to at the beginning of this chapter between a psychological law or laws pertaining to perception and a particular philosophical view of the functioning of scientific language, which holds that scientific terms derive their meanings from prior experiences, beliefs, or theories and possess meaning only in their context.

Or so it seems. If we look closer, we find that even Kuhn admits the possibility of "fundamental novelties of fact"—that is, of genuine discovery that goes *against* a well-established paradigm. Without this possibility, as he himself realizes, science could only develop theoretically and never by adjustment to facts. "Discovery commences with the awareness of anomaly, i.e., with the recognition that nature has somehow violated the paradigm-induced expectations that govern normal science" (Kuhn 1970, 52–53).

We now have to specify exactly where, according to Kuhn, these "paradigm-induced expectations" come from that are violated in discovery: Are they induced by theoretical and abstract structure (à la Duhem) or by the causal properties (à la Hanson) of those elements in question that can be manipulated in experiment? A natural answer for Kuhn would be to say "both!" "[B]oth observation and conceptualization, fact and assimilation to theory, are inseparably linked to discovery" (Kuhn 1970, 55). If we look closer, however, at Kuhn's own examples, we notice that it is almost always the *theoretical interpretation,* the assimilation to theory, that is taken as decisive for discovery and hardly ever any causal experience. We are primarily shown cases where someone identifies a well-known experiment or entrenched phenomenon in a new way. Lavoisier, we are told, for example, was enabled through his new paradigm "to see in experiments like Priestley's a gas that Priestley had been unable to see there himself" and was "to the end of his life" unable to see (Kuhn 1970, 56). Assertions to the contrary notwithstanding, novelty in discovery seems for Kuhn to be the result of a paradigm-induced change in "seeing as" and not in a novel experience or recasting of a causal process.

The only case where Kuhn admits that discovery has been effected by a genuinely novel causal experience seems to be the case of the X rays. "Its story opens on the day that the physicist Roentgen interrupted a normal investigation of cathode rays because he had noticed that a barium platino-cyanide screen at some distance from his shielded apparatus glowed when the discharge was in process" (Kuhn 1970, 57). Although Kuhn seems to consider this observation theory-laden, I maintain that, in Duhem's sense, it is not. If it *were,* Roentgen, by definition of theory-ladenness, would have been able to interpret it in light of the theories of physics he had at his disposal. But here it is exactly the point that his theories deserted him and he could *not* find a place for this new experience in his customary theoretical structure. For this reason he interrupted his investigation and asked himself why the screen had come to glow. Yet it goes without saying that the novel observation is theory-laden in the sense of Hanson, because Roentgen immediately looked for a causal relationship between his apparatus and the glowing of the screen, although this went completely against all his expectations!

A follower of Kuhn might now say that Roentgen would never have paid attention to the glowing screen if he had not disposed of deeply entrenched theories of physics that *prohibited* such a phenomenon. This shows again, as Kuhn's advocate could continue, that observation is governed by expectation—as it happens, a conflicting one this time—and that therefore, at least in this sense, Roentgen's observation was also theory-laden. This might be true, but note that this is now a *third* sense in which the notion of theory-ladenness is used. It says something about the likelihood with which an observation occurs, the ease with which a phenomenon is detected or paid attention to in the light of a paradigm: An observation is theory-laden in this sense if it were improbable that an observer would have made it (that an observer would have noticed it or would have attributed any importance to it) without her holding a particular theory beforehand.

This is *not,* however, a claim about the nature of observation and its relation to theory, as the earlier discussed view of theory-ladenness would require, but about the disposition of a subject to perceptually detect or discriminate a phenomenon in relation to her prior experiences and theoretical belief or disbelief. It is certainly true that we tend to notice or overlook phenomena depending on certain expectations and beliefs. Whatever, though, the relation between the expectations of an observer and her perceptual abilities might actually be, it cannot by itself establish any relevance of an observer's belief to the *meaning* of the observational terms involved. If this were the case we would not be able to detect any anomalies—that is, observations that *contradict* our theoretical expectations. (Since anomalies are prerequisite to scientific revolution in Kuhn's sense, Kuhn cannot renounce them for his own theory.)

In order, therefore, to distinguish this third type of "theory-ladenness" (if this label is still appropriate at all) from the two types associated before with

Hanson and Duhem, respectively, let us call it henceforth "theory-guidance." It refers to how the disposition to make a particular observation depends on the theoretical background of the observer, and it should primarily be associated with Kuhn.[1] As we have seen, however, theory-guidance cannot be taken as genuine theory-ladenness because of its irrelevance to the *meaning* of observation sentences.

Let us step back into Roentgen's laboratory for a moment. What did he do after he noticed the anomaly? He conducted various experiments in order to explore the *cause* of the incident: "Further investigations—they required seven hectic weeks during which Roentgen rarely left the laboratory—indicated that the cause of the glow came in straight lines from the cathode ray tube, that the radiation cast shadows, could not be deflected by a magnet, and much else besides. Before announcing his discovery, Roentgen had convinced himself that his effect was not due to cathode rays but to an agent with at least some similarity to light" (Kuhn 1970, 57). This is perhaps the only place in his book where Kuhn uses the term "cause" (or an equivalent expression) in relation to an experimental investigation. The quotation shows vividly that Roentgen's experiments were not conducted to test a theory but to expand our knowledge of causal connections in relation to the scientific instruments and devices involved. (Steinle [1998, 284–292] investigated this type of experimentation more closely and called it "exploratory.")

What does our discussion suggest as the most adequate description of Roentgen's early investigations? They were certainly theory-guided in the sense of Kuhn and they were causality-laden in the sense of Hanson, but not (or not yet), I claim, theory-laden in the sense of Duhem. The experiments Roentgen conducted during his seven hectic weeks were in the same way a "recital of concrete and obvious facts," as the above-mentioned experiment of decapitating the frog was. To draw an analogy to Duhem's case, we can say that Roentgen could have understood these facts even if he had not known a word of physics. The only knowledge he had to have for conducting his initial experiments was about the *causal power of the instruments* he used. (It goes without saying that looking at an X-ray tube became gradually theory-laden the more the X-ray tube became embedded in a new theory.)

Roentgen's early series of experimentation has to be (and it *can* be!) systematically distinguished and separated from the Kuhnian process of "assimilation to theory." Such an assimilation can of course also proceed by experimentation. Kuhn is right when he says that only after this assimilation has been achieved and the phenomena have received an abstract and symbolic representation can we speak of a "discovery" of X rays. Yet before this interpretation has taken place, we can only say that an anomaly has occurred.

The case of the X rays shows, however, that in an important sense experimentation itself can be, and very often is, autonomous and free from theory.[2] It is wrong to see experiment as nothing more than a test of preconceived ideas gained by a theoretical interpretation. We should therefore learn a lesson from the X-ray case and distinguish between two kinds of experiments: those that are causal but not embedded in a theoretical structure and those that presuppose the knowledge of such a framework. I think that Kuhn's discussion caused much damage by blurring and dissolving this difference and by identifying the concept of a paradigm too much with the Duhemian conception of theoretical interpretation. Sometimes Kuhn seems to realize this when he stresses that "[a]t a level lower or more concrete than that of laws and theories, there is, for example, a multitude of commitments to preferred types of instrumentation and to the ways in which accepted instruments may legitimately be employed (Kuhn 1970, 40). And he explicitly refers to the discovery of X rays as a case in point. So even in Kuhn a sense turns up in which experiment can be independent of the theoretical commitments of a paradigm and dependent only on an entrenched tradition of instrumentation, although Kuhn does not pursue this idea further.

Is my emphasis of an autonomous "lower level" in experimentation a relapse into a positivist spirit? Definitely not. Nowhere in my argument appears an appeal to a neutral experiential authority like "immediate experience," "bare sense-data," "the given," or "pure observation-language" that is to decide conclusively for or against a theory. All that is claimed here is that two types of experimentation should be kept conceptually apart: experimentation at the causal level, where instrumental manipulation is distinguished, and experimentation taking place at the theoretical level, where the results at the causal level are represented in a theoretical superstructure (that can itself also have causal meaning). In this way, all those claims to theory-independence that for Kuhn and Hanson were typical of positivism can be avoided.

Our discussion so far suggests two things: one should first of all distinguish "theory-guidance" from the notion of "theory-ladenness." Even if all experimentation were guided by theory, in the sense found in Kuhn, this alone would not be enough to prove that observations of experimental results are theory-laden. The reason for this is that theory-guidance alone is not able to establish a relation between the *meaning* of observational terms and the *meaning* of the theory that guided it. Second, one should distinguish between "theory-ladenness through appeal to causal understanding" and "theory-ladenness through theoretical interpretation," or, in Kuhn's words, "assimilation to theory." The latter has to reflect the former, but not necessarily the other way around, as Kuhn and Hanson (and many others) usually seem to suggest. It makes much more sense to regard causal understanding and theoretical understanding moving toward each other from separate and independent starting points until they meet at a stable state of equilibrium than to mingle them beforehand.

In order to avoid misunderstandings it is then even better to reserve the term "theory-ladenness" for the cases Duhem had in mind—that is, for "theory-ladenness through theoretical interpretation." When Hanson introduced the term in the course of his own discussion (actually he coined the expression "theory-loaded"), his primary intention was not to claim that all observation is interpreted in the light of a theory but to stress that, contrary to the positivists, observation always presupposes some causal notion that transcends direct experience. One can of course maintain that all causal talk is theoretical talk, and Hanson was perhaps an advocate of this opinion. I think, however, that this goes too far, because there are many cases in our everyday communication where we use causally loaded terms in an explanatory fashion without referring to any theory or theoretical entities. When someone asks me why the light went on, I can use causal words in my answer ("I turned the switch") without invoking any theory about the nature of electricity and electric action whatsoever. The term "theory-laden" in its original sense refers to genuine theoretical interpretation that transcends causal understanding of "common experience." It should be reserved for these cases and not be diluted in the sense advocated by Hanson.

2. Instruments and Their Use in Experimentation

In the light of the foregoing discussion, it seems advantageous to identify as the central and primary constituent of scientific experimentation the causal agents involved—that is, *scientific instruments*. In this way, experimentation can be investigated more closely in its two basic forms: as improvement and expansion of causal knowledge and as adjustment to a theoretical context. In addition, this way of putting things makes it possible to envisage a genuine history of experimentation that is driven by the instruments involved.

If we look at experiment in its *first* form, as causal manipulation by means of instruments, we can distinguish between instruments that are used in order to fulfill a *productive* and those that have a *constructive* function.[3] The goal of productive instruments is to produce phenomena that normally do not appear in the realm of human experience. Roentgen was using his apparatus productively when he tried to accomplish other, hitherto unknown, effects with it besides the glowing of a barium platinocyanide screen. As we have seen, he found out that under certain conditions he could make it cast a shadow. Roentgen's apparatus was, as we might say, *unconditionally productive,* but there are other productive instruments that produce *known* phenomena—although in circumstances where they have not appeared before. I am thinking of instruments, like microscopes or telescopes, used in order to improve human perception. Still another type of a *conditionally productive* instrument is one that tries to analyze or to split a phenomenon into different, previously unknown components. A case in point is the spectroscope.

Roentgen also used his instruments *constructively* when he tried to influence phenomena in order to make them behave in a certain way. The goal of such experiments is to produce an effect in its "pure form," without any complications or additions that could spoil it or that are otherwise alien to it. Another goal is to tame the phenomena in order to be able to manipulate them in a certain desired way. Ernan McMullin spoke of the "causal idealization" of Galileo's experiments in this respect (McMullin 1985, 247–273). We can also refer to another of Kuhn's favorite examples, the Leyden jar, invented around 1745. This instrument was not used at the time to *uncover* the phenomenon of electricity, so to speak, but to collect and store electricity in it. It was developed in order to produce a desired effect in a desired way.

Still another type of experimentation in its first form is experimentation by means of *imitative* instruments. They are used to produce effects in the same way as they appear in nature without human intervention. In biology, for example, we find experiments in which an apparatus is used that closely simulates the production of an enzyme in an organism.

If we look at experiment in its *second* form, as adjustment to a theoretical context or assimilation to a theoretical interpretation with the help of instruments, we see other functions of instruments step to the foreground—above all, what I call the *representative* role of instruments. In this case, the goal is to represent symbolically in an instrument the relations between natural phenomena and thus to better understand how phenomena are ordered and related to each other. Examples of instruments that fulfill such a function are clocks, balances, electrometers, galvanometers, thermometers, etc. These are "information-transforming instruments," as Davis Baird once called them; they transform the input information into a more useful output format while preserving the order of the phenomena vis-à-vis the intensity of the attribute in question (Baird 1987, 328). In a thermometer, for example, the different states of heat accessible to our sense of heat are transformed into different states of the instrument itself (that is, different heights of the mercury column) that are accessible to sight. The order of the heat states is more or less preserved in the order of the heights of the column (cf. Mach 1896). The changes the instrument undergoes can be taken as representative of the changes of the measured phenomena.

The difference between the use of productive and constructive instruments on the first level of experimentation and the use of representative instruments on the theoretical level mirrors to some extent other distinctions that have been proposed: there is the old difference of the seventeenth century between "philosophical" and "mathematical instruments." This is taken up in our time by Jed Buchwald when he suggests to distinguish "discovering experiments" from "measuring experiments" (Buchwald 1993, 200). In a similar way, Willem D. Hackmann makes a difference between "active instruments" that intervene in nature and "passive" ones that try to minimize any effect on the relevant object (Hackmann 1989, 39–40).

In order to illustrate my claim, I would like to have a look at the experiments that led to Georg Simon Ohm's law as a case in point and ask how Ohm's law of electricity theory was discovered and what role instruments played in this discovery.[4] In the series of experiments he conducted between 1825 and 1827, Ohm relied mainly on two instruments: the electroscope to measure what he called the "electroscopic force" or "tension" in the electric circuit, later identified as "potential difference" by Kirchhoff in 1849, and the galvanometer in order to measure the "exciting force" of the current or the "strength of the magnetic effect on the conductor"—that is, the intensity of the current. Both instruments served constructive and productive functions in Ohm's experiments. The electroscope was first of all used as a constructive device to identify electricity in its pure form, in abstraction from any specific situations in which it arises. But later Ohm transposed it from the context of static electricity to the (dynamic) case of electric flow, using it as a conditionally productive device to yield hitherto unknown effects. Consequently, the ensuing usage of "tension" for the dynamic case proved to be very difficult for many of Ohm's contemporaries, and many of them rejected it as unfounded.

The galvanometer had its origin in Oersted's "fundamental experiment," of course, which was conducted in order to produce the magnetic effect of a current carrying wire. In the hands of Ohm it also served as a constructive device when its different states were used as the only aspects relevant to the strength of the dynamic action. This also did not find the acclaim of many of Ohm's contemporaries, because they were convinced that dynamic action of electricity is different in the case where there is some chemical action present. Ohm, however, was a follower of Volta's theory, according to which the electric action depended on the contact of two metals and was not the result of a chemical activity. (The action of the Voltaic pile was explained with the so-called contact theory, which did not see any chemical action present.) Indeed, Ohm thought chemical activities in a "galvanic chain" should be avoided, because they detract from the "natural purity" of the galvanic effect. Following a suggestion of the editor of the *Annalen der Physik,* Ohm used, from 1826 on, a thermoelectric source for his experiments. This thermoelectric apparatus played a double role, both as a productive and a constructive device: a productive role because all other sources of dynamic electricity available at the time were highly unstable, vacillating highly in their electric action, and a constructive role because it produced the action in its pure or idealized form, as Ohm thought, without any chemical "contamination," so to speak.

The constructive and productive usage Ohm made of his instruments as described takes place on a causal level that is theory-free and guided by the causal possibilities available with the instruments in question. The *representative* or *symbolic* level is now superimposed on this causal level. Ohm attained for his experiments a representative and symbolic significance by three means: First, he used his instruments not only as productive and constructive devices but also as representing ones—that is, as measuring instruments—arriving thereby at a "symbolic generalization," as Kuhn called it, which functions as a unifying formula. Second, his approach enabled Ohm to create and define a new theoretical concept, the concept of electric "resistance" or "conductivity." And third, he was able to give a theory of the instruments involved—that is, to "substitute for the concrete objects composing these instruments an abstract and schematic representation," as Duhem (1974, 153) once formulated it.

As already noted, Ohm finally arrived through his measurements at the formula that is known as Ohm's law and which can be written as $I = V/R$. The road to this formula was very winding and tortuous indeed, and Ohm had to make many attempts, both in a practical as well as in a theoretical respect, to obtain his result. It is highly significant that his first theoretical conception of electric activity in a closed circuit was guided by the Coulomb paradigm of static electricity and that he was able to describe this already with some version of his law. This implied a concept of "resistance" similar to the mechanical resistance in friction phenomena. Later Ohm conceived of electric conduction in a new way by establishing an analogy between heat conduction as developed in Fourier's theory of heat and the conduction of electricity. In this sense, "resistance" becomes a truly theoretical or theory-laden term that is not yet present at the causal level of Ohm's experiments; it is reached and formulated only at the symbolic level.[5] Ohm could avoid measuring resistance directly in his first-level experiments by the following reasoning: Let I_n be the intensity of an electric circuit where instead of the outer part of the circuit a short and thick "Standard Conductor" is introduced. ("I_n" stands for "normal" or "standard intensity"). If R_i is the inner resistance of the circuit and R_o the outer one, we can put $I_n = V/R_i$. For a circuit other than with the Standard Conductor, we can thus write $I = (R_i) \times I_n / (R_i + R_o)$. I put the first "$R_i$" in parentheses because Ohm overlooked that this is not a constant factor of proportion but depends on the inner resistance of the circuit. It was only at a later stage that he substituted V for $(R_i) \times I_n$.

It should be obvious that Ohm could also apply his new mathematical formula to the galvanometer and the electrometer and predict their behavior in many different cases. Ultimately it was the practical usability of Ohm's law for the arrangement of all kinds of measurements in the circuit and especially its technical applications, such as in electrical telegraphy, that in the end led to its acceptance. It was soon recognized that Ohm's law was completely neutral with regard to the exact theory of the origin of the electromotive force of electricity; it holds irrespective of "whether that force is regarded as being derived from the contact of dissimilar metals [as its founder himself believed] or as referable to chemical agency," as the Royal Society wrote when it dedicated the Copley medal to Ohm in 1841 (Royal Society 1841, 336).

3. Conclusion

The received view of theory-ladenness in observation and especially in experimentation is too coarse-grained. First of all, we should not mix up theory-ladenness with the concept of "theory guidance" as it appears in Kuhn's work. Second, we should distinguish between theory-ladenness on a primary causal level of scientific experimentation (if it is still appropriate to call it this way) and experimentation on a supervening, secondary level when theory takes possession of the direct causal experience with scientific instruments and when the adjustment of a causal picture to a theoretical and symbolic context is called for. There are many cases where first-level experimentation is and can be pursued without taking into account the secondary level.

It is true that in advanced and mature theories the two levels form an inseparable amalgam, as Duhem and Kuhn have amply demonstrated. Yet it is clear that when a new domain is explored, experimentation is conducted in a theory-free way, only constrained by considerations of the causal power of the instruments used. As I have tried to show, the distinction between a causal and a theoretical level of experimentation also sheds new light on the different roles instruments play in scientific experiment. Last but not least, this way of viewing things enables us to give back to experiment some of the epistemic dignity it used to have when empiricism was still in more esteem. This view also liberates us from extreme modes of constructivism without falling back into naive forms of experientialism.

I am grateful to John Michael for his help in improving my English.

1. I note in passing that theory-guidance as reconstructed here is not the only claim Kuhn takes over from the psychology of perception. We also encounter the *contrary claim* in his book that the more entrenched a paradigm is the more one neglects anomalies and that this disposition weakens only in periods when a paradigm enters a crisis state.

2. This claim was of course first raised by Hacking (1983a, esp. chap. 11). My argument for it as presented here, however, differs from his.

3. The classification of instruments in experimentation as developed in this chapter was first suggested in Heidelberger (1998).

4. For a fuller story, see Heidelberger (1980), or—in German translation, with less misprints—Heidelberger (1983).

5. I have explored in Heidelberger (1979) how Ohm defined "conductivity" in his theory by presupposing the validity of Ohm's law. I also tried to show that this method is *not* circular and that it is followed not only in Ohm's case but that it constitutes a frequently used way to introduce theoretical terms.

Hans Radder

8 / Technology and Theory in Experimental Science

I begin with a point of method. If we want to put forward a specific claim about an issue in the philosophy of scientific experimentation (for instance: "theory-free experiments are impossible"), we need to make explicit, at least in outline, what we take to be an experiment. One suggestion, then, might be that an experiment is what scientists call an experiment. Consistently following this suggestion, however, would imply that we simply take for granted the intuitions and conceptualizations of the scientists. Given that scientists and philosophers pursue different goals, this is—generally speaking—not recommendable. For example, many scientists do not bother to distinguish between experiments in which the object under study is being manipulated and "experiments" in which this is not the case. Such a distinction, however, is philosophically quite relevant. Another approach might be to have scientists' practice, instead of their words, decide on what to call an experiment. Unfortunately, this approach will not work either, since what scientists do does not unambiguously fix what it is that they do. Thus, the question of what counts as an experiment—when and where does it start or end? what to include and what to exclude?—cannot be answered by merely noticing what is going on in scientific practices. We also need to add our own preunderstanding as philosophers, which may then be adjusted and improved in the process of learning more about scientific experimentation. In this way, the resulting account will be informed, yet not fully determined, by the study of experimental practice. That is to say, when studying the practice of scientific experimentation, we will also be confronted with the problem of the hermeneutical circle.

The focus of this chapter is on two central themes in the philosophy of scientific experimentation: the relationship between science and technology, and the role of (existing) theory in experimental practice. In addressing these themes, my own preunderstanding derives from my earlier account of scientific experimentation (see Radder 1986; 1988, chap. 3; 1996, chaps. 2 and 6). According to this account, an experimenter tries to realize an interaction between an object and some apparatus in such a way that a stable correlation between some feature of the object and some feature of the apparatus will be produced. If the experiment succeeds, two aims have been achieved simultaneously. First, a stable experimental (or object-apparatus) system has been materially realized; second, it has proved possible to obtain some knowledge about relevant features of the object by observing and interpreting correlated features of the apparatus. In addition, since scientific practice does not consist of isolated experiments performed by solitary experimenters, we have to examine the ways in which individual experiments are embedded and used in broader experimental and theoretical contexts. Thus, if we want to establish what counts as an experiment, we need to investigate the wider function and significance of stable experimental systems and experimentally acquired knowledge.

1. The Irreducibility of the Theoretical Meaning of Replicable Experimental Results

A subject that arises naturally from the study of scientific experimentation is the relationship between science and technology. This is a vast and complex subject, to which several approaches may be taken (see Staudenmaier 1985, chap. 3; Radder 1987; Joerges and Shinn 2001). One approach involves the study of the interactions between science and technology in the course of their historical development. An issue that has been much debated is to what extent science shapes and has shaped technology and, conversely, what is and has been the impact of technology on science (see Böhme, Van den Daele, and Krohn 1983; Keller 1984). Another approach is more theoretical. It first tries to define both science and technology in terms of certain fundamental characteristics. On this basis it then explains their relationship. A rather common view, for example, claims that there is a fundamental difference between science and technology in that the former is driven by the search for truth, while the latter aims at solving practical, social problems (Bunge 1966). In contrast, quite a few authors who stress the significance of experimentation have developed interpretations in which science is conceptualized as basically similar to technology. In the following sections, two of the latter interpretations—those of Peter Janich and Bruno Latour—will be analyzed and assessed in detail.

1.1. Science as Technology

Peter Janich's account of the relationship between science and technology builds on a broad German tradition, which includes such philosophers as Hugo Dingler and Paul Lorenzen and which has exercised a significant influence on Jürgen Habermas's earlier views of science and technology. This approach is called "methodical" or, more recently, "culturalistic constructivism." It has produced several philosophical studies of the nature and role of scientific experimentation (for reviews, see Janich 1996a, chap. 1; Lange 1999, chap. 3).

In his 1978 article "Physics—Natural Science or Technology?" Janich links the relationship between science and technology to a contrast between observation and experimentation. He makes a distinction between two kinds of activity. Observations, for instance of organic cells, clouds, stars, and lakes, reveal the existence of natural, human-independent phenomena. In contrast, experimentation —and experimental measurement, in particular—leads to artificial results, such as lengths, durations, velocities, and masses, which do not exist independently of human, technological action. "Doing experiments is more an activity to produce *technical effects,* which can be described appropriately as engineering rather than as a scientific activity, properly speaking, as a construction of machines rather than as an inquiry into nature, as an attempt to produce artificial processes or states rather than as a search for true sentences" (Janich 1978, 11). Thus, physics as natural science—that is, observational physics—is here being contrasted to experimental physics, or physics as technology.[1] In the same article, however, the philosophical significance of the distinction between physics as natural science and physics as technology is already qualified, while in his more recent work it is played down even further. To that end, Janich puts forward four different arguments.

First, he notes that scientific observations of natural phenomena rely on instruments that have been developed in the context of an experimental science of artificial phenomena (Janich 1978, 22). This point can be reinforced by noting the following problem that arises from Janich's earlier views. Consider a lake as an example of a natural entity. According to Janich's interpretation, the spatial dimensions of a lake, the time it has existed (its age), the mass of its water, and the velocity of its waves do not exist independently of human action. In what sense, then, can a lake still be said to be a natural entity?

Second, experiment and technology are claimed to be similar in that both are cultural achievements. The point is that experimental or technological action can be successful only in relation to a cultural norm of what counts as success. In particular, a successful experiment requires that the instruments used are free from disturbances (Janich 1978, 17; 1998, 100–101). A time measurement, for instance, cannot be successful if the pendulum of the clock is substantially expanding as a consequence of increasing temperatures. Since the question of what counts as undisturbed operation of instruments cannot be decided on the basis of natural laws, Janich emphasizes the culturalistic character of his account. Accordingly, he criticizes naturalistic interpretations, in which physics is seen as a value-free search for objective, unified laws of nature.

One could add here that the point of the cultural nature of success criteria equally applies to observations. Observational instruments (glasses, telescopes, microscopes) should be screened from disturbing influences just as well. Observations made with dirty or steamy glasses count as inaccurate. On this argument, both experiments and observations qualify as cultural, technological activities. Hence, it leads to a further weakening of the contrast between observation and experimental measurement, and thus it reinforces the arguments for science as technology.

Third, Janich argues that the justification of causal explanations of natural phenomena depends on the possibility of either replicating or simulating these phenomena under artificially produced laboratory conditions (Janich 1996a, 48–49; 1998, 108). Thus, it is laboratory experiments on light emitted by specific materials that justify the claim that the spectral lines in starlight have been caused by specific atoms and molecules, either in the interior of the stars themselves or in the intermediate space between the stars and the observational setup. Or, to cite Janich's own example, it is only experiments on how shadows are formed when light is interrupted by intervening bodies that give us full confidence in the causal explanation of solar eclipses.

In a fourth and final argument, Janich (1996a, 50–51) claims that all scientific theories—even those about "natural" phenomena, such as the theories of astronomy—should be interpreted as prescriptive know-how, as knowledge about how practical states of affair should be brought about by means of technological intervention. Moreover, such know-how is said to be based on the successful performance of language- and theory-independent, instrumental actions. "The foundation of prescriptive know-how sentences . . . can be brought about in a definitive and conclusive manner, to wit by having recourse to the most elementary instructions on how to act. Then, simply following out these elementary instructions, in the sense of carrying them out in a language-free manner, produces the foundational states of affair" (Janich 1996a, 50; my translation).

Physics, for instance, is said to presuppose a methodically prior "protophysics," which provides normative definitions or, better, constructions of the basic physical concepts (length, time, and mass) by means of binding operations with the relevant measuring instruments. Thus, Janich's constructivist account of science as technology implies that the basic theoretical concepts, or the core of scientific theories, rest on language- and theory-independent, instrumental actions. In this sense, science ultimately *is* technology.

I address this science-as-technology approach for two reasons. First, in thinking about the philosophical and sociocultural meaning of science it is crucial to take account of the close connection between science and technology. This connection results, primarily, from the action and production character of experimentation. To be sure, Janich's approach is not the only one that highlights the intimate relationship between science and technology (see, for example, Dingler 1928; Habermas 1970, 1978; Radder 1986, 1996, chaps. 2, 6, 7; Ihde 1991; Lelas 1993, 2000; Lee 1999). Yet the *theoria,* or spectator, view of science is far from extinct, especially in philosophy of science and epistemology (see also Tiles 1993). Hence, it remains important to reiterate the general point of the intrinsic connection between science and technology.[2]

At the same time, however, I want to argue that Janich's science-as-technology view cannot offer an adequate, comprehensive interpretation of science, including experimental science. This deficiency is due to a reductionist tendency that characterizes this view. Janich's constructivist approach aims at the unambiguous definition of the meaning of theoretical concepts by showing how they can and should be constructed on the basis of particular experimental procedures. Against this claim I shall argue that theoretical concepts are open-ended and hence their meaning is affected by a specific kind of ambiguity. Thus, my criticism of Janich's science-as-technology view pertains, primarily, to the fourth argument I reviewed above. More generally, I think that—because of its intuitive and arguable implausibility—any suggestion of a complete reduction of theory and theoretical interpretation of experiments to language-independent action will backfire and thus contribute to the continuation, rather than the demise, of the *theoria* account of science.

1.2. Experimental Replicability and the Ineliminability of Language and Theory

In Janich's account, theory is strongly tied to *particular* experimental setups. He claims, for instance, that the function of theory is to make tradition possible—that is to say, to communicate to others the knowledge acquired through specific experimental interventions (Janich 1978, 19–20; see also Lange 1999, 68). In contrast, I think that an essential step from experiment to theory is to disconnect theoretical concepts from the particular experimental processes in which they have been realized so far.

In order to make this point I need to deal briefly with the issue of the reproduction and reproducibility of experiments.[3] It is useful to distinguish between three types of reproducibility: the reproducibility of the material realization of the experimental process; the reproducibility of the experimental process under a fixed theoretical interpretation; and the reproducibility of the result of an experiment, also called its replicability (Radder 1996, chaps. 2 and 4). For my present purposes, the significant point is the contrast between replicability and the first two types of reproducibility.

The notion of replicability can be explained as follows. In practice, scientists designate one part or aspect of the overall theoretical interpretation of the experimental process as its result. *Which* part or aspect is taken to be the result of the experiment will depend on the particular context. Setting apart a result enables the scientist to focus on a certain part or aspect of the overall experimental process and then to consider this item regardless of the specific process from which it originally resulted. Generally speaking, results will be statements about some outcome of experimental processes. We may, for instance, produce an (approximately) monochromatic yellow light ray by passing white sunlight through a prism, by selecting a yellow ray through screening off the other colors, and by measuring its frequency. If the experiment is reproducible under this theoretical interpretation, then we might designate the statement "this type of yellow light ray is (approximately) monochromatic" to be the result of this experiment.

In experimental practice, however, scientists are often less interested in reproducing the entire experimental process than in replicating an experimental result by means of a quite different experimental setup (see Collins 1985, 19; Hacking 1983a, 231; Rouse 1987, 86–92; Baird, this volume, chap. 3, sect. 3.2). As a matter of fact, many attempted replications prove to be successful. Imagine, for instance, that there is a second experimenter who intends to replicate the result of the above-mentioned prism experiment on the basis of a rather different experimental process. For instance, the yellow light ray might now be emitted by excited atoms produced in a laboratory context, while its frequency might be measured by means of a different frequency-measuring device. Let us consider the second experimenter at the moment when the first experiment has been successfully completed, whereas the second one has been proposed and accepted as being sensible but has not yet been performed. Apparently, in planning and designing the second experiment the experimenter assumes that the statement "this type of yellow light ray is (approximately) monochromatic" might be realized through the new experimental process as well.

More generally, scientists usually suppose that an experimental result remains meaningful *in abstraction of* the specific realization conditions of the original experiment. The sensible attempt to enlarge the scope of a result by applying it to a materially different domain involves abstracting from the original realization context. I call an experimental result *replicable*, if it has been successfully applied to a certain domain and if it *might* be realized in one or more new domains. The notion of a replicable result is a modal notion. It designates some set of realizability conditions that would enable its actual replication if they could and would be realized. The point can be summarized by saying that the theoretical concepts occurring in the description of replicable results possess a *nonlocal meaning,* which transcends the meaning they have as interpretations of the results of the experimental processes that have been realized so far. Thus, the nonlocality of the meaning of such theoretical concepts reflects the "unintended consequences" that might arise from their potential use in novel situations.[4]

As we have seen, on Janich's account science can be founded, ultimately, on language-independent, instrumental action. In contrast, the claim of this section is that the frequent occurrence of replicability claims and the nonlocality of the meaning of theoretical concepts argue against a reduction of the result of an experiment to a fixed set of instrumental actions. Janich assumes that the meaning of experimentally determinable concepts (for example, the measurement of the length of a body) is normatively fixed by the particular experimental procedures by which these concepts can be realized. The above account, however, implies that the theoretical concepts that are linguistically expressed in the description of replicable results do not have a fixed but a nonlocal meaning. Because of this specific "open-endedness," theoretical concepts and their linguistic expressions cannot be reduced to a well-defined set of language-free actions.[5]

1.3. The Role of Replicability in Edison's Construction of the Incandescent Lamp

To demonstrate the practical significance of making claims of replicability, I consider the case of Thomas Edison's invention of the incandescent lamp as it is reviewed and interpreted by Bruno Latour. In Latour's actor-network theory, science is viewed as basically similar to technology (Latour 1987, 131–132, 168–169). More particularly, his view of the theoretical side of science is analogous to that of Janich in claiming that the meaning of theoretical laws and theories does not transcend the sum of the actions (that is, the network interactions) through which they have been produced in the first place. A theoretical equation, such as Ohm's law, "is no different in nature from all the other tools that allow elements to be brought together, mobilised, arrayed and displayed; no different from a table, a questionnaire, a list, a graph, a collection" (Latour 1987, 238). Moreover, laws and theories are not autonomous. In line with the empiricist tendency that characterizes Latour's work, he is at pains not to ascribe an autonomous meaning to "abstract" theories and equations. Their meaning cannot be established in abstraction of the networks (for instance, the production of a graph through a specific experimental process) from which they have been derived. "The equations . . . constitute, literally, the *sum* of all these mobilisations, evaluations, tests and ties. They tell us what is associated with what" (Latour 1987, 240). Laws and theories are merely a special kind of summary of more "empirical" results of science, and hence their meaning remains tied to the meaning of the various local networks, the results of which they summarize.[6] At the same time, Latour emphasizes the basic significance of theoretical and mathematical work for (the study of) science. Mathematical theories are called "the true heart" of scientific networks, and their study and interpretation is deemed to be more important than that of the construction of facts (Latour 1987, 240–241).

Latour illustrates his views on the meaning of laws and theories with the case of Edison's construction of a system of electric lighting during the late 1870s and early 1880s. This large technological system was meant to be competitive with, and eventually to supplant, the system of gas lighting. A major part of the construction work was the invention of the incandescent lamp. Edison's invention resulted from a synthesis of economic, technological, and scientific aspects (Hughes 1979; see also Hughes 1983, chap. 2). On the one hand, in view of the high price of copper, the copper conductors needed to be as thin as possible in order to make the technology competitive. On the other hand, conductors with a large cross section are required if one wants to minimize energy losses through heat production during the distribution of the electricity. The combined laws of Ohm and Joule suggested a way out of this dilemma to Edison and his collaborators. By increasing the resistance of the lamp filaments, the electric currents would be small; hence the energy loss, even in thin conductors, would be limited. Thus, the task was to find a high-resistance filament that also met the other technical requirements for use in an incandescent lamp, such as durability, infusibility, and the possibility of parallel wiring.

In this way, Edison's reasoning combined a theoretical-mathematical argumentation with economic and technological considerations. Led by results derived from the laws of Ohm and Joule, he started a systematic experimental search for an appropriate filament. Thomas Hughes describes the episode as follows: "Edison stated that in the fall of 1878 he had experimented with carbon filaments but that the major problem with these was their low resistance. . . . An apparently remote consideration (the amount of copper used for conductors), was really the commercial crux of the problem. . . . Edison then said that he turned from carbon to various metals in order to obtain a filament of high resistance, continuing along these lines until about April 1879 when he had a platinum of great promise because the occluded gases had been driven out of it, thereby increasing its infusibility" (Hughes 1979, 137). Latour presents the case of Edison as supporting his actor-network approach in general and his "operationalist" interpretation of the meaning of theoretical concepts—such as those occurring in the laws of Ohm and Joule—in particular.

The point of my review of this case is, however, to show that Edison's invention of the incandescent lamp does not suit Latour's framework but instead confirms the significance of the nonlocality of the meaning of replicable experimental results. To see this, consider the following account of the events. The experimental results at issue were phrased in terms of the concepts from the nineteenth-century theories of electricity and heat, such as electric current, voltage, resistance, power, and energy. At the time Edison started his searches, these results had been realized in a number of particular experimental processes by Ohm and Joule, and they had been shown to be related in a specific manner. Edison's search for a high-resistance filament and his intention to experiment on "various metals," including platinum, was based on the assumption that the experiments of Ohm and Joule are replicable. That is to say, Edison's plans make sense only if he presupposed that the relevant experimental results might be replicated in the novel context of his own experiments. In other words, when Edison started his experimental search he assumed that these results have a nonlocal meaning that transcends the meaning they have as a consequence of their use in the experiments by Ohm and Joule.

Edison searched for a new metal with specific properties. He did not try to make his experiments as similar as possible to those of Ohm and Joule. In a specific way, he "severed" the results obtained by Ohm and Joule from the networks in which they had been realized thus far. In other words, the implied claim about the nonlocality of the meaning of the relevant concepts and laws structured Edison's search process and guided him in a particular direction. The crucial point is that this structuring and guiding took place *before* the replications of the experiments of Ohm and Joule had been actually realized. Latour, in contrast, claims that theoretical laws and mathematical concepts or equations can make a difference only "once the networks are in place" (Latour 1987, 239). Hence, he is unable to explain this important episode in Edison's construction of a system of electric lighting. The underlying reason for this shortcoming is that an interpretation of experimental results as a mere sum of past actions cannot do justice to the future-oriented, heuristic significance of the idea of experimental replicability.

2. The Impossibility of Theory-Free Experiments

The second theme of this chapter is the role of (existing) theory in experimental practice. Several philosophers have emphasized the relative autonomy of experimentation from theory. In his 1978 article, Peter Janich phrases the point as follows: "A high respect for academic traditions on the one hand and a certain disdain for technical application on the other led historically to the belief that experiments are *nothing but* a means for gaining true knowledge of nature although experiment played an important role from the very beginning of men's use of tools and caused the development of a technology independent of any theoretical claims" (Janich 1978, 18). Ian Hacking has developed a similar line of argumentation. He discusses the claim that "an experiment must always be preceded by a theory" and proposes to distinguish between a weak and a strong version of that claim. "The *weak version* says only that you must have some ideas about nature and your apparatus before you conduct an experiment. . . . There is however a *strong version*. . . . It says that your experiment is significant only if you are testing a theory about the phenomena under scrutiny" (Hacking 1983a, 153–154). Next, Hacking argues that the weak version is quite plausible or even trivial. According to his understanding of experimental practice, "experiments" without ideas are meaningless and hence not experiments at all. In contrast, the strong version of the claim—advocated by Justus von Liebig and Karl Popper, among others—is said to be unwarranted on the basis of examples from the history of science. Thus, many noteworthy phenomena would have been experimentally investigated independently of any theoretical interpretation.[7] In particular, Hacking mentions a number of experiments from the early history of optics, between 1600 and 1800.

More recently, the issue of the theory (in)dependence of experimentation has been examined in somewhat more detail. Friedrich Steinle discusses the question of whether experiments are always "guided" by theories. An experiment is theory-guided if it is being explicitly planned, designed, performed, and used from the perspective of particular theoretical claims about the objects under scrutiny (Steinle 1998, 286, 289). Theory guidance in this sense lies somewhere between having an idea and testing a theory. It is much more specific than the former but includes more than the latter. Next, Steinle introduces the notion of exploratory experiments and he argues that this type of experimentation is not theory-guided in the above sense, even if it does require reflection and deliberation. He illustrates his arguments with a discussion of some early electromagnetic experiments carried out in 1820 by André-Marie Ampère. In addition, he claims that exploratory experimentation occurs in the practice of other scientists and other sciences as well.

A further recent contribution has been made by Michael Heidelberger. He uses the term "theory-ladenness" and makes a distinction between "theory-ladenness through theoretical interpretation" and "theory-ladenness through reference to a prior understanding" (Heidelberger 1998, 85–86; see also Heidelberger, this volume, chap. 7). In the same spirit as Hacking and Steinle, he admits that experiments do depend on prior understanding, which entails certain expectations about the performance of the experiments. But he also claims that certain types of experiments are possible that depend neither on a theory of the relevant phenomena nor on a prior theoretical interpretation of the instruments. By adding the latter, his views seem to be stronger than those of Hacking and Steinle, who merely claim that sensible experimentation does not always presuppose a theory *of the phenomena* under scrutiny. Heidelberger develops this view as follows. He proposes a classification of instruments into three categories: productive instruments (for example, an air pump), constructive instruments (for example, a wind tunnel), and representative instruments (for example, a thermometer). On this basis, he claims that experiments that employ only productive or constructive instruments are fully theory-free.

The views advocated by Hacking, Steinle, and Heidelberger share two basic claims. First, they assume that there is a philosophically significant contrast between "ideas" (or "understanding"), on the one hand, and "theories" of nature and apparatus, on the other. Second, they adduce examples from the history of science that are meant to exemplify experiments led by ideas or understanding but not by theories.

2.1. The Relationship between Experiment and Theory: Four Different Claims

The issue of the relationship between experiment and theory is complex, and the claim of this chapter is that it demands a more differentiated treatment than has been advanced so far. The challenge is to take account of both the specific features of experimental activities and the various roles played by theoretical interpretations. In this section, I discuss four different claims about how experiment and theory relate to each other. The first three are relatively uncontroversial. Hence I will deal with these rather quickly and focus on the claim that involves a substantial disagreement with the views summarized above.

All Experiments Are Explicit Tests of Existing Theories About the Objects in Question

This is the strongest version of the claim that experimenting is theory dependent. It completely subordinates experimental inquiry to theoretical research. It can be safely concluded, though, that this claim is most certainly false on any reasonable understanding of what is taken to be an experiment (see the above-mentioned studies, among many others). Yet, the fact that experimentation involves much more than theory testing does not, of course, mean that testing a theory may not be an important goal in particular scientific settings. Thus, Gerd Lüer (1998, 200–204) argues that so far psychological experiments have had too much of a life of their own. In psychology, an exclusive focus on experimental methodology has been at the expense of a theoretical understanding of underlying psychological mechanisms. Accordingly, Lüer claims that psychological experiments, if they are to become more informative, should be developed in closer connection to explanatory theories.

All Experiments Are Theory-Guided

This claim says that experiments are planned, designed, performed, and used from the perspective of one or more theories about the objects in question. Since "guidance" is meant here in the rather strong sense of a specific and continuing impact of theories of the objects under study on various activities during the successive stages of the experimental process, this claim is too extreme to be plausible. Steinle is certainly right in arguing that not all experiments are theory-guided in this sense. But again, this does not preclude the case of theory guidance being found in scientific practice. Steinle himself (1998, 282–284) discusses the example of Ampère's experiments to test his hypothesis that all magnetism is caused by electric currents circulating in material bodies. Modern particle physics is another area where theory-guided experiments occur regularly.

The (Immediate or Later) Significance of Experiments Is Affected by the Theoretical Context in Which They Are Situated

This claim involves a broader role for theories than the first two. The significance of an experiment is said to depend, to a lesser or greater degree, on its theoretical relevance. Consider again the early experiments on the interaction between galvanic electricity and magnetized needles performed by Hans Christian Oersted and, among others, Ampère. The immediately acknowledged significance of Oersted's original experiments largely derived from the fact that they realized a kind of phenomenon that appeared to be at odds with the dominant theoretical discourses. An important point was that there seemed to be a non-Newtonian—that is, a noncentral—force at work (Gooding 1990, 29–36). In particular, as Steinle (1998, 274–277) notes, the existence of an interaction between electricity and magnetism also did not fit in Siméon Poisson's theoretical account of electricity. Another illustration of the above claim is provided by Evangelista Torricelli's well-known experiment, performed in 1644. Torricelli poured mercury into a glass tube and then inverted the tube in a dish filled with the same substance. A crucial question concerned the nature of the space left at the top of the mercury column: is there a vacuum or not? Thus, the experiment derived its theoretical significance from its direct bearing on the long-standing dispute over the possibility of a vacuum (Shapin and Schaffer 1985, 41). In these two cases, the theories in question were neither being tested nor guiding the realization of the experiments in all sorts of ways. Instead, they increased the theoretical relevance of these new kinds of phenomena and thus strongly stimulated their continuing experimental exploration.

The significance of an experiment may also change over the course of time as a consequence of further theoretical developments. In this case, the same material realization of an experiment gives rise to different theories of the experimental process and different theoretical results (cf. Radder 1988, chap. 3; 1996, chap. 2). An illustration can be found in Carrier (1998, 186–189). He discusses the experiment by which Hippolyte Fizeau measured the shift in the interference pattern of light when it passes through moving water. Carrier shows that, over the course of time, three conceptually disparate interpretations of this experiment were proposed: Fizeau's own ether drag account (1851), Hendrik Lorentz's approach in terms of the interaction between light and the charged constituents of matter (1892), and Albert Einstein's explanation by means of the relativistic addition of velocities (1905). Hence, the significance of Fizeau's original experiment changed fundamentally as a consequence of reinterpretations through radically new theories.

On the basis of such cases, the claim that the immediate or later significance of experiments is affected by the theoretical context is plausible enough. The present claim, however, is not necessarily incompatible with the accounts of Hacking, Steinle, and Heidelberger, because the relationship between experiment and theory is less tight than in the case of the first two claims. As we have seen, it even applies to Steinle's example of Ampère's exploratory experiments. Unwittingly, Hacking (1983a, 156) provides another example. He notes that the phenomenon of double refraction in Iceland spar surprised seventeenth-century physicists and that this surprise was, in part, caused by an apparent conflict with the theoretical laws of refraction.

Performing and Understanding Experiments Depends on a Theoretical Interpretation of What Happens in Materially Realizing the Experimental Process

This is a view I have held myself for a long time through claiming that experimentation involves both material realization and theoretical interpretation (Radder 1988, 59–76; 1996, 11–12). As I explained at the beginning of this chapter, experimenting involves, at least, the material realization of an interaction between an object and some apparatus in such a way that a stable correlation between some feature of the object and some feature of the apparatus will be produced. In these terms, the claim is that materially realizing a stable correlation and knowing what can be learned about the object from inspecting the apparatus depends on theoretical insights about the experimental system and its environment. Thus, these insights pertain to those aspects of the experimental process that are relevant to obtaining a stable correlation. It is not necessary, and in practice it will usually not be the case, that the theoretical interpretation offers a full understanding of any detail of the experimental process.

To avoid misunderstanding, it should be clear that this account of scientific experimentation does not imply that each and every activity that occurs in experimental practice involves theoretical interpretation. On the contrary, this account explicitly admits the possibility of a description of the material realization of experiments in "common," nontheoretical language. What it denies is that an adequate philosophical understanding of scientific experiments can be obtained without including the role of theoretical interpretation.

This particular form of theory dependence, however, is at odds with the account advocated by the authors whose views I discussed at the beginning of this section. Hacking, for instance, writes: "It remains the case . . . that much truly fundamental research precedes any relevant theory whatsoever" (Hacking 1983a, 158). And Heidelberger states: "Normally, the experimental use of productive and constructive instruments does not presuppose a theoretical interpretation" (Heidelberger 1998, 87; my translation). As I noted before, the view that much experimental inquiry precedes, or does not presuppose, theoretical interpretation is claimed to be supported by examples from the history of science. Hence, to assess the plausibility of this view I will examine two of the mentioned cases in some detail.

I start with some experiments from the early history of optics. Hacking (1983a, 156) tells us that Newton's experimental observations of the dispersion of light preceded any theoretical interpretation, and Heidelberger (1998, 82) includes the prism in his category of theory-free, productive instruments. In many of Newton's experiments on the dispersion of light, carried out between the 1660s and the 1720s, the prism was his major instrument. One series of experimental trials involved the use of a single prism to produce colored light projected on a screen. A further, very important, series of experiments investigated what happens when light of a specific color that goes out from a first prism is sent through a second one. The following account of Newton's prism experiments shows that they depended in at least three ways on a theoretical interpretation of what was going on in the material realization of the experimental processes (see Schaffer 1989; see also Hakfoort 1986, chap. 2).

A first question concerns what may be learned from inspecting the apparatus (looking at the colors on the screen) about the object under study (the light before it impinges on the prism). Only if the outgoing light is basically of the same nature as the incoming light are we able to learn something about the latter by examining the former. Scholastic theories of light, however, usually distinguished two kinds of color: apparent colors produced from modified light (for example, in prisms) and real colors disclosed, but not produced, by light. Hence, on the basis of these theories, prisms were inappropriate for probing the nature of real light and real colors. That is to say, on this view prismatic colors are mere artifacts of the instrument and not intrinsic constituents of light. In his studies on optics, however, René Descartes had criticized the distinction between real and apparent colors and claimed that all colors were apparent, or secondary. In this respect Newton followed him. As Schaffer (1989, 73–74) concludes, it was only because of this changed theoretical context that prisms could acquire a key new role in the experimental study of light and color.

Second, there is the notion of a light ray. The idea of rays traveling in straight lines and capable of being refracted upon entering another medium was employed routinely in analyzing experiments by Newton and many others before him. It played a role in the performance of the experiments as well. In many experiments a "single ray" was needed. For this purpose, light should be passed through a slit of the right dimensions, while the room should otherwise be as dark as possible. According to Newton, such conditions were crucial for realizing a stable experiment that could be reproduced by others (see Schaffer 1989, 88, 93). Thus, the theoretical model of light as consisting of rays, traveling along straight lines, and refrangible on entering a different substance structured the way in which Newton performed and understood his experiments on the dispersion of light.[8]

Third, there was the more specific notion of primitive rays, each possessing a particular refrangibility. This notion was very important to Newton. He stated that the existence of such rays could be demonstrated by means of the two-prism experiment: a ray that comes out of the first prism is primitive (or uncompounded) if its color does not change when it is passed through the second prism. It proved quite difficult, however, to realize this experiment in a stable manner. Several experimenters from different countries failed to reproduce it. In response, Newton claimed exclusive authority for his own theoretical interpretation by arguing that his critics had failed to produce a primitive ray in the first place. Thus, Newton stuck to his theory that light is a heterogeneous mixture of differently refrangible, primitive rays, which he said had informed all his prism experiments since about 1666 (Schaffer 1989, 71). This, however, was not the only possible theoretical interpretation. Robert Hooke, for instance, did succeed in reproducing Newton's experiments, but he interpreted them on the basis of his own vibrational theory of light and color (Schaffer 1989, 86–87).

Let us now turn to a second case of claimed theory-free experimentation. As we have seen, Heidelberger distinguishes between productive, constructive, and representative instruments. Both productive and constructive instruments aim primarily at the creation of stable phenomena. Air pumps, prisms, and particle accelerators are examples of productive instruments, while Leyden jars and wind tunnels exemplify constructive instruments. Experiments employing only productive or constructive instruments are claimed to be theory-free. That is to say, neither is it the proper aim of such experiments to test a presupposed theory or even the relevant prior understanding, nor does this prior understanding imply a theoretical interpretation of the experiments (Heidelberger 1998, 86–87). I will leave aside the issue of the adequacy of the proposed typology of instruments (for a discussion of this issue, see Buchwald 1998, 384–391). Instead, I will examine a case that is put forward by Heidelberger as a prime example of a theory-free, productive instrument, to wit, the air pump.

Consider, more specifically, the case of the air-pump experiments performed by Robert Boyle during the late 1650s and early 1660s (see Shapin and Schaffer 1985, chaps. 3 and 4). This case shows that both the working and the possible outcomes of Boyle's productive experiments with the air pump were directly dependent on theoretical assumptions. These assumptions were made explicit by Thomas Hobbes in his criticisms of these experiments. Boyle's central claim was that his air pump produced a space that was (almost) totally devoid of air. This claim required that the routine leakage of air was "negligible." Hobbes, however, argued on the basis of his theory of the nature and composition of air that the pump leaked, not just slightly and incidentally—as was conceded by Boyle—but in a significant and consistent way (Shapin and Schaffer 1985, 115–125). According to this theory, common air is a mixture of different substances: earthy or aqueous fluids and pure or ethereal air. Because of the presence of the latter, air is infinitely divisible. Consequently, an absolutely impermeable seal is impossible, and thus the air pump is bound to leak. In addition, Hobbes claimed that his theory was able to account for the results of Boyle's experiments just as well. In one trial, for instance, a candle placed in the receiver of the air pump went out after some time of pumping. And while Boyle accounted for this on the basis of the absence of air, Hobbes simply claimed that the candle had been extinguished as a consequence of the violent circulation of the air caused by the pumping. In this way, Hobbes intended to show that (the results of) Boyle's experiments depended on questionable theoretical assumptions.

What does this episode from experimental science tell us with respect to Heidelberger's claims? Heidelberger is right in as far as Boyle did not intend his air-pump experiments to be an explicit test of either an already available theory or his prior understanding. Thus, the case confirms the earlier point that not all experiments are meant to test theories or other knowledge claims.

There is more to this case, though. Consider the claim that "the air pump produces a vacuum, which human beings could not experience in nature without this instrument" (Heidelberger 1998, 81; my translation). In fact, this claim has not always been evident. In Boyle's days, the conclusion that the space in the receiver was a vacuum was highly contestable because of the big theoretical controversy between plenists and vacuists about the possibility of a vacuum in nature. Hence Boyle, who wished not to go beyond "matters of fact," intentionally refrained from drawing that conclusion. Thus, he understood a vacuum to be a space (almost) totally devoid of air and not a space without any bodies at all (Shapin and Schaffer 1985, 46).

However, the claim that the air-pump experiments demonstrated the existence of a vacuum in Boyle's sense depended on theoretical interpretation as well. The conclusion that he had produced a space devoid of air required that Boyle took a stance in the ongoing theoretical debate on the nature and constitution of air. In his reply to Hobbes's criticism, Boyle conceded that it was possible that air had an ethereal part and that this part might always be left in the receiver. But he denied that the presence of this ether could have an impact on his experimental results, since the ether was not experimentally "sensible" (see Shapin and Schaffer 1985, 178–185). The general point is that the correlation between features of the apparatus (for example, the candle going out) and corresponding features of the object under study (for example, the space devoid of air) is stable only if the experimental setup constitutes a closed system.[9] To know whether or not a system is closed requires a theoretical interpretation of the possible interactions between the (object-apparatus) system and its environment. In the case under discussion, Hobbes argued that there was such an interaction (the intrusion of outside air into the receiver) *and* that it disqualified the claimed experimental results. Boyle, in contrast, admitted that there might be such an interaction, but he argued that it was *irrelevant* anyway for the production of stable experimental matters of fact.

In the above discussion, the examples have been deliberately chosen from early experimental science. After all, here the view that experimentation may be theory-free seems to have an initial plausibility. Surely, such a view will be far less likely when the objects under study are further removed from ordinary experience and the apparatus is more advanced and complex (see the example of the absorption spectrometer discussed in Rothbart and Slayden 1994). Heidelberger claims, in particular, that the particle accelerators of modern physics enable a theory-free experimental production of new constituents of matter. It is, however, not hard to show that performing and understanding these experiments does require a lot of theory, both about the particles themselves and about the operation of the instruments (see, for example, Morrison 1990, 6–14).

2.2. The Relationship between Experiment and Theory: Systematic Arguments

Thus far, my analysis has focused on the role of theory in a number of cases from the history of experimental science. At the same time, this analysis was informed by a theoretical account of scientific experimentation. In this section I will discuss some further, more systematic, arguments relating to the claim that performing and understanding scientific experiments depends on theoretical interpretation.

First, there is a general and, I think, quite forceful argument for the theory dependence of experimentation. It does apply to the historical cases discussed so far, even if it did not come up explicitly. The argument is most clear in the case of quantitative experiments, but it holds for qualitative ones as well. It has to do with the stability of the results of different runs of the same experiment. As Rainer Lange (this volume, chap. 6) emphasizes, a single experimental run is not enough to establish a stable result. A set of different runs, however, will almost always produce values that are, more or less, variable. The question then is this: What does this fact tell us about the nature of the property that has been measured? Does the property vary within the fixed interval? Is it a probabilistic property? Or is its real value constant and are the variations due to random fluctuations? In experimental practice, answers to such questions are based on an antecedent theoretical interpretation of the nature of the property that has been measured. In the case of qualitative results, the variations will be qualitative in nature. Yet, similar questions may be asked about the meaning of the differences between the various runs of a qualitative experiment.

We may, for example, perform a number of experimental runs to measure the boiling temperature of a fluid under conditions that are taken to be the same. Of course, the measured values can be averaged in a certain way, but the claim that this average represents a fixed boiling *point* is clearly theory-dependent (see Radder 1988, 68). Similarly, David Sohn (1999) discusses the significance of the variations in the results of psychological experiments with different individuals.

He rightly argues that the claim that experimental effects are variable (or, alternatively, constant) across individuals depends on a prior theoretical interpretation. In both cases, the theoretical interpretation fixes a specific and significant aspect of the object(s) under study.

Next, it is important to note that, in experimental practice, the theoretical interpretation will not always be explicit and the experimenters will not always be aware of its use and significance. Once the performance of a particular experiment becomes routine, the theoretical assumptions "drop out of consciousness" and they become like an (invisible) "window to the world" (cf. Heelan 1983, chap. 11). Yet, in a context of learning to perform and understand the experiment or in a situation where its result is very consequential or controversial, the implicit interpretation will be made explicit and subjected to empirical and theoretical scrutiny. This means that the primary locus of the theoretical interpretation is the relevant scientific community and not the individual experimenter.[10] Thus, in his discussion of the inference from a state of the apparatus to a state of the object, Peter Kosso is right to claim that "the inference we speak of here is that justification which is available to the epistemic community should an explicit account of the informational content be called for" (Kosso 1989, 141). Hence, the philosophical claim that the performance and understanding of an experiment depends on theoretical interpretation is itself a theoretical claim. It does not primarily denote an empirically ascertainable process taking place in individual observers, but it refers to a logical dimension of how experiments function in epistemic communities. At the same time, as I indicated above, under specific conditions the theoretical claim of the theory dependence of experiments can be empirically confirmed.

Finally, there is the question of the distinction between "ideas" (or "understanding") and "theories." As we have seen, such a distinction is used by Hacking, Steinle, and Heidelberger to support their view that experiments may be guided by ideas or prior understanding and still be essentially theory-free. Since a full assessment of the philosophical merits of the theory-idea distinction is clearly beyond the scope of this chapter, I will limit myself to indicating briefly some major problems. Even so, it will become clear that the philosophical significance of the contrast between theories and ideas is severely limited.

First, making a distinction between ideas and theories is part of a more general strategy of differentiating the notion of theory.[11] Thus, Hacking (1983a) introduces a (further?) contrast between high-level and low-level theories. On the basis of this contrast, it is claimed, two important philosophical consequences of the doctrine of theory-ladenness can be avoided: the vicious circularity that would arise if all experimental tests of a theory were laden with that theory itself, and the antirealist conclusion that scientific knowledge cannot be about a human-independent reality if it were completely dependent on ever-changing high-level theories. I think that the first claim is plausible, although cases of quite direct circularities do occur. Examples are certain attempts at experimentally detecting free quarks (see Pickering 1981) or the experimental observation of single atoms by means of electron microscopes (see Kosso 1989, 87–94). In contrast, the second claim—employing the distinction between theories and ideas to argue for an entity realist view—cannot stand critical scrutiny.[12]

Apart from this, there is a second reason for the restricted philosophical significance of the contrast between ideas and theories. The main purpose of the early proponents of theory-ladenness—such as Karl Popper, Norwood Hanson, Thomas Kuhn, and Paul Feyerabend—was to demonstrate the inadequacy of empiricism. That is to say, they opposed all forms of epistemology according to which scientific knowledge is founded on theory-free observations or experiments. Since their criticism works as well in the case of ideas (or prior understanding), as in the case of presupposed theories, any appeal to this distinction for empiricist purposes is unwarranted. Thus, whatever the further merits of the "new empiricism,"[13] it cannot include the classical empiricist claim that scientific knowledge is, or should be, justified on the basis of "given" experience.

Furthermore, it will be clear that a strict definition of what a theory is, and hence how it contrasts with a mere idea about nature and apparatus, is hard to come by. Yet, some attempts have been made. Thus, Hacking (1983a, 175) sees a theory as "a fairly specific body of speculation or propositions with a definite subject matter." And Steinle (1998, 285) conceives of theories as systems that aim to account for entire domains of unobservable entities. On these views, however, Newton's interpretation of the nature of light and color and Boyle's account of air surely qualify as theories. Thus, the claims that Newton's experiments on the dispersion of light and Boyle's air-pump experiments were led by ideas but not by theories prove to be historically wrong.

A final drawback of the contrast between ideas and theories is that it may induce a certain presentist bias. Simply put, using this distinction may engender a tendency to interpret yesterday's theories as mere ideas, while current—and especially microscopic—theories are seen as genuinely theoretical. If the philosophical claims about theory (in)dependence are to be historically adequate, this presentist pitfall should be avoided.

Thus, the cases and arguments discussed in sections 2.1 and 2.2 allow me to conclude that the significance of experiments is affected by their contemporary or subsequent theoretical context and that performing and understanding scientific experiments depends on a theoretical interpretation of what happens in materially realizing the experimental process. This conclusion is supported by the philosophical analysis of experiments as attempts to realize stable correlations between specific features of objects and apparatus. Moreover, Newton's prism and Boyle's air-pump experiments show that even cases that appear to be theory-free at first sight prove, on closer inspection, to rely on theoretical interpretation in several ways.

3. Conclusion

In this chapter, the following claims have been vindicated. First, we have seen that it is primarily through the material realization of experiments that science is intimately related to technology. Yet, the (theoretically interpreted) results of experiments cannot be reduced to language-independent actions, as Peter Janich claims, or to a sum of past network interactions, as Bruno Latour has it. The reason for this irreducibility is the nonlocal character of the meaning of replicable experimental results.

Next, I have endorsed the view that experiments are not subordinate to theories of the objects under study. At the same time, I have put forward a number of arguments that show that and show why performing and understanding experiments require a theoretical interpretation of (at least some aspects of) what is going on in the experimental process. Hence, against Peter Janich, Ian Hacking, Friedrich Steinle, and Michael Heidelberger, I would insist on the impossibility of theory-free experiments.

In this way, two different dimensions of the relationship between experiment and theory have been dealt with. On the one hand, the impossibility of theory-free experiments implies that specific experimental practices are always structured by certain theoretical interpretations. On the other hand, the abstraction of replicable results from the contexts in which they have been realized originally constitutes a first but crucial step from specific experimental practices to a scientific discourse in terms of nonlocally meaningful, theoretical concepts.

In spite of this, it should be clear that the claims in this chapter do not imply a return to a theory-first approach, neither in a logical nor in a temporal sense. An interactive view of scientific experimentation, in which the mutual dependencies between material realization and theoretical interpretation are systematically taken into account, proves to be the most adequate.

NOTES

1. In much of Janich's work the focus is on physics, but his more recent studies (for example, Janich 1996a) deal with other sciences as well. For a detailed account of experimental biology, which builds on Janich's work, see Lange (1999).

2. To appreciate the significance of the general approach, one does not need to endorse every aspect of Janich's account. Thus, a less convincing aspect of this account (besides the points to be discussed in the main text) is its view of the scope of experimental intervention. According to Janich (1996a, 33–36; 1998, 102–107), the purpose of experimental action and production is to prepare and start an experimental process, which then needs to be left to its own course. Against this point of view—originally put forward in Von Wright (1971)—I have argued that experimental processes cannot be left alone: the conditions for closing the experimental system need to be realized and maintained during the entire course of the experimental process (see Radder 1988, 63–69; 1996, chap. 6).

3. Some students of experimentation have claimed that in actual scientific practice experimental reproduction and the norm of reproducibility are insignificant. It is, however, not hard to show the opposite, in part with the help of the work of those same scholars (see Radder 1996, 20–28). Latour and Woolgar (1979, 169–170), for instance, describe an episode in immunological research in which the guiding role and effectiveness of the norm of reproducibility is very obvious. Yet these authors fail to notice this fact and instead they interpret the episode as a series of "accidentally related events."

4. The issues surrounding meaning and abstraction are philosophically complicated. Here, I can offer no more than a brief summary of my views. For more, see Radder (1996, chap. 4) and, in much more detail, Radder (2002); the latter also includes a discussion and criticism of operationist theories of meaning and a comparison with Popper's three-worlds ontology.

5. At a more general level this conclusion is connected to the fact that concepts would be useless for communication across varying situations if their meaning would be fully fixed by specific local contexts.

6. Thus, Latour's (descriptive) operationalism advocated in the above quotation parallels Janich's (normative) operativism. Cf. Janich (1996a, 26–31).

7. This, at least, is Hacking's 1983a view. The principal claim of his 1992 essay, however, is that the stability of the laboratory sciences results from a process of mutual adjustment of a variety of heterogeneous elements, including theories and experimental interventions. In particular, he now asserts that theories "are among the intellectual components of experiments" (Hacking 1992, 44). It is not immediately clear whether—and, if so, how—these two views can be reconciled (see also Harré 1998, 369–374).

8. For an insightful account of the theoretical nature and the practical uses of this model, see Toulmin (1967, esp. 16–28). Note also that this model is theory-relative: from the perspective of the theory of general relativity, light does *not* propagate along straight, Euclidean paths.

9. For this notion, see Radder (1988, chap. 3; 1996, chap. 6). See also Boumans's (1999) account of measurement in economics. To his claim that external influences should be either constant, negligible, or absent, I would like to add "or irrelevant."

10. In Radder (1988, chap. 3) I did not sufficiently acknowledge this point. Those critics of the theory-dependence claim who base their criticism on separate case studies of individual scientists have missed the point as well.

11. I would like to thank Francesco Guala for a helpful discussion about this subject.

12. See Radder (1988, 119–120); Morrison (1990).

13. For this label, see Rouse (1987, 9–12); see also Carrier (1998, 179–181) on the similarities with logical empiricism.

Giora Hon

9 / The Idols of Experiment

TRANSCENDING THE "ETC. LIST"

> Our logic instructs the understanding and trains it, not (as common logic
> does) to grope and clutch at abstracts with feeble mental tendrils, but to dis-
> sect nature truly, and to discover the powers and actions of bodies and their
> laws limned in matter. Hence this science takes its origin not only from the
> nature of the mind but from the nature of things; and therefore it is no won-
> der if it is strewn and illustrated throughout with observations and experi-
> ments of nature as samples of our art.
>
> Francis Bacon, *Novum Organum,* 1620

1. Introduction

In the concluding session of the workshop "Experiments: Their Meaning and
Variety" (Bielefeld, Germany, March 1996; see Heidelberger and Steinle 1998),
it became apparent that a divide separates the historians of science from the
philosophers of science as to scientific experimentation. It transpired that the
philosophy of experiment is lagging behind the extensive historical studies of
experimentation and has not yet incorporated the many facets (technological,
cultural, sociological, and anthropological) that historians have addressed. It was
clear that a stronger case for the philosophy of experiment should have been
made. To be sure, there have been attempts at such philosophy, and I shall outline a few of them shortly. However, these attempts have not cohered into a

forceful and cohesive philosophical analysis of experiment, incisive for episte-mology as well as for the historiography of experimentation.

Hans Radder concurred with this view. In his paper "Issues for a Well-Developed Philosophy of Scientific Experimentation," he took stock of the Bielefeld workshop from the perspective of philosophy of experimentation. Radder explicitly acknowledged that the philosophy of experimentation is still underdeveloped, especially in comparison to historical and sociological studies of experiment (Radder 1998). This observation applies, according to Radder, to both the Bielefeld meeting and more generally to the position of the philosophy of experiment within the wider field of philosophical studies of science (Radder, this volume, chap. 1). I fully share with Radder this impression. My objective is therefore to contribute toward a more developed philosophy of experiment. I grope then to bridge the divide between history and philosophy of scientific ex-perimentation by helping consolidate the philosophical bank of the divide. It is conducive to the project, as a preliminary step, to identify and characterize the principal obstacles to the construction of a philosophy of experiment, obstacles that have proved quite recalcitrant. An outline of the tension between history and philosophy of experiment will serve as a background.

2. History vs. Philosophy of Experiment

The position of the historians of science may well be represented by the view of Jed Buchwald. He claims succinctly and bluntly that "living sciences cannot be corralled with exact generalizations and definitions. Attempting to capture a vibrant science in a precise, logical structure produces much the same kind of information about it that dissection of corpses does about animal behavior; many things that depend upon activity are lost" (Buchwald 1993, 170–171). Indeed, according to Buchwald, "axiomatics and definitions are the logical mausoleums of physics" (Buchwald 1993, 170–171). The position of the contemporary histo-rian of science is then to regard science as an activity, not an end result but a process, a "living" and "vibrant" process. The historian's claim is that any gen-erality in the form of, say, logical structure, simply kills this lively activity. The metaphor of the living and the dead appears to be crucial to Buchwald and to historians of science at large. They follow Kuhn's directive, which he formulated right at the beginning of his *Structure of Scientific Revolutions*. According to Kuhn, the aim of history of science "is a sketch of the . . . concept of science that can emerge from the historical record of the research activity itself." "Activity" appears to be the key feature as distinct from "finished scientific achievements" (Kuhn 1970, 1; see also Latour 1987).

The historian may well be happy, therefore, with a detailed description and a thorough analysis of the activity—the "living" particular; but the philosopher must strive, as Hacking put it simply and directly, for "both the particular and the general" (Hacking 1992, 29). There is no escape. If we want to do philosophy— that is, if we believe that philosophy has a bearing on a certain kind of activity— we have then to seek its general features, its underlying principles. In other words, we have to uncover logical structures and characterize methodological principles that govern this activity without, however, losing sight of its particulars, namely, its "living" execution. Now, as to experimentation, it is unquestionable that philosophy ought to have a bearing on this activity—it being one of the chief methods of producing knowledge. We have then no choice but to analyze experiment in vitro, as it were, keeping an open wide eye on its features as an activity in vivo. Buchwald's claim should serve as a warning rather than a condemnation. We should give heed to this warning and follow Whitehead's cautious dictum: "Too large a generalisation leads to a mere barrenness. It is the large generalisation, limited by a happy particularity, which is the fruitful conception" (Whitehead 1929, 39). Thus, a well-developed philosophy of scientific experimentation should bring together in a consistent fashion both the normative aspect of the experimental activity—its descriptive as well as prescriptive dimensions—and a comprehensive theoretical conception of experiment that throws light on its central features, features that underwrite the reliability of the knowledge thus obtained.

I propose the notion of experimental error as an efficient vehicle for attaining this objective. I seek generalizations of the experimental activity that emerge through a study of the notion of experimental error. I claim that while capturing the nature of the experimental activity, the notion of experimental error also reflects, albeit negatively, central conceptual features of experiment. Put differently, the thesis exploits types of experimental errors as constraints by which one may uncover general features of experiment. The articulation of the notion of experimental error originates in the normative dimension—how to rectify and indeed avoid errors in the execution of experiment. However, this articulation reflects at the same time structures and principles of experimentation. The attempt then is to capture at once, via the notion of experimental error, both the normative aspect and the theoretical conception of experiment.

3. Setting the Philosophical Scene: Two Clusters of Problems

To set the philosophical scene, it is useful first to identify the obstacles that obstruct the way to a viable philosophy of experiment. I discern two principal clusters of obstacles to the construction of such philosophy. Not surprisingly, both clusters have to do with the transition from the particular to the general. For reasons that will shortly become clear, I wish to call the first cluster "epistemological" and the second "methodological." As it happened, right at the beginning of the last century, two physicists cum philosophers published pioneering, influential works that bear on these issues. Ernst Mach published in 1905 his *Knowledge and Error.* In this collection of essays, he addresses problems pertaining, in his words, to "scientific methodology and the psychology of knowledge" (Mach 1976, xxxii). Mach dedicated one essay (chapter 12) to the analysis of physical experiment and to identifying its leading features. A year later, in 1906, Pierre Duhem published his book entitled *The Aim and Structure of Physical Theory.* In this book Duhem raised explicitly the question as to "What exactly is an experiment in physics?" (Duhem 1974, 144). While Duhem focuses on the epistemological problem, Mach is concerned with methodological issues.

3.1. The Epistemological Cluster: The Transition from Matter to Argument

The first cluster of obstacles to a philosophy of experiment is in my view the transition from the material process, which is the very essence of experiment, to propositional knowledge—the very essence of scientific knowledge. As Duhem sees it, the experimental physicist is engaged in "the formulation of a judgment interrelating certain abstract and symbolic ideas which theories alone correlate with the facts really observed." The conclusions of any experiment in physics, and for that matter in science, are indeed "abstract propositions to which you can attach no meaning if you do not know the physical theories admitted by the author" (Duhem 1974, 147–148). The end result of an experiment is not, to refer once again to Duhem, "simply the observation of a group of facts but also the translation of these facts into a symbolic language with the aid of rules borrowed from physical theories" (Duhem 1974, 156). In other words, the obstacle is the problematic passage from matter that is being manipulated and undergoes some processes, via observations to propositions whose meaning is provided by some theory—a language expressed in symbols.

Andrew Pickering, to turn to a contemporary author, addresses this problem as a substantial element of the issue of realism. Pickering writes that he is concerned with the process of "finding out about" and "making sense of"; that is, he inquires into the relation between articulated scientific knowledge and its object —the material world (Pickering 1989, 275). He conceives of a three-stage development in the production of any experimental fact: a material procedure, an instrumental model, and a phenomenal model (Pickering 1989, 276–277). These three stages span, according to Pickering, the material and conceptual dimensions of the experimental practice. It is in the arching of these two dimensions that the passage from matter to knowledge should be forged. Pickering is of the opinion that this passage "is one of made coherence, not natural correspondence." In other words, the coherence between material procedures and conceptual models is an artificial product due to the successful achievements of actors in accommodating the resistances arising in the material world (Pickering 1989, 279).

A different attempt at addressing the epistemological issue is that of Davis Baird, who approaches the problem head-on by developing a materialist conception of knowledge (Baird, this volume, chap. 3). The metaphysics of material knowledge may point to a new way toward a philosophy of experiment by avoiding altogether the need to ascend (or arch, in Pickering's terms) from matter to proposition. For Baird, ascent is the wrong metaphor. In his view, the material products of science such as instruments and drugs constitute knowledge, to be sure knowledge of a different kind than propositional knowledge, yet knowledge all the same based indeed, so Baird claims, on equal footings. Baird gropes for a metaphysics in which both the material world and the world of sign could be constitutive of knowledge on equal terms.

In a similar vein, I had recourse elsewhere to a concept that I termed "material argument" (Hon 1998a). With this concept I tried to bring together in a philosophical context all the elements that are involved in experimentation: the theoretical context and the scheme of manipulation, the material processes, and the resulting scientific knowledge that is essentially propositional. I introduced the notion of "material argument" precisely for the purpose of rendering intelligible the transition from the manipulation of matter to the propositions that characterize experimental knowledge—the declared end result of experiment. The transition from matter to proposition presents then the first set of difficulties for a philosophy of experiment. I refer to this cluster of obstacles as the "epistemological issue."

3.2. The Methodological Cluster: Transcending the List of Strategies, Methods, Procedures, Etc.

The second cluster of obstacles is at the level of manipulation of matter; we may refer to this cluster as "methodological." Here we are concerned with the transition from myriad strategies, methods, procedures, conceptions, styles, and so on to some general, cohesive, and coherent view of experiment as a method of extracting knowledge from nature. From a philosophical perspective it would have been fruitful had we obtained a general yet fundamental scheme of experiment that captures in a tight economic fashion these myriad facets and features.

A convincing historical account that exhibits the enormous variety of facets and features that experiment possesses is Darrigol's notion of "transverse principles," which he applied to nineteenth-century electrodynamics. These principles are not general rules of scientific method; rather, they are methodological precepts that regulate at once theory and experiment, hence "transverse." Guided by tradition or one's own ingenuity, the physicist follows a transverse principle that links one's theoretical conception of the physics that one studies to actual experimentation. Clearly, the application of the principle contributes much to the formation and definition of the physicist's methodology (Darrigol 1999, 308, 335).

For an example, consider Faraday. According to Darrigol, Faraday's theories "were rules for the distribution and the interplay of various kinds of forces." Faraday dispensed with the Newtonian distinction between force and its agent. In Faraday's view, "an agent could only be known through actions emanating from it" (Darrigol 1999, 310–311). Thus, the best course to take in the study of a body acting on another body consisted in mapping the various positions and configurations of the body acted upon. This position called for a principle of contiguity. It is this principle that regulated, according to Darrigol, both the theoretical and experimental practice of Faraday: "On the theoretical side, this principle entailed his concept of the lines of forces as chains of contiguous actions and his rejection of the dichotomy between force and agent. On the experimental side, it determined the emphasis on the intermediate space between sources and the exploratory, open character of his investigations" (Darrigol 1999, 312). When Darrigol juxtaposes this approach of Faraday to the studies of other nineteenth-century electrodynamicists, the variety and richness of conceptions of theory and experimental practices become apparent. Darrigol argues persuasively for a close connection between theory and experiment in nineteenth-century electrodynamics. As it is so tightly connected to theory, the conception of experiment and its actual procedures become, at least in this historical episode, enormously varied and complex. The question immediately presents itself as to how should one, as a philosopher, capture in general terms this enormous variety of conceptions of experiment and the practices of their material procedures?

In his essay on the leading features of physical experiment, Mach realizes that these features may not be exhausted. It seems then that a generalization may not be attained. The formative features of experiment, which Mach describes, have been abstracted, so he writes, "from experiments actually carried out. The list is not complete, for ingenious enquirers go on adding new items to it; neither is it a classification, since different features do not in general exclude one another, so that several of them may be united in the experiment" (Mach 1976, 157). Is the list indeed open or is it in fact in the final analysis constrained? If no constraints were to be imposed on this method of inquiry, then no classification and indeed no generalization would be obtained. The approach would be eclectic and ad hoc.

A good illustration of an elaborated list that goes beyond Mach's preliminary list and yet remains ad hoc is Allan Franklin's list of "epistemological strategies," which he convincingly buttresses with elaborated case studies. Here is the list of strategies Franklin has drawn:

1. Experimental checks and calibration, in which the apparatus reproduces known phenomena.

2. Reproducing artifacts that are known in advance to be present.

3. Intervention, in which the experimenter manipulates the object under observation.

4. Independent confirmation using different experiments.

5. Elimination of plausible sources of error and alternative explanations of the result.

6. Using the results themselves to argue for their validity.

7. Using an independently well-corroborated theory of the phenomena to explain the results.

8. Using an apparatus based on a well-corroborated theory.

9. Using statistical arguments. (Franklin 1990, 104; cf. 1986, chaps. 6, 7; 1989)

Franklin argues that these strategies have been designed to convince experimenters that experimental results are reliable and reflect genuine features of nature. The list of strategies demonstrates, according to Franklin, the different ways experiments gain credibility. Practicing scientists pursue such strategies to provide grounds for rational belief in experimental results (Franklin 1989, 437, 458). For Franklin, the use of these strategies has then the "hallmark of rationality" (Gooding, Pinch, and Schaffer 1989, 23), and in that sense he is seeking to contribute to a philosophy of experiment.

However elaborated and complex, the list of strategies Franklin puts forward is essentially similar to the list Mach presents in his essay on the leading features of experiment. Like Mach, Franklin is aware of the limitation of this approach —the account is ad hoc. Franklin indeed states that the strategies he documented are neither exclusive nor exhaustive. Furthermore, these strategies or any subset of them do not provide necessary or sufficient conditions for rational belief. "I do not believe," he states, that "such a general method exists" (Franklin 1989, 459). Nevertheless, Franklin is convinced that scientists act rationally. According to Franklin's unfailing optimism, scientists use, as Gooding, Pinch, and Schaffer (1989, 22–23) aptly put it, "epistemological rules which can be applied straightforwardly in the field to separate the wheat of a genuine result from the chaff of error."

Franklin is much concerned with the working scientist, or rather the practicing experimenter, and it appears that the strategies he lists have been in fact abstracted from actual experiments, precisely as Mach did a century earlier. As such, his list, although rich and varied, remains eclectic and ad hoc. While each item on the list provides a specific illustration of an experimental procedure that is designed to give grounds for rational belief, there appears to be no overall guiding principle to govern the list itself. Most importantly, the list does not throw light on the inner processes, the epistemic dynamic processes, inherent in experiment that connect propositions with matter—for example, the link between the background theory that governs the experiment and the actual functioning of the setup; or, for another link, the connection between matter and propositions in obtaining an experimental result from observations and measurements. Each item on the list addresses a certain procedure, a certain experimental methodology, external, as it were, to what in effect takes place within experiment, namely the translation of a controlled and confined process into some proposition that expresses the result of the process. It is not surprising, therefore, that the items do not form a comprehensive view of experiment. Such a list cannot be completed since no constraint is being imposed. This approach would not result in a coherent generalization of experiment.

This is then another problem that is posed to the philosopher of experiment, namely, how to transcend "the list"? How to generalize the various items that comprise the list? In attempting an answer to this issue we should give heed to Hacking's warning and be careful not "to slip back into the old ways and suppose there are just a few kinds of things, theory, data, or whatever" (Hacking 1992, 32; cf. 43).

4. The "Etc. List"

Following Hacking, I call this problem the "etc. list." In his "Self-Vindication" paper, Hacking refers to several authors and in particular to Pickering and Gooding, identifying in their writings lists of items. So, for example, what Pickering calls "pragmatic realism" is the coproduction of "facts, phenomena, material procedures, interpretations, theories, social relations etc." (Hacking 1992, 31). Similarly, Hacking portrays Gooding as having another "etc. list." According to Hacking, Gooding "speaks of an 'experimental sequence' which appears as the 'production of models, phenomena, bits of apparatus, and representations of these things'" (Hacking 1992, 32). We agree, Hacking continues, "that the interplay of items in such a list brings about the stability of laboratory science" (Hacking 1992, 32). On his part, Hacking gives the *matériel* of an experiment a crucial role to play in the stabilization process of experimental science. By the *matériel* he means "the apparatus, the instruments, the substances or objects investigated. The matériel is flanked on the one side by ideas (theories, questions, hypotheses, intellectual models of apparatus) and on the other by marks and manipulations of marks (inscriptions, data, calculations, data reduction, interpretation)" (Hacking 1992, 32).

It looks then as if Hacking presents us with an "etc. list" of his own. Hacking however is not content with "lists and etc.'s" (Hacking 1992, 32), and he ventures a taxonomy of elements of experiment that takes him further afield, beyond Mach and Franklin.

The conception that in experiment the *matériel* is flanked on one side by ideas and on the other by marks is the clue to Hacking's proposal for making the open list converge onto three groups of elements of experiment, namely, "ideas, things, and marks" (Hacking 1992, 44). "Ideas" are the intellectual components of experiment; "things" represent the instruments and apparatus, and finally "marks" comprise the recording of the outcomes of experiment. Hacking is not worried by Mach's claim that classification will not do, "since different features do not in general exclude one another, so that several of them may be united in the experiment" (Mach 1976, 157). In fact, Hacking delights in constructing a flexible taxonomy, since in his view the stability of experimental results arises from precisely the very interplay of elements—on all accounts, the taxonomy should not be rigid (Hacking 1992, 44). With this taxonomy Hacking seeks at once to demonstrate the "motley of experimental science" and to contribute toward a philosophy of experiment so that one would not meander, as he puts it, "from fascinating case to fascinating case" (Hacking 1992, 31–32).

As I have indicated, I was concerned elsewhere with the first cluster of problems; namely, I analyzed the "material argument" of experiment as a crucial aspect of the epistemological issue of experimentation (Hon 1998a). In the present chapter I wish to tackle the second cluster—that is, the methodological issue, the "etc. list." My objective is to transcend the list, reach the taxonomic stage, and aim beyond it to experimental principles.

5. The Guiding Idea: Approaching Knowledge from the Perspective of Error

My guiding idea is to study experiment by the nature of its possible faults. I suggest that by examining possible failures of experiment, light may be shed on this method of inquiry. My approach takes then a different route altogether from that of Franklin. I am not seeking epistemological strategies that are designed to secure reliable outcomes, strategies that may in turn provide bases for rational belief. As we have seen, this approach results in an open, ad hoc list. I am looking rather for general characterizations of classes of possible faults. We shall see that in many respects the emerging typology of classes of experimental errors reflects, albeit from a negative perspective, Hacking's typology. There will be, however, some crucial differences. It is hoped further that the resultant typology would serve as a framework for developing a theory of experiment out of which general principles may emerge.

For clarification, here is a brief illustration of how the method works. Consider the standard approach to experimental error—that is, the dichotomy of systematic and random error. Clearly, this dichotomy reflects an interest in the mathematical aspect of error: does a deterministic rule govern the error? Or is it a statistical law? In the former case, as is well known, the error is systematic, and in the latter it is random. The dichotomy is very useful and much in use in the practice of experimentation, especially in the analysis of the results by introducing correction terms and reducing the data. The dichotomy could therefore be included in the list of strategies. However, the distinction throws no light on the source of the error; in other words, philosophically it is not useful. Error that may originate in the assumption of, say, incorrect instrumental theory is classified together with an error that has originated in a faulty calibration—both being systematic. For another example, small error in judgment on the part of the observer in estimating the scale division and unpredictable fluctuations in conditions, such as temperature or mechanical vibrations of the apparatus, are classified together since these errors are all random in nature (for a detailed analysis, see Hon 1989b, 474–479).

I maintain that for philosophical purposes clear distinctions should be introduced among possible sources of error. I follow here Kant's dictum: "to avoid errors . . . one must seek to disclose and to explain their source, illusion. Very few philosophers have done that, however. They have only sought to refute the errors themselves, without indicating the illusion from which they arise. This disclosure and breaking up of illusion is a far greater service to truth, however, than the direct refutation of errors, whereby one does not block their source and cannot guard against the same illusion misleading one into errors again in other cases because one is not acquainted with it" (Kant 1992, 562). From an epistemological perspective, it is worth to inquire into the source of the error and not so much to examine the mathematical features of the error and the means of calculating it away—the causal feature being of a higher interest here than the pragmatic one. Thus, errors that have originated in the use of the apparatus should be set apart from errors that pertain to the interpretation of the data. Once distinctions between the different kinds of sources are being introduced, retained, and elaborated, we may see how knowledge of the scheme of possible sources of error in experiment throws light on the very structure of the method at stake. Specifically, as we shall see, the features of the different kinds of source of error reflect the various elements that are involved in experimentation.

The approach to knowledge from its negative perspective—that is, from errors and faults—is not new. In fact, "the first and almost last philosopher of experiments"—to use Hacking's characterization of Francis Bacon (Hacking 1984, 159)—employed a similar methodology. Bacon was philosophically aware of the problem of error and explicitly addressed it. Indeed, Bacon deployed the notion of error as a lever with which he hoisted his new program for the sciences. As expounded in the *Novum Organum* (Bacon 1859, 1960, 2000), his programmatic philosophy consists of two principal moves: first, the recognition of error and its rebuke if not elimination and, second, the commencing anew of the true science based on experiment and induction. I shall presently argue that Bacon's conception is found wanting, especially when experiment, the very instrument of his research, is in question. The shortcomings of his approach would be the key to my move. So here is a précis of Bacon's theory of error.

6. Bacon's Typology of Errors: The Four Idols of the Mind

Bacon argues in his celebrated *Novum Organum* that Aristotle "has corrupted Natural Philosophy with his Logic; . . . he has made the Universe out of Categories" (Bacon 1859, 39). In Bacon's view, the application of Aristotle's doctrine has more the effect of confirming and rendering permanent errors that are founded on vulgar conceptions than of promoting the investigation of truth (Bacon 1859, 13–14).

Bacon builds his program on the doctrine that truth is manifest through plain facts, but for this claim to be valid the student of nature has to get rid of all prejudices and preconceived ideas. As Bacon instructs, "the whole work of the mind should be recommenced anew" (Bacon 1859, 4); only then would the student experience things as they are. "Our plan," he explains, "consists in laying down degrees of certainty, in guarding the sense from error by a process of correction . . . and then in opening and constructing a new and certain way for the mind from the very perceptions of the senses" (Bacon 1859, 3). In this way, Bacon concludes, "we are building in the human Intellect a copy of the universe such as it is discovered to be, and not as man's own reason would have ordered it" (Bacon 1859, 120). Thus, the first task of the scientist is to eliminate errors from his or her cognition by the "expiation and purgation of the mind"; only then can the scientist enter "the true way of interpreting Nature" (Bacon 1859, 51).

Bacon finds it necessary, therefore, to expound in considerable detail the subject of the obstacles to the true interpretation of nature before proceeding to unfold his positive program: the method of inductive inquiry based on experimentation. He devotes nearly the whole of the first book of *Novum Organum* to the examination of these obstacles which he calls idols "idols of the mind." This name reflects the Platonic concept of *eidolon,* which refers to a fleeting, transient image of reality, in contrast to the concept of *idea,* which represents reality in the Platonic sense (Bacon 1859, 16–17 n). Although Bacon claims that "to draw out conceptions and axioms by a true induction is certainly the proper remedy for repelling and removing idola" (Bacon 1859, 21), he still finds it of great advantage to explicitly indicate the idols and expound them in detail. For, as he explains, "the doctrine of idola holds the same position in the interpretation of Nature, as that of the confutation of sophisms does in common Logic" (Bacon 1859, 21). In other words, to use Jardine's formulation, "the idols . . . bear a relation to the inductive method analogous to that which cautionary lists of fallacious arguments bear to syllogistic" (Jardine 1974, 83). However, as I have indicated, I wish to go further from mere "cautionary lists" and to obtain a conceptual scheme of experiment based on a typology of sources of error. Bacon's theory of error, his typology of idols and its critique, serves as a philosophical illustration of the approach I am taking.

Bacon classifies four types of idol that, as he puts it, "beset the minds of men" (Bacon 1859, 21): idols of the tribe, the cave, the marketplace, and the theater.

Idols of the Tribe

The first type of idols, idols of the tribe, are errors incidental to human nature in general. The most prominent of these errors are the tendency to support a preconceived opinion by affirmative instances while neglecting all counterexamples; the tendency to generalize from a few observations; and the tendency to consider mere abstractions as reality. Errors of this type may also originate in the weakness of the senses, which affords scope for mere conjectures (Bacon 1859, 21, 24–29). Bacon warns the student of natural philosophy against the belief that the human sense is the measure of things. For Bacon, "the human intellect is like an uneven mirror on which the rays of objects fall, and which mixes up its own nature with that of the object, and distorts and destroys it" (Bacon 1859, 21). To obtain the true interpretation of nature, the human mind should function, according to Bacon, like an even mirror.

Idols of the Cave

The second kind of idols, idols of the cave, are errors incidental to the peculiar mental and bodily constitution of each individual (the cave is a direct reference to Plato's simile in the Republic). These errors may be either of internal origin, arising from the peculiar physiology of the individual, or of external origin, arising from the social circumstances in which one is placed by education, custom, and society in general (Bacon 1859, 22, 29–30, 32–33).

Idols of the Marketplace

The third class of idols, idols of the marketplace, are errors arising from the nature of language—the vehicle, as Bacon puts it, for the association of men, their commerce and consort (Bacon 1859, 22–23, 33–35). Language, according to Bacon, introduces two fallacious modes of observing the world. First, there are some words that are merely "the names of things which have no existence (as there are things without names through want of observation, so there are also names without things through fanciful supposition)." Secondly, there are "names of things which do exist, but are confused and ill defined" (Bacon 1859, 34). Bacon is aware then that language may lead the researcher astray by being opaque to nature. He therefore cautions the researcher of the faults of language and advises to regain its transparency.

Idols of the Theater

Finally, the fourth class of idols, idols of the theater, are errors that arise from received "dogmas of philosophical systems, and even from perverted laws of demonstrations" (Bacon 1859, 23; cf. 35–49). Here Bacon refers mainly to three kinds of error: sophistical, empirical, and superstitious. The first error corresponds to Aristotle, who has, according to Bacon, "made his Natural Philosophy so completely subservient to his Logic as to render it nearly useless, and a mere vehicle for controversy" (Bacon 1859, 30). The second error, the empirical, refers to leaping from "narrow and obscure experiments" to general conclusions. Bacon has in mind particularly the chemists of his time and Gilbert and his experiments on the magnet (Bacon 1859, 41–42). The third error, the superstitious, represents the corruption of philosophy by the introduction of poetical and theological notions, as is the case, according to Bacon, with the Pythagorean system (Bacon 1859, 42–44).

CONCLUDING HIS DISCUSSION of the idols, Bacon demands that all of them "must be renounced and abjured with a constant and solemn determination" (Bacon 1859, 49). He insists upon purging and freeing the intellect from them so that "the approach to the Kingdom of Man, which," as Bacon conceived of his quest, "is founded on the Sciences, may be like that to the Kingdom of Heaven" (Bacon 1859, 49). Thus, having performed these "expiations and purgations of the mind," one "may come to set forth the true way of interpreting Nature" (Bacon 1859, 51). The religious connotation is explicit and should be underlined.

Clearly, Bacon's doctrine of the idols is systematic and methodical if somewhat contrived. He neatly classifies the idols as "either adventitious or innate. The adventitious," Bacon explains, "come into the mind from without—namely, either from the doctrines and sects of philosophers or from perverse rules of demonstration. But the innate are inherent in the very nature of the intellect, which is far more prone to error than the sense is" (Bacon 1960, 22). The classes of idols proceed progressively from the innate to the adventitious, respectively from the most persistent to the easiest to discard. They commence with the general character of human beings—the tribe—move on through the features of individuals that comprise the tribe—that is, the cave—further on to the interactions, negotiations, and commerce between individuals—the marketplace— and reach finally the doctrines that individuals conceive—the theater. Bacon is aware of the fact that the innate features are hard to eradicate, so that these idols may not be eliminated. "All that can be done," he instructs, "is to point them out, so that this insidious action of the mind may be marked and reproved (else . . . we shall have but a change of errors, and not clearance)" (Bacon 1960, 23). The adventitious idols, by contrast, could and should be eliminated. Having undergone these epistemological ablutions, one is ready, according to Bacon, to commence anew the true interpretation of nature.

Bacon designed the typology to shed light on the nature of sources of error. The scheme of idols presents a systematic and methodical view of the elements involved in the obstruction of knowledge: the interplay of sources of error pertaining to the nature of the mind in general, to individuals and their community, to language and doctrines. The scheme may appear somewhat artificial and contrived, but it constitutes an essential element of Bacon's comprehensive conception of the emergence of new knowledge and its impediments. In many respects the scheme of idols anticipated new disciplines, namely, the study of anthropology, ethnology, psychology, and linguistic and cultural, political and religious ideologies (Coquillette 1992, 233–234; for references, see 300 n. 24).

7. A Critique of Bacon's Scheme

The question naturally arises whether or not this all-embracing typology of errors is applicable to the very method of research that Bacon advocates for use—that is, experimentation. "It will doubtless occur to some," Bacon acknowledges the question, that "there is in the Experiments themselves some uncertainty or error; and it will therefore, perhaps, be thought that our discoveries rest on false and doubtful principles for their foundation" (Bacon 1859, 111–112). This appears to be a surprising remark. Could it be that Bacon's proposed method of research is open to objections and that all the cleansing and ablutions were for nothing? No! Bacon dismisses the threat right away: "this is nothing," he exclaims, "for it is necessary that such should be the case in the beginning." By way of an analogy he explains that "it is just as if, in writing or printing, one or two letters should be wrongly separated or combined, which does not usually hinder the reader much, since the errors are easily corrected from the sense itself. And so men should reflect that many Experiments may erroneously be believed and received in Natural History, which are soon afterwards easily expunged and rejected by the discovery of Causes and Axioms" (Bacon 1859, 112). Bacon assures us that we should not be disturbed by these objections, and he reiterates this confidence in his outline for experimental history (Bacon 1960, 280). However, he admits that "it is true, that if the mistakes made in Natural History and in Experiments be important, frequent, and continuous, no felicity of wit or Art can avail to correct or amend them" (Bacon 1859, 112).

Thus, if there lurked at times "something false or erroneous" in Bacon's natural history that has been proved with "so great diligence, strictness, and," Bacon adds, "religious care," what then must be said, he asks rhetorically, "of the ordinary Natural History, which, compared with ours, is so careless and slipshod? or of the Philosophy and Sciences built on ... quicksands"? (Bacon 1859, 112).

Notwithstanding Bacon's resolute assurance, the objections are disturbing. Bacon appears to be waving his hands, so to speak, rather than providing convincing arguments in defense of his position. He would have us believe that the analogy between a printer's error and an experimental error is faithful. This is, however, not the case. According to Bacon's analogy, the sense of the experimental context, like the meaning of a text, is given; but it is precisely this sense, the physical meaning of the experimental arrangement, that the experimental sciences lack and in fact seek to discover and establish. The blackening of photographic plates, for example, in close vicinity to functioning cathode-ray tubes had been a well-known nuisance before Roentgen apprehended the meaning of this bothering phenomenon. Given the correct physical meaning, the sense of this physical context (the discharging tubes emitted a new form of radiation), Roentgen was able to transform this nuisance into a great discovery that revolutionized experimental physics. The two types of error—namely, the printer's and the experimental—are categorically different. (I distinguish elsewhere between these two possible faults. I call a mistake a fault that is due to avoidable ignorance, while error arises in unavoidable ignorance. Thus, a printer's error, like a miscalculation, is in fact a mistake [Hon 1995]).

Surprisingly, it appears that Bacon did not apply consistently his critical scheme of errors to the very instrument of his inquiry—experiment. Admittedly, he was concerned with errors that beset the mind: once one had purged one's mind from the idols and, to use Bacon's mirror metaphor, smoothed away with religious fervor every protrusion and cavity in one's intellect so that it became an even surface reflecting genuinely the rays of things (Bacon 1960, 22), one was ready to embark on the true way of interpreting nature. At issue here is not whether this instruction to cleanse one's mind is practicable or not, but rather can the instrument of one's inquiry be itself an object of critical scrutiny. Indeed, as we have seen, it had taken some time before the question "What exactly is an experiment in physics?" was explicitly raised and addressed (Duhem 1974, 144). The persistent impediment that the occurrence of errors poses for knowledge one claims to have obtained from experiment may not be covered by Bacon's scheme of idols of the mind. Bacon's trust in his instrument of inquiry, which he expressed with his offhand dismissal of experimental errors, is objectionable. I respond to this criticism and propose to examine the different idols that beset experiment.

8. The Idols of Experiment: Script, Stage, Spectator, and Moral

The construction of a scheme of idols that beset experiment has a similar objective to the scheme of Bacon, but the analysis goes further in that it explicitly argues that the scheme reflects underlying principles of experimentation. My intent, to repeat, is not to seek strategies in an ad hoc fashion following Mach and Franklin. That is, to refer once again to Gooding, Pinch, and Schaffer's well-phrased remark (1989, 22–23) that, in Franklin's view "there are epistemological rules which can be applied straightforwardly in the field to separate the wheat of a genuine result from the chaff of error." The objective is not to list such rules in an eclectic way but to construct a scheme of "the chaff of error" that reflects the structure of experiment as an instrument of inquiry that results in knowledge. The motivating principle is thus to classify different possible sources of error in such a way that each idol reflects an element in the experimental procedure, so that together the scheme of idols covers the elements that comprise experiment.

In the spirit of the metaphoric language of Bacon and following his idols of the theater, I suggest to discern four kinds of idol that beset experiment: idols of the script, the stage, the spectator, and the moral. The image of theatrical play constitutes a convenient and useful metaphorical setting for experiment since, like a play, an experiment is the result of an activity that truly has "a show" at its center (on experiment and drama, see Cantor 1989, 173–176). In an experiment, nature is made, if you will, to display a show on a stage conceived and designed in some script. The show is observed and registered by a human or automated spectator and, finally, interpretation is proposed with a view to providing a moral —that is, the outcome of the experiment as knowledge of the physical world. These four idols reflect elements of the experimental process: the background theory of the experiment—the script; the assumptions concerning the apparatus and its actual working—the stage; making observations and taking measurements—the spectator; and, finally, theoretical conclusions—the moral.

Error is a multifarious epistemological phenomenon. It is an expression of divergence whose mark is discrepancy—a discrepancy that emerges from a procedure of evaluation against a chosen standard. The nature of this discrepancy, the reason for its occurrence, how to treat it, and what can be learned from it once it has been perceived and comprehended constitute the vast subject of the problem of error. The four idols that comprise the proposed scheme depict different kinds of discrepancy that may arise at different stages of the process of experiment.

Experiments proceed essentially in two stages: preparation and test. In the preparation stage the experimenter sets up the initial conditions of the apparatus and the system within which the experiment is designed to evolve—this is the theoretical and the material framework of the experiment. Once the experimenter sets the framework, the experiment may commence its runs: the testing—the evolution of the system within the designed framework (for a detailed analysis, see Hon 1998a). It should be underlined that the terms "preparation" and "test" are being used here in a very loose sense. Experiment is not necessarily a test, nor does it have by definition a set of known initial conditions. However, the dichotomy between these two distinct stages, the preparation and the test, is crucial in the sense that experiment always exhibits the evolution of a (prepared) system.

It may be seen immediately that the idols of the script and of the stage are associated with the first stage, the preparation, whereas the idols of the spectator and of the moral pertain to the latter stage, the testing. In this way the idols cover all possible faults in terms of the different contexts in which sources of error may crop up in experiment. The idols that comprise this typology of sources of error reflect, therefore, the roles that faulty elements would play in the overall structure of experiment. The claim is that the possible faults in the different contexts of the process of experiment illuminate the epistemic structure of experiment. The scheme provides an overview as well as a detailed analysis of the four stages of experiment: the background theory, the apparatus, observation and measurement, and finally interpretation.

A distinct characteristic of the proposed taxonomy is then its focus on the source rather than on the resultant error. By concentrating on the definitions of different classes of source of error, the typology illuminates from a negative perspective the elements that are involved in experiment and their interrelations. Thus an incorrect or ill-suited background theory (for example, the application of Stokes's law to the minute and jagged metal dust particles in Ehrenhaft's alleged discovery of subelectrons), an idol of the script, is different from assuming erroneously that certain physical conditions prevail in the setup (for example, technical difficulties in establishing the physical conditions for the determination of the Hall effect)—the latter error being an idol of the stage. Physical, physiological, and psychological elements interfering with the display of the phenomenon under study or with the reading of a measuring device (for example, Blondlot's autosuggestive perception of N rays), an idol of the spectator, is different from conferring an erroneous interpretation on experimental results (for example, Franck and Hertz's interpretation that the first critical potential they measured was an ionization and not excitation potential)—an idol of the moral. However we conceive of errors—that is, experimental errors— they would be covered, I submit, by one of the four idols. (For a detailed account of the four classes of experimental error illustrated with historical examples, and for references for the above cases, see Hon 1989b).

This typology of experimental error addresses the issue of experiment and error in an entirely different way than the analysis Deborah Mayo offers in her book *Error and the Growth of Experimental Knowledge* (Mayo 1996). Mayo's reference to experimental rather than to scientific knowledge has to do with her claim that experimental knowledge is knowledge grounded on argument from error (Mayo 1996, 7). Mayo sides with Peirce, Neyman, and Pearson and stands against their common opponents—the Bayesians. She christens her position "error-statistical philosophy of experiment" (Mayo 1996, 410, 442, 457, 464), because the chief feature that her approach retains from the Neyman-Pearson statistical methods is the centrality of error probabilities (Mayo 1996, x–xi). The demand that it is necessary to take into account the error probabilities of experimental procedures in order to determine what inferences are licensed by data is the principal element that fundamentally distinguishes, according to Mayo, her approach from others (Mayo 1996, 442). She concludes that "we make progress in experimental knowledge—experimental knowledge grows—because we have methods that are manifestly adequate for learning from errors" (Mayo 1996, 464). On this account, errors and the statistical methods for treating them have become the tools for building the body of knowledge we call science.

There is no doubt that error-statistical analysis is a powerful tool much needed in the technical realm of the reduction of data. After all, the analysis of error probabilities permits quantifying trustworthiness (Mayo 1996, 424, 425). Indeed, the analysis can throw light in this way on methodological issues and as such can be added to Franklin's list of strategies. However, it may not do philosophical justice, at least not on its own, to such complex concepts as error and experimental knowledge. Mayo's book is not about error but about error probabilities, and the notion of experimental knowledge it develops is rather the knowledge of the probabilities of specified outcomes in some series of experiments (Mayo 1996, 12).

Mayo's philosophy of experiment relies neither on scientific theories nor on a theory of experiment; it relies instead on methods—statistical methods—for producing experimental effects (Mayo 1996, 15). This observation is crucial. It explains the limited view of experiment exhibited in this study. Notwithstanding Mayo's talk about the need to address the actual practice of experimentation, she focuses her attention *solely* on statistical calculations. This is not what one would expect of, say, a Faraday, a Helmholtz, a Pasteur, a Hertz, a Rutherford, a Gibson, a Rabi, or a Kapitza. Consider Peirce's observations on experimental style: "Of all men of the century Faraday had the greatest power of drawing ideas straight out of his experiments and making his physical apparatus do his thinking, so that experimentation and inference were not two proceedings, but one. To understand what this means, read his *Researches on Electricity*. His genius was thus higher than that of Helmholtz, who fitted a phenomenon with an appropriate conception out of his store, as one might fit a bottle with a stopper" (Peirce 1966, 272). Mayo's "full-bodied experimental philosophy" (Mayo 1996, 444) is not attuned to the act of experimenting; it focuses rather on the end result: data and their statistical tests. For example, questions as to the interpretation of the experimental result do not arise in this framework. For another example, no theory of experiment is forthcoming in this approach (see by way of comparison Radder 1995).

In Mayo's philosophical framework, experimental knowledge becomes completely statistical. Success in obtaining experimental knowledge is explained in this philosophy by the properties of the statistical methods applied. The errors upon which Mayo builds her error-statistical philosophy of experiment are not error at large but rather a specific and indeed limited kind of error, namely, error probabilities. Error probabilities are not probabilities of hypotheses, but the probabilities that certain experimental results would occur were one or another hypothesis true about the experimental system (Mayo 1996, 367). What we have then for the theme of the book is *Error Probabilities and the Statistical Assessment of Experimental Data.* Mayo's study constitutes a contribution to experimental design in the traditional sense of the term as well as to the analysis of error probabilities. It does not, however, illuminate the inner epistemic processes of experiment that the idols of experiment—the typology I offer—seek to bring out by focusing on sources of error and not on their statistical behavior. (For an essay review of Mayo's book, see Hon 1998b).

Another important feature of the typology is that it characterizes "the script" —the conceptual, theoretical guiding lines of apparatus and instruments—that is, the background theories—as analytically distinct from "the moral," theories that provide the basis for the interpretation of the outcome of experiment. This distinction is logically crucial since it keeps apart the theories that constitute the conceptual framework of experiments and the theories that provide the outcome of experiments with physical meaning. One of the essential features of the method of experimentation, namely, procedures of correction, was recognized early on by Galileo. The experimenter should be, amongst other things, a good accountant: "Just as the computer who wants his calculations to deal with sugar, silk, and wool must discount the boxes, bales, and other packings, so the mathematical scientist *(filosofo geometra),* when he wants to recognize in the concrete the effects which he has proved in the abstract, must deduct the material hindrances, and if he is able to do so, I assure you that things are in no less agreement than arithmetical computations. The errors, then, lie not in the abstractness or concreteness, not in geometry or physics, but in a calculator who does not know how to make a true accounting" (Galileo 1974, 207–208). Clearly, to conduct successfully this true accounting the experimenter would need to resort to a theory. This theory should be provided by "the script" and not by "the moral," lest the argument be circular.

Duhem's insightful logical analysis of the correction procedure of systematic error, a crucial feature of modern experimentation, is rightly based on theories that belong to "the script" and not to those that belong to "the moral" of experiment. Duhem observes that a physical experiment is not merely the observation of a group of facts produced under some controlled constraints. If it were so, it would have been absurd to bring in corrections, "for it would be ridiculous to tell an observer who had looked attentively, carefully, and minutely: 'What you have seen is not what you should have seen; permit me to make some calculations which will teach you what you should have observed'" (Duhem 1974, 156).

Following Duhem, observations in experiment have to be capable of translation into a symbolic language—for example, an equation—and it is physical theories that provide the required rules of translation. The experimenter has constantly to compare, to continue Duhem's line of argumentation, two objects: on the one hand, the real, concrete object that is being physically manipulated—the apparatus—and on the other hand the abstract, symbolic object upon which one reasons (Duhem 1974, 156). This crucial comparative activity in experimentation, which allows for the introduction of necessary correction terms, depends entirely on "the script." By contrast, the theories that provide the basis for interpretation, that is, "the moral," are brought, as it were, from without; they are not involved in the process of correcting systematic errors. They are, however, crucial for correcting errors of interpretation (for a case study of error of interpretation, see Hon 1989a).

Hacking, it may be recalled, has grouped experimental elements into three classes: "ideas, things, and marks" (Hacking 1992, 44). As I have indicated, my proposed scheme of idols that beset experiment reflects, albeit negatively, Hacking's typology. The scheme of idols diverges, however, from the typology of Hacking on two important points. Roughly, "ideas" correspond to "idols of the script," "things" to "idols of the stage," and finally "marks" relate to elements of "idols of the spectator." There remains the class of "idols of the moral," which Hacking's typology appears not to cover; or, alternatively, in his typology elements of "ideas" cover both the background and the outcome of experiment without distinguishing between these two sets of elements. I agree with Hacking that flexibility and interplay of elements are crucial to the stability of experimental results, and so one may cover the fourth set of idols, "idols of the moral," by "ideas." This is a realistic view of experimental practice since "the script"—"ideas" in Hacking's terms—often inform the interpretations of results.

Nevertheless, I do hold strongly that for analytical, logical reasons there should be a clear separation between "the script" and "the moral." Hacking's taxonomy eliminates the crucial difference between these two sets of idols. Again, the "script" consists of theories that are presupposed to govern and shape the experiment: both the working of the apparatus and the application of instruments. The experimenter does not put these theories to the test; they are presupposed and considered correct. These theories provide the framework for the execution of experiment. By contrast, theories that belong to the "idols of the moral" are being tested and may be dispensed with, replaced, or rejected and indeed proved false without affecting at all the overall experiment, its argument, and the body of its accumulated data. (The Franck-Hertz experiment is a case in point; see Hon 1989a.)

Furthermore, the alternative of grouping together "the spectator" and "the moral" under Hacking's class of "marks" should also be avoided. Again, the sources of error and procedures of correction that take place in the reading of data are distinct from analyzing, reducing, and interpreting the data. Thus, from the negative perspective—that is, from the perspective of error—it is instructive to split Hacking's "marks" into two different, distinct classes I call "spectator" and "moral."

9. Concluding Remarks

Against the background of collapse and decline of Scholastic epistemology, a breakdown that led to the proliferation of often conflicting views of knowing (Solomon 1998, xv), Bacon conceived of a science in which one seeks "to discover the powers and actions of bodies and their laws limned in matter. Hence this science," according to Bacon, "takes its origin not only from the nature of the mind but from the nature of things"; Bacon offered a new logic that he had designed "to dissect nature truly" (see the motto above). This new logic leads to the true "Interpretation of Nature" (see also Martin 1992, 147). It consists essentially of two moves. The first, as Bacon puts it, is the "expurgation of the intellect to qualify it for dealing with truth" (quoted by Martin 1992, 147), and the move to follow is "the display of a manner of demonstration for natural philosophy superior to traditional logic" (Martin 1992, 147). Bacon developed the scheme of idols that besets the mind to facilitate the first move; the second move proceeds by founding philosophy on natural and experimental history—the furnishing of the material of knowledge itself (Martin 1992, 146–147).

By way of analogy, I addressed the second Baconian move, the move by which the material of knowledge is furnished. My proposed scheme of the idols of experiment takes its cue from the Baconian two-tier approach to the true way of interpreting nature. However, the point of the scheme is not epistemological but rather methodological—it is here that the analogy to Bacon's approach ends. The proposed scheme carries the critical, Baconian program over to experimentation itself.

The scheme focuses on the different kinds of possible sources of error that may crop up in experiment. In that sense, the scheme reflects the normative aspect of experiment. However, once the typology is set up, it may be seen that the different kinds of source of error comprise four different contexts in experimentation that exhaust, so the claim goes, all possible sources of error. In other words, the four idols (the script, the stage, the spectator, and the moral) cover all sources of error—each idol characterizing a class of similar sources of error, that is, discrepancies of similar nature. The constraints imposed by the scheme, with its clear delineation of the classes, provide a comprehensive overview of experiment from a negative perspective that does not depend on open lists—the "etc. list" has been thus transcended. I submit that the study of the relations between the elements that comprise each idol could provide an insight into the dynamic epistemic processes that are inherent in experiment—the conceptual underpinnings of experimentation. By transcending the list, the set of idols of experiment provides us with both a normative and a comprehensive, conceptual view of experimentation.

10. Coda

Projecting via the idols the understanding of the general concept of error unto the various possible occurrences of error in experiment, it may become clear that it is problematic to render an experiment erroneous. Since experiment is a physical process that exhibits some phenomena as its outcome, it is philosophically misguided to consider the results spurious. After all, in its modern conception nature takes its natural course and never errs. We do expect nature to submit, as it were, to its laws even under stringent and rigorous physical constraints of the experimental setup. Indeed, this is the essence of the analytic method. As Cartwright explains, "to understand what happens in the world, we take things apart into their fundamental pieces; to control a situation we reassemble the pieces, we reorder them so they will work together to make things happen as we will. You carry the pieces from place to place, assembling them together in new ways and new contexts. But you always assume that they will try to behave in new arrangements as they have tried to behave in others. They will, in each case, act in accordance with their nature" (Cartwright 1992, 49; see also Hintikka 1988). The errors, it may be recalled, "lie not in the abstractness or concreteness, not in geometry or physics, but in a calculator who does not know how to make a true accounting" (Galileo 1974, 208). The task of the experimenter is to avoid error in the quest for knowledge. "*Shun error!*" as James put it. "We may regard the chase for truth as paramount, and the avoidance of error as secondary; or we may, on the other hand, treat the avoidance of error as more imperative, and let truth take its chance" (James 1897, 17–18).

By focusing on sources of error in experiment the scheme of idols facilitates the characterization of the location—that is, the specific context in experiment in which possible errors may arise and, further, the grasp of the features of the arising discrepancies. Clearly, correction follows in accordance with the kind of discrepancy that has been characterized. However, experimental error, as I have tried to show, is not just a matter to shun and calculate away. Its recognition and the accompanied methods of correction may provide an insight into the elements that comprise experiment and the way they cohere to produce knowledge.

NOTE

I am grateful to Willem B. Drees, Hans Radder, and an anonymous reader for incisive and valuable comments on an earlier draft of this chapter.

Evelyn Fox Keller

10 / Models, Simulation, and "Computer Experiments"

As soon as an Analytical Engine exists, it will necessarily guide the future course of the science.

Charles Babbage, *Passages from the Life of a Philosopher*, 1864

1. Introduction

As Nelson Goodman famously observed, "Few terms are used in popular and scientific discourse more promiscuously than 'model'" (Goodman 1968, 171). Writing more than thirty years later, much the same might be said of the term "simulation." Yet this was not always the case. Both words have ancient histories, but until very recently, the meaning of "simulation," at least, was manifestly stable: it invariably implied deceit. Usages offered by the Oxford English Dictionary (OED) prior to 1947 include "false pretence"; "A Deceiving by Actions, Gestures, or Behaviour" (1692); "a Pretence of what is not" (1711). Evidence provided by the OED, in short, suggests that it was only after World War II that the word took on the meaning that brings it into its current proximity with models: "The technique of imitating the behaviour of some situation or process . . . by means of a suitably analogous situation or apparatus, especially for the purpose of study or personnel training." Here, the valence of the term changes decisively: now productive rather than merely deceptive,[1] and, in particular, designating a technique for the promotion of scientific understanding. The shift

198

reflects a crucial change not only in the perceived value of simulation but also, as others have already noted, in the means of production of scientific knowledge (see, for example, Rohrlich 1991; Humphreys 1991; Galison 1996; Winsberg 1999). Furthermore, it is this new sense of the term that encourages its use in much of the current historical and philosophical literature as either interchangeable with the term "model," or as one part of a single composite noun (as in "models and simulations"). An obvious question arises, however, and it is this: Do the actual uses of simulation in contemporary scientific practice in fact warrant such facile assimilation? Or, to pose the question somewhat differently, does the use of simulation in post–World War II science add significantly new features to the range of practices that had earlier been subsumed under the term "modeling"? My answer is yes, but I argue that the novelty has been multilayered and cumulative in its effects, requiring a more nuanced history than has yet been made available.

The rise of simulation in post–World War II science is not exclusively associated with the advent of the computer—in fact, the earliest invocations of the term relied primarily on the use of electrical and electronic analogue devices designed to mimic the behavior of real-world phenomena.[2] However, it was the introduction of the digital computer that provided the major impetus for the adoption of simulation techniques in scientific research, and for that reason, my discussion will be confined to what has come to be known as "computer simulation." Very crudely, it might be said that the immediate effect of these new techniques on scientific practice was to radically extend the range of problems amenable to quantitative analysis. They did so in a variety of ways, however, and with widely varying implications. Indeed, even the term "computer simulation" covers so complex a range of activities that some sort of taxonomy would seem to be in order. What kind of taxonomy? We might start with a division along disciplinary lines—distinguishing the uses of simulation in the physical sciences from those in the biological sciences, in cognitive science, in economics or management. But proceeding with a canonical evolutionary tree will clearly not serve, for such a structure misses the cross-structures needed for and resulting from ongoing hybridization. On the other hand, differences in aims, interests, and tradition are plainly evident, and they bear critically on subsequent historical developments. I suggest, therefore, that it is useful to follow the history of simulation along quasi-disciplinary lines while at the same time remaining alert to the extensive cross-disciplinary traffic of technical innovations that has been so much a part of this history. In this chapter, I focus primarily on the physical sciences, and I argue that, even within such a major disciplinary category, important subdivisions need to be demarcated. Furthermore, each of these subdivisions pushes the boundaries of the initial disciplinary divide in distinctive ways.[3]

2. Computer Simulation in the Physical Sciences

In one of the first attempts to bring the novel features of computer simulation to the attention of philosophers of science, the physicist Fritz Rohrlich put forth the claim for "a qualitatively new and different methodology" lying "somewhere intermediate between traditional theoretical physical science and its empirical methods of experimentation and observation. In many cases," he wrote, "it involves a *new syntax* which gradually replaces the old, and it involves *theoretical model experimentation* in a qualitatively new and interesting way. Scientific activity has thus reached a new milestone somewhat comparable to the milestones that started the empirical approach (Galileo) and the deterministic mathematical approach to dynamics (the old syntax of Newton and Laplace)" (Rohrlich 1991, 907). Others have argued in a similar vein. Peter Galison (1996), for example, draws a sharp distinction between the new computer simulations and the earlier analogue simulations. He suggests that, while the latter can be readily assimilated into a long history of analogue models (including not only ship models and wind tunnels but also nineteenth-century models built out of "pulleys, springs, and rotors to recreate the relations embodied in electromagnetism" [Galison 1996, 121]), the new techniques of computer simulation effected a radical epistemological transformation in the physical sciences, ushering "physics into a place paradoxically dislocated from the traditional reality that borrowed from both experimental and theoretical domains" and creating a "netherland that was at once nowhere and everywhere on the methodological map" (Galison 1996, 120). Such claims have become familiar, and they are generally taken to be uncontroversial. Over the last half a century, a new domain of physical science has come into being that is widely recognized as different from the older domains of both theoretical and experimental physics, and that has accordingly warranted a new designation, namely "computational physics." "Computational physics" is simply a term referring to the use of computer simulation in the analysis of complex physical systems, and, as such, it is unquestionably both new and distinctive.

Controversy arises only in response to the question, What exactly is it that is so distinctive about this new endeavor? In one sense, the answer is obvious: computer simulation opened up the study of complex systems—that is, it brought a range of phenomena that had hitherto been mathematically intractable into analytic reach. Until the advent of computers, the primary tool physicists had at their disposal for representing their theoretical understanding of the mechanics and dynamics of material systems had been the differential equation, and their principal task was to relate the solutions of these equations to observed experimental effects. But differential equations are notoriously difficult to solve once they depart from the linear domain, and especially so when representing the interactions of many bodies. Thus, prior to the computer, the study of complex, nonlinear phenomena by physicists had been limited to what could be achieved by perturbation methods, simplifying models/approximations (for example, "effective-field" approximations), or paper-and-pencil schemes for numerical approximation,[4] and the first and most obvious use of "computer simulations" (in the widest sense of the term) was to provide mechanized schemes of calculation that vastly expanded the reach of available methods of analysis.[5] While computers were not capable of giving exact solutions of the equations already provided—either by established theoretical principles or by the various models that had been developed to make these principles more tractable—they could give approximate solutions to high degrees of accuracy, and with astonishing rapidity, and this capacity in itself clearly transformed both the domain and the practice of physical science.

Yet even so, there remains the question of epistemological novelty. How does the availability of high-speed computation qualitatively alter the epistemic character of what numerical analysts had already been doing, albeit on a manifestly smaller and slower scale? Indeed, in what sense can these early computational schemes be said to be *simulations*? What are they simulations *of*? Finally, although it might be easy to see how numerical analysis falls outside the range of what is conventionally regarded as "theory," what is it about these techniques that brings them into the domain of "experiment"? Taking these questions in order, I will argue that what we have now come to see as the epistemological novelty of computer simulation in fact emerged only gradually—not as a consequence of the introduction of any single technique, but as the cumulative effect of an ever-expanding and conspicuously malleable new technology: a technology that may originally have been designed to meet existing needs but that was, from its inception, already generating new opportunities and new needs. Just as with all the other ways in which the computer has changed and continues to change our lives, so too, in the use of computer simulations—and probably in the very meaning of science—we can but dimly see, and certainly only begin to describe, the ways in which exploitation of and growing reliance on these opportunities changes our experience, our sciences, our very minds.

Computer simulation may have started out as little more than a mechanical extension of conventional methods of numerical analysis, where what was being "simulated" were the precomputer, handwritten equations and where the early deprecatory sense of the term was still very much in place, but such methods rapidly grew so effective that they began to challenge the status of the original, soon threatening to displace the very equations they were designed to simulate. Over the course of time, evolving practices of computer simulation generated qualitatively different ways of doing science in which the meaning as well as the site of "theory," of "modeling," and eventually of "experiment" and "data" all came in for similar dislocations: Simulation came to lose its earlier sense of ontological inferiority, its status of "pretender," but also its sense of epistemological inferiority, at first nothing more than a mechanization of the lowliest form of scientific work, numerical computation. Paraphrasing Galison (1996, 119), we might agree that simulation eventually came to constitute "an alternate reality." Yet no single technical innovation can be held responsible. The transformation to which Galison and others refer emerged out of the collective and cumulative successes of many different effects in which new technical developments built on older ones in ways that might look seamless from afar but that, upon closer inspection, reveal a number of several more-or-less distinct stages (or branches), each bearing its own marks of epistemological novelty and its own disturbances to traditional notions of "theory," "experiment," and "data."

Provisionally, I suggest three such stages: (1) the use of the computer to extract solutions from prespecified but mathematically intractable sets of equations by means of either conventional or novel methods of numerical analysis; (2) the use of the computer to follow the dynamics of systems of idealized particles ("computer experiments") in order to identify the salient features required for physically realistic approximations (or models); (3) the construction of models (theoretical and/or "practical") of phenomena for which no general theory exists and for which only rudimentary indications of the underlying dynamics of interaction are available. With the growing success of these practices, use of the new techniques (as well as reliance upon them) increased steadily, inevitably enhancing the perceived epistemological and even ontological value of the simulation in question. But the originals whose privileged status was thereby threatened, and that may even have been put at risk of being supplanted by the simulation, were of widely different kinds in these three different practices. What is most directly called into question by the first case is the traditional status of the differential equation as the primary tool of theoretical physics; in the second, it is the nature of modeling and its relation to the construction of theory; in the third, it is both the meaning and the goals of explanation that come in for transformation. Interestingly, the roots of all these practices, as well as the first invocation of computers as "experimental" tools, can be found in the work of the mathematician Stanislaw Ulam (1909–1984) at the Los Alamos National Laboratories.

2.1. Computers as Heuristic Aid: "Experiments in Theory"

The immediate impetus for the development of computer simulation techniques came from research at Los Alamos from 1946 to 1952 on the feasibility of various proposals for building effective thermonuclear weapons.[6] Here, the first and foremost need was to bypass the mathematical intractability of equations conventionally used to describe the highly nonlinear phenomena that were involved (for example, neutron diffusion, shock waves, and "multiplicative" or branching reactions). Ulam, working with Von Neumann, Fermi, and others, originated a number of novel approaches to existing computational procedures that have since become staples in the analysis of complex systems. Ulam's main contributions were not, however, dependent on the computer, but rather on the deployment of methods of combinatorial analysis and statistical sampling for exploring the solution space of conventional differential equations.

The most famous of these, the Monte Carlo method, has been extensively discussed by Galison (1996), and its introduction (Richtmyer and Von Neumann 1947; Metropolis and Ulam 1949) is sometimes taken as synonymous with the origin of simulation (see, for example, Mize and Cox 1968, 1). In point of fact, however, the epistemological novelty of Monte Carlo (at least as it was first introduced) had little to do with the computer. Its application to differential equations depended on the formal isomorphism of such equations with certain equations in probability theory. The first novelty of the method lay in inverting the customary use of that relation (that is, in exploiting the probability relations to solve the differential equations rather than using the differential equations to analyze the probability relations); the second lay in replacing the computation of the combinatorial possibilities for all sequences of events (or individual trajectories) by estimates of successful outcomes obtained by sampling a number of different "experimental" trajectories, trials, or "games." As Ulam wrote, "Given a partial differential equation, we construct models of suitable games, and obtain distributions on solutions of the corresponding equations by playing those games, i.e., by experiment" (Ulam 1952, 267). The method's first published use was in "solving" prespecified (Boltzmann-type) equations for neutron diffusion. Ulam explained the procedure as follows: "[D]ata are chosen at random to represent a number of neutrons in a chain-reacting system. The history of these neutrons and their progeny is determined by detailed calculations of the motions and collisions of these neutrons, randomly chosen variables being introduced at certain points in such a way as to represent the occurrence of various processes with the correct probabilities. If the history is followed far enough, the chain reaction thus represented may be regarded as a representative sample of a chain reaction in the system in question. The results may be analyzed statistically to obtain various average quantities of interest for comparison with experiments or for design problems" (Ulam 1990, 17).

What lent the computer its importance to this application was simply its ability to perform the required "detailed calculations" on a scale and at a speed exceeding anything that could have been done by hand or by other mechanical devices. In other words, the difference made by the computer in the application of this method was identical to that which it made to more conventional kinds of numerical analysis. It was because of its speed (leading in turn to its versatility) that Ulam described the electronic computer as enabling "one to make, as it were, 'experiments in theory' by computations, either of the classical or of the 'Monte Carlo' type computations" (1990, 122). What did he mean by "experiments in theory"? I submit that, for Ulam, computer simulations were "experimental" in the same sense in which a thought experiment was "experimental" (or in which repeated games of chance were "experimental"), different only in that the computer permitted the working out of the implications of a hypothesis so rapidly as to rival the speed of thought and was certainly vastly faster than any of the traditional means of computation that had been available. They extended the mathematician's powers of analysis and, as such, ought to have been as valuable for solving problems in pure mathematics as in mathematical physics. In no sense were they to be confused with actual experiments ("experiments in practice"?) on which confirmation of theory depended. Nor were such simulations to be confused with the "design problems" posed by the physical materials with which the engineers were working—problems that had perforce to be solved before a device with real explosive power could be detonated. When it came to building actual bombs, nowhere is there any evidence of either confusion or slippage between simulation and the real thing.[7]

As is perhaps inevitable, such "experiments in theory"—especially as they proved ever more successful—began to take on a life of their own. Indeed, their very success brought conspicuous pressure to bear on the primacy of (or need for) more conventional mathematical tools, most notably, the differential equation. Similarly, that success added new legitimacy to the practice of numerical analysis that had been around for so long. But, while both these effects can claim certain kinds of epistemological novelty (for example, challenging both the hegemony and the realism of continuous variable representations and demanding "an empirical epistemology, and not merely a mathematico-deductive one" [Winsberg 1999, 290]),[8] neither, it seems to me, yet alters the perception of the basic aim of matching theoretical predictions with experimental findings. Computer simulation is at this stage still directed toward eliciting the implications of well-formulated theoretical models.

2.2. "Computer Experiments" in Molecular Dynamics

A rather different sense of "experiment"—one that is noticeably closer to physicists' understanding of the term than to that of a mathematician—enters the literature on computer simulations in the mid to late 1950s. Where, in the earlier sense of the term, what was to be simulated were the equations of the traditional theoretical physicist, and the aim was to obtain approximate solutions of these equations, in the new practice of simulation it was an idealized version of the physical system that was to be simulated, the aim of which was to produce equations (or models) that would be both physically realistic and computationally tractable.[9] In fact, the practice of "computer experiments" (as such techniques soon came to be called) deployed two levels of simulation: first, substitution of the actual physical system by an "artificial" system, and, second, replacement of the equations to which the first level of simulation gave rise by computationally manageable schemes for numerical analysis. Thus, they were "experimental" in two senses of the term—not only in Ulam's sense of "experiments in theory" but also in the sense we might call "experiments in modeling"—and they were aimed at redressing theoretical intractability on two corresponding levels: descriptive and computational. As two of the early advocates of this approach described it, "This half-way house between elegant theory and experimental hardware, our programmed version of the physical laws and boundary conditions, we call a "computer experiment." It differs from a typical computation of a theoretical result in that we do not evaluate mathematical expressions derived from the laws of nature, but we make the computer simulate the physical system" (Buneman and Dunn 1965, 4).

To understand the need out of which this new use of simulation arose, we need to recall the state of "theory" in the mid-1950s for the macroscopic (thermodynamic) properties of liquids, gases, and solids. "Theory," in this context, means statistical mechanics, and its aim is to derive the equilibrium and nonequilibrium behavior of many-body systems from the molecular dynamics of the component particles, and the obvious problem is how to deal with so large a number of particles. Clearly, some form of simplification and successive approximation is required. For example, one might begin by ignoring all interactions between particles (the ideal gas approximation); a next step, and slight improvement, would be to treat the particles as quasi-independent, each moving in some average potential due to all the other particles in the system. But neither of these approximations is adequate at high density, nor for describing the phenomena of greatest interest, namely phase transitions; for this, one needs a more realistic representation of the effects of molecular interactions. The crucial next step (often said to be the origin of modern liquid theory) was taken in 1935 when J. G.

Kirkwood rewrote the equations in terms of pairs of particles moving in an effective potential (due to all other particles). Thus rewritten, one needed only two functions: the intermolecular pair potential and the radial (or pair) distribution function; yet, without knowledge of the interatomic or intermolecular forces, neither of these functions could be specified. Various models for the pair potential were available (for example, hard spheres; hard spheres plus square well; Lennard-Jones), but until the mid-1950s, the only available access to the radial distribution function (representing the distribution of distances between atoms or molecules) was from X-ray or neutron diffraction patterns in simple fluids. Such measurements were not only cumbersome (having to be redone for each change in density or temperature) but also limited to finite ranges of frequency, and even in those ranges dependent on data that were often fraught with ambiguities. Finally, and perhaps most important, such empirical derivations were theoretically unsatisfying in that they offered no insight at all into the molecular dynamics responsible for the shape or behavior of the function.

Computers enter the history of this field with the development of an alternative approach to the problem by two physicists working at the Livermore National Labs, Berni Alder and Ted Wainwright, in the late 1950s. Building on the Monte Carlo computations of N-body systems that the Los Alamos group had pioneered, and using the high-speed computers available at Livermore, Alder and Wainwright were able to follow the behavior of systems of a finite number of particles (ranging between 32 and 500) idealized as hard spheres under conditions of varying density and temperature. As they wrote, "With fast electronic computers it is possible to set up artificial many-body systems with interactions which are both simple and exactly known. Experiments with such a system can yield not only the equilibrium and transport properties at any arbitrary density and temperature of the system, but also any much more detailed information desired. With these 'controlled' experiments in simple systems it is then possible to narrow down the problem as to what analytical scheme best approximates the many-body correlations" (Wainwright and Alder 1958, 116). Others—both in studies of classical fluids and in plasma physics—soon picked up on the method, as well as on the nomenclature, and ran with it. In the first of a series of papers on computer experiments, in which Alder and Wainwright's approach was extended to systems of particles interacting through a Lennard-Jones potential, Loup Verlet explained, "The 'exact' machine computations relative to classical fluids have several aims: It is possible to realize 'experiments' in which the intermolecular forces are known; approximate theories can thus be unambiguously tested and some guidelines are provided to build such theories whenever they do not exist. The comparison of the results of such computations with real experiments is the best way to obtain insight into the interaction between molecules in the high-density states" (Verlet 1967, 98).

Here, the aim of computer experiments is clearly stated: to test "approximate theories" where they exist and to provide guidelines for building such theories where they do not. Here, also, use of the word "exact" highlights the difference between the Monte Carlo simulations initiated at Los Alamos for the purpose of computation and the new methods: in contrast to the "exact" machine computations, it is the theories (or models) to be computed by these methods that are now acknowledged as "approximate." In this shift of usage, the purpose of computation (conventionally associated with the application of a theory) is tacitly subordinated to another kind of aim—namely, that of building theory—and it is in pursuit of this latter aim that the simulation (or "artificial system") serves as an "experimental" probe. The simulation is a trial theory, and the role of machine computation is to render the test of that theory "unambiguous." As the final sentence (in the passage quoted above) makes clear, however, the ultimate power of arbitration was still seen as residing in "real experiments," especially in the real experiments that provided measurements of macroscopic properties.

That physicists were well aware of the irregularity of this use of the term "experiment"—and equally of the threats such usage raised for traditional understandings of "theory" and "experiment"—is well attested to by the recurrence of discussions of the matter to be found in the literature throughout the 1960s. In one of the earliest such discussions, the authors, Buneman and Dunn, focus their attention on the relation between "computer experiments" and "theory." They begin by observing, "we are at the threshold of a new era of research." Computer experiments "yield surprising and significant answers"; they permit the deduction of "a qualitative or even an analytic theory" and allow one to "guess what are the significant effects and what is the correct way of looking at a problem" (Buneman and Dunn 1965, 56). In just the brief time they'd been around, use of this new mode of analysis had already increased dramatically. From a retrospective search of the literature performed in 1966, another observer, Charles Birdsall, estimated the growth rate of articles on computer experiments over the years since 1956 by an exponential factor of 1/3. The rapid increase, Birdsall wrote, "shows the strong entry of computer experiments into at least junior partnership with theory, analysis and laboratory experiment" (Birdsall 1966, 4). It was obvious that the trend would continue and equally obvious that questions —both about their epistemological and their professional status—would arise and had in fact already arisen. In response to such questions, Buneman and Dunn rose to the defense of their methods, and they did so in a way that makes manifest a certain already existing heterogeneity regarding the meaning of "theory." They wrote: "One encounters, at times, a prejudice against computer experiments. Partly, such prejudice is based on mathematical snobbery (the formal description of the skin effect in Bessel-functions of complex argument enjoys higher prestige than a few graphs showing how it actually goes!). But often one hears the complaint that a computer can at best say "this is *how* it happens" and never "this is *why* it happens." The examples produced here should suffice to answer this complaint. The mere fact that the computer was able to produce the 'how' has, many times, told us the 'why'" (Buneman and Dunn 1965, 56).

By the early 1970s, however, contestation had spread (if not shifted) to the relation between at least some real experiments and computer experiments. Computations of the radial distribution function based on X-ray scattering remained fraught with technical difficulties, but, by contrast, the prowess of the simulators, their machines, and those who programmed the machines increased rapidly. As a consequence, confidence in the reliability of computer "observations" soon came to rival (if not overtake) confidence in the reliability of observations based on experimental measurements. Not only were the former easier to obtain (that is, more economical), but also the repertoire of internal consistency checks available to the "simulators" soon granted their "observations" a trustworthiness that the methods of the experimentalists were unable to inspire.[10]

Nevertheless, it remains the case that it is primarily in the domain generally referred to as "theory" that "computer experiments" have had their major triumph. Today, with the accumulation of four decades of experience, their value as theoretical tools—that is, for building theory—has been amply vindicated, and defense of the sort that Buneman and Dunn had earlier felt called upon to make seems no longer to be required. Or so, at least, one might conclude from their prominence in the literature. For a contemporary assessment, I quote from the contribution to the current *Encyclopaedia Britannica* on the molecular structure of liquids. Here John M. Prausnitz and Bruce E. Poling write: "Since 1958 such computer experiments have added more to the knowledge of the molecular structure of simple liquids than all the theoretical work of the previous century and continue to be an active area of research for not only pure liquids but liquid mixtures as well" (Prausnitz and Poling 1999).

2.3. Cellular Automata and Artificial Life

The third class of computer simulation I want to discuss departs from the first two in at least one crucial respect: It is employed to model phenomena that lack a theoretical underpinning in any sense of the term familiar to physicists—phenomena for which no equations, either exact or approximate, exist (as, for example, in biological development), or for which the equations that do exist simply fall short (as, for example, in turbulence). Here, what is to be simulated is neither a well-established set of differential equations (as in Ulam's "experiments in theory") nor the fundamental physical constituents (or particles) of the system (as in "computer experiments"), but rather the phenomenon itself. In contrast to conventional modeling practices, it might be described as modeling from above.

Perhaps the most conspicuous example of this use of simulation is to be found in A-Life studies, officially christened as such at a conference held at Los Alamos in 1987 and organized by Christopher Langton, at that time a member of the Theoretical Division of the Los Alamos National Laboratory (Langton 1989). I include A-Life under the category of "simulation in the physical sciences" for the simple reason that, despite its explicitly biological allusion, it was developed by—and for the most part has remained in the province of—physical scientists. Furthermore, Langton may have been responsible for introducing (as well as for popularizing) the term "Artificial Life" as a label for computer simulations of biological evolution,[11] but the basic project of simulating biological processes of reproduction and natural selection on the computer is in fact of much longer standing: indeed, it has its origins in the same context (and in the work of the same people) from which the first use of computer simulation for numerical analysis arose.

Von Neumann is the man most frequently credited as the "father of Artificial Life," and his contributions to the field arose directly from his preoccupations with a question that might be regarded as the oldest and most fundamental of all questions about simulation, namely, how closely can a mechanical simulacra be made to resemble an organism? More specifically, he asked: Is it possible to construct an automaton capable of reproducing itself? Beginning in the 1940s, Von Neumann worked with a kinematic model of automata afloat in a sea of raw materials but never fully succeeded in capturing the essential logic of self-reproduction. The breakthrough came with the suggestion of his close colleague, Stanislaw Ulam, that a cellular perspective (similar to what Ulam was using in his Monte Carlo computations)—in which the continuous physical motion required in the kinematic model would be replaced by discrete transfers of information—might provide a more effective approach. "Cellular automata," as they have since come to be called, have no relation to biological cells (and, indeed, from the beginning they were also invoked for the analysis of complex hydrodynamic problems), but they did suggest to Von Neumann a way of bypassing the problems posed by his kinematic model. Here, all variables (space, time, and dynamical variables) are taken to be discrete: An abstract space is represented as a lattice with a cellular automaton (a mathematical object—that is, "a finite-state" machine) located at each node of the lattice. Each such automaton is connected to its nearest neighbors, and it evolves in time by reading the states of its neighbors at time t_n and, according to prespecified and simple rules, moving to a new state at time t_{n+1}. Ulam and Von Neumann reasoned, and indeed soon proved, that the collective dynamics resulting from such simple rules might bear a formal resemblance to the biological process of self-reproduction and evolution.

Von Neumann's initial construction in the early 1950s was cumbersome (requiring 200,000 cells with twenty-nine states for each automaton), but it made the point. The story of its subsequent development (and dramatic simplification)—from John Conway's "Game of Life" (see Gardner 1970) to Chris Langton's even simpler self-reproducing "loops" (1984)—has been recounted many times and hardly needs repeating here.[12] Somewhat less well known is the history of the use of cellular automata in the modeling of complex physical phenomena (for example, turbulence, crystallization, etc.)[13]—an activity that, like "Artificial Life," also exploded with the appearance of "super-computers" in the 1980s. Indeed, the very first conference on "Cellular Automata" was also held at Los Alamos (preceding the A-Life conference by four years), and while it provided the occasion for Langton's initial foray into artificial life, the primary focus of the earlier conference was on the physical sciences (Farmer, Toffoli, and Wolfram 1984).[14] A proper account of this part of the history of cellular automata remains to be written by historians, but my focus here is not so much historical as it is conceptual: that is, to try to identify what is distinctive about this new kind of simulation and to capture its epistemological novelty. What follows is at best a very rough (and necessarily brief) characterization.

Cellular automata are simulations par excellence: they are artificial universes that evolve according to local rules of interaction that have been prespecified. Change the initial conditions, and you change the history; change the rules of interaction, and you change the dynamics. In this sense, the analogy with differential equations is obvious. Also obvious are many of the differences between CA and DE's: the universe of CA is discrete rather than continuous; its rules generally describe interactions that are local (for example, nearest neighbor) rather than long range, and uniform rather than spatially variable; the temporal evolution of CA systems is exactly computable for any specified interactions (given enough time) while DE's are rarely susceptible to exact analytic solutions and only approximately computable when they are not.[15] But more important by far are the differences in the uses to which they are put, in the processes by which they are crafted, and in the criteria by which they are judged.

CA have a home in A-Life studies precisely because of the unavailability of differential equations for the processes they simulate; similarly, they lend themselves to the simulation of excitable media, turbulence, and earthquakes because the equations that do exist are not adequate to describe the phenomena of interest. And indeed, in some of their uses, CA models might be viewed simply as an alternative to DE's in which exact computability enables unambiguous tests of approximate theories—that is, just as it was claimed to do in molecular dynamics, only this time around without the quotation marks. More often, however, they are employed in a radically different spirit, aimed more at producing recognizable patterns of "interesting" behavior in their global or macrodynamics than in their microdynamics. As Stephen Wolfram writes,

> Science has traditionally concentrated on analyzing systems by breaking them down into simple constituent parts. A new form of science is now developing which addresses the problem of how those parts act together to produce the complexity of the whole.
> Fundamental to the approach is the investigation of models which are as simple as possible in construction, yet capture the essential mathematical features necessary to reproduce the complexity that is seen. CA provide probably the best examples of such models. (Wolfram 1986, v)

Several points bear emphasizing here, and they are related: one has to do with the process by which CA models are constructed, another with their synthetic capacities (in both senses of the word), and a third, with the focus on formal similarity between the outcomes they yield and the "overall behavior" of the processes they are designed to mimic (physical, biological, economic, or other).[16] Toffoli and Margolus's introduction to the subject is instructive, and I quote it at length:

> In Greek mythology, the machinery of the universe was the gods themselves. . . . In more recent conceptions, the universe is created complete with its operating mechanism: once set in motion, it runs by itself. God sits outside of it and can take delight in watching it.
> Cellular automata are stylized, synthetic universes. . . . They have their own kind of matter which whirls around in a space and a time of their own. One can think of an astounding variety of them. One can actually construct them, and watch them evolve. As inexperienced creators, we are not likely to get a very interesting universe on our first try; as individuals we may have different ideas of what makes a universe interesting, or of what we might want to do with it. In any case, once we've been shown a cellular-automaton universe we'll want to make one ourselves; once we've made one, we will want to try another one. After having made a few, we'll be able to custom-tailor one for a particular purpose with a certain confidence.
> A cellular automata machine is a universe synthesizer. Like an organ, it has keys and stops by which the resources of the instrument can be called into action, combined, and reconfigured. Its color screen is a window through which one can watch the universe that is being "played." (Toffoli and Margolus 1987, 1)

The seductive powers of CA are obvious, and many (including Toffoli) have been seduced. Their successes at simulating global effects have encouraged them to shift not only the meaning of simulation (and of model), but, at least in some writings, the status (or even locus) of what had earlier been taken to be the original, the real thing. Thus, for example, G. Y. Vichniac (1984) proposes "[c]ellular automata as original models of physics," and he suggests the possibility that the physical world really is a discrete space-time lattice of information bits evolving according to simple rules, an enormous CA running with one of many possible sets of rules. This view was represented in a number of presentations at the 1983 conference, and it has since gained considerable legitimacy in the world of computational physics (occasionally referred to as "synthetic physics"); in fact, however, it had already been advocated by some (notably, Ed Fredkin) ever since the 1960s.[17] Moreover, from such claims about the physical universe to Langton's (1989) arguments for "A-Life" ("we expect the synthetic approach to lead us not only to, but quite often *beyond,* known biological phenomena; beyond *life-as-we-know-it* into the realm of *life-as-it-could-be*") is a short step. The point to be noted is that, in both Vichniac's and Langton's proposals, the very cellular automata that had originally been invoked as explanatory crutch, as simulation of something prior, of features of a world assumed to be simultaneously more fundamental and more "real," have somehow metamorphosed into entities with ontological primacy in and of themselves.

So radical an inversion of conventional understandings of the relation between simulation and reality are not yet widespread—either in the physical or the biological sciences (indeed, they have yet to make any noticeable impact on the majority of biologists)—but the very fact that they have become thinkable, and in certain circles even acceptable, is surely worth noting. Minimally, it provides an indication of the power of CA models to subvert conventional distinctions between real and virtual, or between real and synthetic, and hence of their efficacy in establishing an "alternate reality." The epistemological novelty of CA modeling is in this sense quite different both from that of the Monte Carlo techniques first introduced by Ulam and Von Neumann (however much it may owe these early inventions for its technical development) and from that novelty associated with the computer experiments of molecular dynamics. Where one extended the meaning of "mathematical," and the other the range of "theory," the primary novelty of CA modeling may well lie in extending the range of the "real." I would argue, however, that the principal route by which CA modeling achieves this extension is to be found less in its capacity to present visually compelling images of synthetic (that is, artificial or virtual) objects than in its synthetic powers in the other sense of that term, namely, through its utility in synthesizing new kinds of objects that are unambiguously real. This, I suggest, is especially evident in Artificial Life studies.

Despite initial hopes in the value of CA modeling in promoting better theory —in particular, a better understanding of biological principles—Artificial Life studies have made little impression on practicing biologists. Far more significant has been their influence on engineers. In a recent book entitled *Creation: Life and How to Make It,* Steve Grand writes, "Research into artificial life is inspiring a new engineering discipline whose aim is to put life back into technology. Using A-life as an approach to artificial intelligence, we are beginning to put souls into previously lifeless machines. . . . The third great age of technology is about to start. This is the Biological Age, in which machine and *synthetic* organism merge" (Grand 2002, 7–8).

Synthetic life forms that are real objects in the sense that they are made from material components and assembled in real space and time are clearly being built, and in ways that draw directly from work on "lifelike" simulations in cyberspace. Engineering is a science that specializes in negotiating the gap between symbol and matter, and robotic engineers, like their colleagues in allied disciplines, have well-developed techniques for translating from one domain to the other, for realizing the metaphors of simulation in the construction of material objects. Computer simulations of biological organisms may well be "metaphorical representations," but they are also models in the time-honored sense of guides or blueprints: in the hands of skillful engineers, they can be, and are, used as guides to construction in an altogether different medium. Here, the simulated organisms of cyberspace are used to guide the synthesis of material objects mimicking the behavior of biological organisms in real space. Without doubt, these entities are real. But another question immediately arises: are they "alive"? This is a question that worries many philosophers, but, as I argue elsewhere (Keller 2002, chap. 9), it may well be a question that belongs more properly in the realm of history than in that of philosophy.

NOTES

1. While the shift from deception to instruction is undoubtedly worthy of study in itself, my aim in this chapter is merely to examine its impact rather than the process by which it occurred.

2. See, e.g., discussion of echo simulators developed to train AI operators of an aircraft interception radar set in Garman (1942).

3. The use of computer simulations in the biological sciences—how it both draws from and differs from its uses in the physical sciences—is of particular interest to me, but I refer the reader to chapters 8 and 9 of Keller (2002) for discussion of this topic.

4. Past computations may have employed large organized systems of human computers, yet the material medium of computation consisted of paper and pencil (and later, calculators).

5. This is the sense that conforms to Paul Humphreys's "working definition" of computer simulation as "any computer-implemented method for exploring the properties of mathematical models where analytic methods are unavailable" (1991, 501).

6. The heading "Experiments in Theory" should not to be confused with Deborah Dowling's notion of "Experimenting on Theories" (1999), a notion intended to capture a broad range of functions of computer simulation.

7. And yet it is surely both curious and notable that a new discourse of simulation, with all the slippage invited by that discourse, arose precisely at the place and the time in which no such slippage could be permitted in actual practice, precisely in the context of a project aimed at annihilation. Such a coincidence cries out for explanation, but it may be that its accounting will need to be sought in the psychological adjustment required for men and women who, in ordinary life, would likely have found such weapons of mass destruction abhorrent. (Thanks to Loup Verlet for this suggestion.)

8. Especially important is the attention Winsberg calls to the processes by which the computational results produced by simulation acquire authority and reliability.

9. R. I. G. Hughes draws this distinction as well. In a paper that came to my attention only after my own was written, Hughes refers to the earlier uses as "computer techniques to perform calculations" (1999, 128) and reserves the term "computer simulation" for simulations of physical systems and not of equations per se.

10. Verlet recalls an encounter between simulators and experimentalists from the early 1970s that is illustrative: the "theorists" had been greatly puzzled by apparent anomalies in an earlier report of a radial distribution function for Argon obtained from X-ray scattering, and many of them had suspected it was based on some error. At a conference held two years later, the author of the original report announced his new, improved results, claiming, "At last, molecular dynamics and X-ray experiment have come into agreement." Whereupon Verlet interjected (to the general amusement of the audience), "For the record I would say, but molecular dynamics has not changed!" (personal communication, March 15, 2000).

The anecdote highlights the ambiguity between "models," "theory," and "experiment" that has become so prominent a feature of simulation in the physical sciences. But it might also be argued that that same ambiguity has always been characteristic of the practice of theory in physics (see, e.g., Buchwald 1995). It also highlights the importance, as Winsberg (1999, 288) has emphasized, of attending to the nitty-gritty of "the process by which these results get sanctioned."

11. In Langton's first use of the term, he wrote: "The ultimate goal of the study of artificial life would be to create 'life' in some other medium, ideally a *virtual* medium where the essence of life has been abstracted from the details of its implementation in any particular model. We would like to build *models* that are so life-like that they cease to become models of life and become *examples* of life themselves" (1986, 147).

12. Although a fuller account of this history—one that also includes the work of Ulam, Barricelli, Holland, and many others—would certainly be welcome.

13. For a good overview of the use of CA in the fluid dynamics and statistical mechanics, see Rothman and Zaleski (1997).

14. That so much of this work has come out of Los Alamos is no accident, for Los Alamos was one of the few laboratories at this time to have the parallel-processing "super-computers" that made the execution of CA systems practical. As Toffoli and Margolus write, "In this context, ordinary computers are of no use. . . . On the other hand, the structure of a cellular automaton is ideally suited for realization on a machine having a high degree of parallelism

and local and uniform interconnections" (1987, 8). Conversely, however, it must also be said that the design of such machines (at least as envisioned by Hillis) was "based on cellular automata" (Hillis 1984).

15. CA advocates see exact computability as a major advantage over DE's. As Toffoli writes, "any properties that one discovers through simulation are guaranteed to be properties of the model itself rather than a simulation artifact" (1984, 120).

16. A good part of the appeal of CA models derives from the presentation of forms that are striking for their visual similitude, but such visual presentations are in an important sense artifacts—i.e., they result from self-conscious efforts to translate formal resemblance into visual resemblance.

17. Today, Stephen Wolfram is the leading advocate of a digitally based physics, but on numerous occasions, Richard Feynman also expressed support for the idea, hypothesizing early on "that ultimately physics will not require a mathematical statement, that in the end the machinery will be revealed, and the laws will turn out to be simple, like the chequer board with all its apparent complexities" (1967, 57).

Mary S. Morgan

II / Experiments without Material Intervention

MODEL EXPERIMENTS, VIRTUAL EXPERIMENTS,
AND VIRTUALLY EXPERIMENTS

1. Introduction

Experiments may be portrayed as involving manipulations of elements in the material world under conditions of control. A simple material experimental manipulation or intervention, such as adding a certain amount of a substance to an amount of liquid in a test tube and directly observing (and perhaps measuring) the results, provides a stereotype example of the idea of experiment gleaned from high-school chemistry experience. Such an experiment incorporates areas of control in both the circumstances and in the procedures of the manipulation.

Recent science studies have emphasized both how very different from this stereotype most experiments really are and how much more is involved.[1] Even in those "simple" cases of experiment, theoretical beliefs about the way the world works, personal knowledge of how the experiment can be made to work, and networks of trust that indulge the credibility of results and subsequent inferences to parallel situations all figure as essential elements of the experimental experience. The ways of modern science are often even further from this vision, for they rely on a huge technological interface between scientist and nature, both in making the intervention and in capturing and assessing outcomes of the intervention. This interface consists of tools to carry out the intervention, instruments to detect the resulting changes in phenomena and yet more instruments to clarify, assess, and present the evidence in a form that can be read by humans.

The instrumentation and technology of experiment is now well rehearsed, as are the roles of tacit knowledge and the interdependence of theoretical guidance and experimental action. All this seems to make it much more problematic to characterize experiments as things that can be studied independently of their place in specific contexts and of the theories and natural phenomena they relate to.

Strangely, perhaps, at the same time as science studies were writing these more integrated accounts of science based on "practice," the "new experimentalism" suggested both that we can treat experiments as having a life of their own and that the old image of experiment as a controlled manipulation or intervention in nature—in something material—still remains on the table.[2] Despite the considerable recent exploration of scientific practice, the conventional methodological category of "experiment" remains largely intact. More striking, it does so even in the face of the various kinds of nonmaterial experiments that have now invaded scientific investigations in many fields. And while the philosophical analysis of such "simulations" has recognized the presence of something new in these epistemological instruments, this work is only slowly feeding back into accounts of experiments.[3]

In this chapter, I am concerned with the kinds of "vicarious" experiments that involve elements of nonmateriality either in their objects or in their interventions and that arise from combining the use of models and experiments, a combination that has created a number of interesting hybrid forms.[4] Using cases from economics,[5] mechanics, and biology, I explore two issues that seem pertinent to a philosophical account of these modern hybrids:

Question 1: What counts as a material intervention?

The use of computers in experiments focuses attention on this problem in a way that leads me to ask, How far may we stretch the notion of what counts as "material" in experiments? I will outline the character of two types of experiments: "virtually" experiments, ones in which we have nonmaterial experiments on (or with) semimaterial objects, and "virtual" experiments, ones in which we have nonmaterial experiments but which may involve some kind of mimicking of material objects.

Question 2: How do the results of these vicarious experiments relate to the world?

The issues of experimental validity are many, but I suggest in this context that the validity of experimental results also hinges on how the objects and models used in experiments relate to the things they are supposed to tell us about in the world. Models used in experiments represent the material world in different ways, and this has implications for the range of inference that such vicarious experiments can support.[6]

2. What Counts as a "Material" Intervention?

2.1. Laboratory Experiments and Experiments with Mathematical Models

The archetype of experiment assumes that however much the experimental situation is constrained, controlled, and even constructed, it is nevertheless an experiment on (or in, or with) a material system. From the use of a vacuum pump to supercolliders, however artificial the environment that is created, however artificial the outcome, the experimental intervention itself involves an action upon or the creation of a material object or phenomenon. In contrast, modern economics tends, in the main, to function by using extended thought experiments in which the material domain remains absent. Since I will use this as a limiting case in opposition to material experiments, let me characterize this practice further.

In the post-1950s period, economists have become avid users of mathematical models. In a recent paper on the way in which such models are used in economics, I suggested that their usage involved being able to trace through deductively the answers to "what if" or "let us assume" type questions about the economic world represented in the model (Morgan 2001). For example, in the early days of modeling, economists built small mathematical models (between three and eight or so equations) to represent the essential elements of Keynes's macroeconomic theory. Such models allowed those economists to trace through in a coherent way the answers to a number of likely real-world questions about such a system: What happens if government expenditure increases? What happens if investment declines? Such questions provided the external "intervention" or change to the model system, and the models were manipulated to answer such questions by exploring the internal behavior and limits of the economic world represented in the mathematical model.

We can portray this modern use of mathematical models as extending economists' verbal thought experiments of earlier times that were limited by the capacity of the mind to follow the paths of more than two or three variables in a system. In characterizing such model usage in terms of glorified thought experiments, we can see how asking questions and exploring the answers with mathematical models have allowed economists to think through in a consistent and logically deductive way how a larger number of variables may interrelate and find the solutions to systems with a larger number of units. Often the questions used in mathematical model exploration are questions about theories; other times, the questions enable analysis of policy problems from the world. In this way, models may function both as tools for theory development and as tools for understanding the world (see Morrison and Morgan 1999). But to treat this extension of the powers of the mind in using mathematical models as experiments of the same form or kind as material experiments requires more analysis of what is involved. There are a number of other ways in which doing experiments with these mathematical models in economics is different from doing laboratory experiments in biology or chemistry or in economics for that matter.

First, there is a difference in how experimental control is achieved. In experiments, isolating the relevant elements of interest and carrying out a controlled manipulation (or intervention) on them is partly achieved through the experimenter's choice of site and particular material process. But however carefully the site choice is made, it is likely that the particular process of interest cannot be isolated without further work by the experimenter, namely by rigorous attention to the control of circumstances. Here it is useful to draw upon Marcel Boumans's (1999) exemplary dissection of ceteris paribus conditions in his work on the functioning of economic models as measuring instruments.[7] Just as in making good measurement instruments, the experimental scientist must work hard to take account of all conditions and factors that are likely to interfere with the process of interest. Boumans extends the analysis of such factors by distinguishing between three sorts of conditions: *ceteris paribus, ceteris neglectis,* and *ceteris absentibus*. Some disturbing causes may be declared absent if the experimenter can physically rule them out of the setup (ceteris absentibus). Of those causes that are present but are not the subject of experiment, some may be thought to be so minor in effect that they can be neglected (ceteris neglectis). The others that are present have to be controlled for by procedures that hold them constant during the experiment (ceteris paribus). These control conditions make the material setup of the laboratory experiment somewhat (more or less) artificial, but it remains material and of the real world for all that, because however ingenious the scientist, the material world can only be controlled and manipulated to an extent.

In contrast, the control requirements for experiments with a mathematical model are achieved by simplifying assumptions: it is assumed that minor causes can be neglected, it is assumed that certain things are zero, it is assumed that certain things are unchanging. In fact, economists state the phrase "ceteris paribus" as a catch-all to imply that all three of Boumans's conditions hold by assumption without discriminating between them. In addition, in the mathematical model experiments, the user can impose, by assumption, a total independence between two or more elements in the model that might not be achievable in the equivalent material system. Such confounding causes may prevent experimental isolation and demonstration in the laboratory experiment, whereas they can so easily be assumed away in the mathematical model experiment: everything else *may* not be the same or may not be ruled out when you manipulate the material system whereas it *can* be held the same or set at zero when manipulating the mathematical model. The agency of nature creates boundaries and constraints for the experimenter. There are constraints in the mathematics of the model, too, of course, but the critical point is whether the assumptions that are made there happen to be the same as those of the situation being represented and there is nothing in the mathematics itself to ensure that they are.[8]

The second difference lies in the production of experimental results. In laboratory experiments we intervene in the material system to produce material results for the particular situation found in the experimental setup. In the mathematical model experiments, as I have just argued, the "intervention" into the model begins with a question that prompts the deductive or logical reasoning power of mathematics to derive the results (see Morgan 2001). This difference between producing or deriving results is the contrast between experimental demonstration and mathematical demonstration.

It is tempting to see this contrast as one between a system in which the outcome to the question is already built into the model that we construct and is merely revealed by mathematical reasoning, whereas however much faith we have in our theory of how the material world works, we still (at some stage) need to make the experimental demonstration. That material world experimental demonstration is necessary precisely because the resources for the result that *we expect to find* are not necessarily present in the experiment setup: we might have the wrong account or theory about what will happen or our knowledge of the world might be seriously incomplete. In the mathematical model world, of course, we know the resources for the result that *we do find* because we built those resources into the model that constitutes the experimental setup. We might initially state this difference in outcomes as follows: the possibilities of the material world are such that experiments in that domain may surprise us, but those with mathematics should not. However, the point of using models is to find out things that we do not already know about how those structures behave when the parts of the model are put together or when we vary certain things in the model. These are the kinds of things we do not know in advance of our experiments with models, so that in using models, we should expect them sometimes to surprise us. In principle, though, having been surprised, we can go back through the model experiment and understand why such surprising results occurred. That possibility may not be open to us with material experiments where our ignorance may prevent us from explaining why a particular set of results occur. Perhaps, then, we might better restate the contrast by saying that in experiments on our mathematical models of physical, biological, or economic

TABLE 11.1. Types of Experiment: Ideal Laboratory and Mathematical Model

	Ideal lab experiment	**Mathematical model experiment**
Controls on inputs, intervention, and environment	experimental	assumed
Demonstration method	experimental in laboratory	deductive in model
Degree of materiality of inputs, intervention, and outputs	material	mathematical

systems, we may be surprised, but in material experiments directly on those systems, we may be confounded.[9] This suggests that material experiments have a potentially greater epistemological power than nonmaterial ones.

Table 11.1 records these two distinctions—the means of control and the method of demonstration—between the pen-and-paper experiments with mathematical models compared to the experiments with material processes or objects. There is a third important difference between these two types of experiments, namely, in their range of potential inference. In the laboratory case, we use the experimental results to argue (or perhaps infer) from this process and results to the same kind of material process and results in other (possibly) nonlaboratory setups. In the mathematical model experiments also, we use the results to argue, or infer, to other systems we think have similar characteristics to those of the model, but in this case those systems might be mathematical or material. These differences in inferential range will be discussed more fully in section 3 of this chapter (see also Morgan 2002). Meanwhile, let me look at two kinds of intermediate cases. The first kind are cases where we use model resources to experiment on semimaterial objects, the second kind are cases where we experiment with models to create non- or pseudo-material objects.

2.2. Experiments on Semimaterial Objects: Computer Bones

The internal structure of our bones looks extraordinarily complicated to the layperson, and it seems that even the experts find it a challenge. Tony Keaveny, director of the Orthopaedic Biomechanics Laboratory at the University of California, Berkeley, conducts laboratory experiments to investigate the strength of bones (see Keaveny et al. 1994; Niebur et al. 2000) and particularly that part due to their architecture. Assessing the strength of the structure is problematic in real material experiments in which force is exerted mechanically on samples of bone because it does not allow the investigator to distinguish between the strength of the material and of its structure, and because the process is a destructive testing regime in which it is difficult to see and analyze how the detailed internal structure responds to increasing force.

In a 1999 seminar paper at University of California, Berkeley,[10] Keaveny outlined two different experiments to overcome this problem. In the first type of experiment, his team converts a real cow hipbone into a computerized image (see Beck et al. 1997). This procedure involves cutting very thin slices of the bone sample, preparing them in a way that allows the complicated bone structure to stand out clearly from the nonbone spaces, and making a photographic digital image of the slice. These digital-image slices are reassembled in the computer to provide a high-quality 3-D image of that particular real hipbone. In the second type of experiment, the team creates a computerized 3-D image of a stylized bone, giving it a structure that begins as a simple 3-D grid of internal squares. The individual side elements within the grid are given assorted widths based on averages of measurements of internal strut widths (taken from a number of real cow bones) and are gently angled in relation to each other by use of a random-assignment process.

In both cases, the sliced-bone case and the grid image case, the experiment consists of the "application" of a conventionally accepted (tried and tested in the applied domain) mathematical version of the laws of mechanics in which the strength of the bone material is assumed to be that already gathered from other tests. The computer experiment calculates the effect of the "force" on individual elements in the grid and assembles the individual effects into an overall measure of the strength due to structure. The process also allows a visual display of how the individual parts of the internal bone structure, as represented in the two different images, bend and fracture as they are "compressed." Both experiments are exploratory rather than theory testing in aim: they are designed to investigate how the architecture of bone behaves under stress and thus to foster understanding and learning about how the architecture responds in real accidents and ultimately perhaps how breakages in human bones might best be repaired.[11]

Consider first the manipulation taking place. It is the same in both cases: the application of a mathematical model of mechanical forces to see how the "bone" behaves under such an intervention. And, as with much computer usage of mathematical models, the resources of the computer are used to show the process visually and to calculate out the effects for the specific cases shown, that is to produce results for the particular case used, not to deduce or derive general solutions. So, the focus in both experiments is on a process of demonstration that is more nearly that of a conventional experimental intervention than of a mathematical model manipulation. Nevertheless, the demonstration relies on mathematics for its conduct, for the mathematical resources replace the mechanical resources as the agent of intervention or manipulation. This mixed method of demonstration is symptomatic of computer model and mathematical model simulations and may well be one of the defining methodological features of such simulations.[12] These are experiments *using* a mathematical model as an experimental instrument, not experiments *on* the mathematical model to find out how it behaves. Yet the demonstration in both cases also requires a computer-model representation of the bone, the material to which the mathematical model was applied in conducting the experiment. Thus, models come in twice: there is a model of the intervention and a model of the object, and it is the behavior of the object that is of interest and that forms the subject of experiment.

Despite the role of mathematics in the intervention, and the twofold appearance of models, my instinct remains that I want to describe at least the first of these two interventions as close to a regular experiment. I do so primarily because of the different ways in which material features are maintained in the two model objects, for that difference occurs precisely at the point of interest of the experiment, namely in the structural characteristics of the bones. Let me recap the differences in the way the experimental model of the bone is constructed to make this clear. The process of creating the 3-D computer image in the bone-slicing case retains a high degree of verisimilitude of structure for each particular bone sample. The scientist starts with a particular physical sample of bone and follows traditional laboratory preparation procedures such as cleaning away disturbing elements, making sure only the bone structure is visible, etc. As little as possible is added on, filtered out, or translated in the process of preparing the bone sample and turning it into a computer model. For these reasons, I label these particular bone images as having a semimaterial status.

We could similarly interpret the model of the bone in the second, grid image case in terms of control factors; that is, we could say that it was made by starting with a bone and then omitting many complicating factors and attempting to hold some other factors that are thought to be relevant present and constant at certain average values. But such an interpretation ignores the actual process of constructing the model. This second bone model was constructed anew by the scientist first visualizing the bone structure along radically idealized lines: a simple grid structure is hypothesized. Only then does the scientist add back certain material features based on the average of a number of material bone samples to create an idealized and simplified abstract structure.[13] Here we remain close to the construction techniques of mathematical models.

Thus, the construction of the first bone model involved procedures more like those of an experimental setup and the construction of the second bone model involved something more like a mathematical modeling exercise. By reason of construction, all the input features of the second structure are chosen by, and so "known" (in a sense) to, the scientist, but this is not necessarily so for the first structure. In this respect, the first experiment is more like an experiment on a material object and the second experiment more like an experiment on a mathematical model. The behavior of the second object may surprise the experimenter (as indeed it did), but the first type of object, because it may contain things in the structure the scientist did not "know" about, may yet confound the experimenter. Extending Tony Keaveny's own terminology in his Berkeley seminar, we might say when we look at the quality of the object being experimented upon that the first is "virtually an experiment" and the second more like a "virtual experiment." The terms "virtual" and "virtually" fit well here because the medium of experiment is computer based, but they may have a wider writ to cover different kinds of non-computer-based experiments. The characteristics of these two types of hybrids, virtual and virtually experiments, are shown in the middle two columns of table 11.2 between the two original types of experiments.

Both of Keaveny's bone experiments might be contrasted with several experiments on joints observed at the Nuffield Orthopaedic Lab in Oxford: in one case we saw physical experiments on physical models of joints made from non-bone material (a material experiment on a material object), and in the other we saw computer experiments on highly stylized (stickman) computer models or representations of the bones and joints (nonmaterial experiments on nonmaterial objects).[14] Note that neither of these experiments were concerned with bone structural strength, and they are mentioned here merely to provide further examples in a closely related domain. We might describe the first as a material experiment and the second as a virtual experiment, but to do so reflects again how difficult it is to make sensible judgments without having much more detailed knowledge of how far the material qualities of joints are embedded in the "joints" involved in either the real experiment or the computer experiment in that Oxford laboratory *and* knowledge of what features of the joints those experiments were designed to explore.

2.3. Experiments to Create Pseudo-Material Outputs: Stock Market Prices

An alternative hybrid tradition, which mixes material and nonmaterial in a different way, are experiments with mathematical models with the explicit intention to produce outputs that mimic, if not the world directly, certainly the relatively unfiltered observations generated by the world. These are the outputs I label "pseudo-material."

This kind of experimental activity has a comparatively long tradition in economics, predating the computer simulations of the type so familiar nowadays.

TABLE 11.2. Types of Experiment: Ideal Laboratory, Hybrids, and Mathematical Model

	Ideal lab experiment	Hybrid experiments		Mathematical model experiment
		Virtually	*Virtual*	
Controls on:				
Inputs	experimental	experimental on	assumed	assumed
Intervention	experimental	inputs; assumed	assumed	assumed
Environment	experimental	on intervention and environment	assumed	assumed
Demonstration method	experimental in laboratory	simulation: experimental/ mathematical using model object		deductive in model
Degree of materiality of:				
Inputs	material	semimaterial	nonmaterial	mathematical
Intervention	material	nonmaterial	nonmaterial	mathematical
Outputs	material	nonmaterial	non- or pseudo-material	mathematical

It consists of statistical or mathematical models that are simulated, or "run," to generate output series with the aim of mimicking observed economic time-series data. For example, one of the most commonly available, but least understood, sets of economic data is that of stock market prices. Mathematical models built to understand their properties date back at least to a Brownian motion model of the stock market in the early twentieth century. In the 1920s and 1930s, economists used stochastic process models in searching for ways to describe economic data and tested their descriptions by the extent to which they could mimic such data. Nowadays, mathematical models of the stock market are also explored numerically in ways that have become common to many fields where analytical solutions for nonlinear and dynamic equations remain infeasible.

Since such activities involve experiments on mathematical or statistical models, all the assumed controls that go along with mathematical model usage can be adduced at work here. But I characterize such experiments as virtual experiments because the method of demonstration involves the hybrid "simulation" form, which mixes mathematical and experimental modes. Yet, from out of these experiments, we may generate something very similar to the material we gain from the natural system; so, when economists use such simulations of mathematical models to produce "observations" of the same form and same type (that is, they share the same statistical characteristics) as the observations produced by real markets, I label them as producing "pseudo-material outputs" (see table 11.2 again). When this happens, such experiments are said to provide "evidence" on the possible economic structures and mechanisms that may inhabit the world.

The experiments do this because they may rule out certain structures as impossible and may rule in others as possible. And they do this in spite of the fact that they make no recourse to material inputs, nor are the models taken to represent the material inputs with any accuracy, in the experiment itself.

Although such virtual experiments aim to mimic the observations produced by real market processes—that is, to produce pseudo-material outputs—some of these stock market experiments are virtually experiments because they use input objects that have the same kind of semimaterial status as in Keaveny's sliced-bone case (for a further example of such hybrids in economics, see Morgan 2002). Experiments conducted by my colleague at the University of Amsterdam, Cars Hommes, in conjunction with William Brock, epitomize the range of elements involved (see, for example, Brock and Hommes 1997). Such studies use, as inputs, mathematical decision rules appropriate for different kinds of behavior, labeled with classifications that separate "fundamentalists" (those who believe that stock prices reflect fundamental values of the companies concerned) from "technical traders" or "chartists" (those who trade on observed patterns of price changes) and "trend followers" (those who follow trends and may overreact in doing so). Hommes and Brock (among others) use model-experiments to explore what happens when various kinds of mathematically described "traders" are put together in mathematically characterized "markets."

In attempting to be descriptively accurate about stock market buyers and sellers at the individual behavior level, these mathematical decision rules may succeed in capturing the well-observed or long-reported trading strategies used in the market. To the extent that traders in the stock market trade on the decision rules as proposed in the model, or that the mathematical rules traders use are those used in the model experiments, we might accord these model experiments the status of experimenting on semimaterial objects. There is some evidence that this claim might be reasonable, for at least with the "chartist" or "technical" traders we do have evidence of the analysis that underpins these trading actions, while we also know that institutional traders rely on mathematical whiz kids to design systems on which trading is carried out. It appears, therefore, that rather than the mathematics offering a nonmaterial description, the mathematics provides, directly, a semimaterial input.

Both these types of simulation experiment in economics present a strong contrast to the exploratory, analytical tradition in experimenting with mathematical models of the economy already mentioned at the beginning of the chapter (see section 2.1). What we discussed there were economic systems and situations represented in abstract models of the economy. Experiments consist of manipulations of the model enabling the economist to explore deductively what happens in the model when specified events, policy interventions, or structural changes occur affecting certain variables. That branch of model-experiment activity relies on highly unrealistic or general assumptions and has very limited intentions to connect with real-world data or to produce anything equivalent to pseudo-economic observations. Such model-based experiments, as I already noted, satisfy other aims than those of mimicry. Rather, they are designed to explore the solutions for rather general economic hypotheses or deduce how complicated elements of the model fit together.

We have now filled in the cells in the four-way contrast shown in table 11.2. At one extreme, we have laboratory experiments in the material world. At the other, we have experiments on mathematical models based on idealized or simplified assumptions and so lacking strong material connections. In between we have virtually experiments on models using semimaterial inputs, and virtual experiments using models on nonmaterial inputs that create nonmaterial outputs (as in the case of the second kind of bone experiments) or pseudo-material (mimicking) outputs (as in the economic cases). It is these two mixed cases, hybrid in the way that models and experiments come together and hybrid in the way that material and nonmaterial elements are involved, that are of particular interest, though I shall continue to use the comparison to the ideal type cases in the next part of the chapter.

3. How Do Vicarious Experimental Results Relate to the World?

The question of whether an experimental situation can be taken to be similar enough to the nonexperimental world that the results from the experiment can be taken as valid for the world is an important and complicated problem.[15] Rather than attempting a general analysis of this question, I will concentrate on how the hybrid model/experimental work I have discussed above might be compared with other kinds of experimental work in this respect. The dimension of interest remains materiality and, in particular, how materiality intersects with the way different kinds of models are taken to represent things in the world. This might aptly be termed the nature of the representing relation.

3.1. Representing Relations and Experimental Validity: Model Organisms

Model organisms used in biology, such as the mouse or the fruit fly, raise prospects about how experiments with models support inferences both within and beyond a particular range that are directly relevant for our hybrid cases. If we can establish experimental results for our lab mice, we may be able to use the fact that they are mice to make valid inferences to the population of mice. But the extent to which we can infer from the experimental organism to other possible organisms depends not only on material qualities but also on the extent to which we can label our experimental subject, situation, and manipulation somehow typical of those others in the world. Model organisms carry out their representing roles in two different ways that provide an interesting benchmark for our hybrid cases.

The mice, fruit flies, and other model organisms that are used for experimental manipulation are bred to a standardized strain so that experiments on them can rely on certain aspects of the organism in which variation has been controlled. The experimental results on a batch or sample of flies or mice may be inferred to the population of flies or mice of that same particular type. At the same time, model organisms remain material systems: however standardized and unnatural lab animals are, they remain, materially, fruit flies and mice. Lab flies remain *representatives of* fruit flies, and lab mice remain *representatives of* mice. This may be used to validate a range of relatively unproblematic inferences beyond the particular strain but within the species on the grounds of strong similarity.[16] However, much of the interest in model organism research lies in how far experimental results established on them can be taken as *representative for* other organisms. Can results about alcoholism or cancer in mice be taken as typical for those conditions in humans?

Both these notions, *representative of* and *representative for,* are appropriate for material models such as model organisms and their representing relation to other objects to which they might refer and therefore to which inferences might be made. When we ask if the lab mouse acts as a representative *of* all mice, and if the mouse acts as a representative *for* other classes of mammals, we are implying two different relations.[17] While inferring to other objects in the "representative of " case relies on establishing representativeness in the same sense as the sample/population relation (that is, they are of the same case), inferring in the "representative for" case depends on establishing similarity relations between two different organisms: a similarity that something that is generic to mice is also generic to humans. In the first case, we have some ways of checking that the laboratory mouse is representative of other mice (for example, the use of arguments based on statistical theory and evidence can form the basis for such inferences). Here, both the particularity and the materiality of the case help underwrite possible inferences from the lab mouse to other mice—they are the same type and the same stuff. But that same particularity and materiality makes inference to other organisms more problematic and so inference has to be separately justified by other evidence using case-based analogical or similarity reasoning (see Ankeny 2001).

This appears rather different from the situation that Nancy Cartwright (2000) characterizes from experiments in physics, namely that once we have established the validity of results from experiments on isolated balls on planes, we can take that process to teach us something about canon balls, planets, and telecommunication satellites (her example). In other words, the behavior of a particular such experimental object in physics is taken as typical of a process that "occurs across a variety of circumstances and in a variety of different kind of individuals" (Cartwright 2000, 7). Establishing such typical behavior in the experimental world may enable us to move from the simple controlled case to the complicated uncontrolled case and even to interventions in the world using that knowledge. Here we are looking at a process taken to be the same throughout many domains, for, as Cartwright suggests, in these physics experiments we have "an underlying ontology of specific processes that can operate in much the same way across a (possibly big) difference in the kind of systems in which they occur" (Cartwright 2000, 8).[18]

Of course, there are no doubt things that are typical, in the sense of being the same, across biological organisms, as in Cartwright's physics example, so that results from one field carry validity to another. A model organism example that appeared in the science column of my newspaper while I was writing this chapter concerned the worm *c. elegans* and its reaction to continuous microwave emissions, the kind of emissions from mobile phones (paper by David de Pomerai, reported in the *Financial Times*, May 18, 2000). Two possible reasons for valid inference of the results from worms to humans were mentioned: one was a generally shared process of the type Cartwright discusses, namely, that the worm released a particular kind of protein released by *all* organisms when threatened. The other was a case-based similarity argument resting on a particularity shared between the two organisms, namely, that humans and worms share many common genes.

In contrast to these model organism examples, I find both terms, *representative of* and *representative for,* uncongenial as ways to describe the way mathematical models represent the things they relate to in the world. The problem may be clarified if we make use of R. I. G. Hughes's (1997, 1999) terminology, which takes the representing idea for mathematical models to be one of denoting: a mathematical model denotes some phenomenon in the world. We can usefully say that the mathematical model denotes a material economic system, but we cannot sensibly say that the mouse denotes the human. Although the model organism (mouse) may stand in for the organism of interest (human), as a mathematical model does for the physical or economic system, it does not represent it in the same way. Mathematical models in economics, such as those I discussed in section 2.1 of this chapter, are best understood as providing *representations of* processes or objects or relations in the world.[19]

3.2. Inference in Hybrid Cases

Tony Keaveny's bone experiments once again provide the example that allows us to distinguish what is at issue here for experimental inference in our hybrid cases. As shown in table 11.3, the two hybrid forms of experiment become separated and align with the two end cases. Recall that the second bone experiment, the one with the grid image, relied, from the start, on a specially constructed *representation of* bone structure. This *representation of* relation creates an immediate inferential gap between the object of experiment and the object of reference, a gap that is independent of other qualities of the model, for the experimental and reference object are no longer made of the same stuff. As is often the case with an abstract model, not only was the medium of expression different, but the object had also been simplified and idealized in certain respects in the model, and its empirical (material) features were no longer dependent on any particular case, rather it incorporated certain average or typical features of bones. Experiments with this model representation of bones might suggest interesting possibilities about the behavior of bones under stress. Indeed, in this grid image case, the investigators were surprised that only a few of the model bone struts bent during the experiment. The result was suggestive but of limited validity for understanding the detailed architecture of real bones. The nature of the model construction and the representing relation—the qualities of the original object having become abstract, simplified, and idealized—creates huge barriers for valid inference.

In comparison, Keaveny's first experiment, with the sliced-bone model, shares features with model organism inference. It relied on a particular hipbone to produce a model object that maintained (I suggest) enough material qualities necessary to establish valid experimental results about that one real bone. Its validity as a *representative* hipbone had to be established by other methods[20]—that is, for example, by testing whether certain characteristics for the particular bone matched those gained from the average of such samples used in destructive testing. Once the particular bone was established as a *representative of* such bones, it became possible to use the experiment to make inferences about the behavior of hipbones in general. This example fits into the model organism genre quite closely for another reason: recall that the experimentalists used individual cow hipbones, and these functioned as *representatives of* cow hipbones in general but were also intended to be taken as *representative for* human bones, for the ultimate aim was to learn something about human bones.

All inference from experiments to the world is subject to many conditions. My discussion here suggests that the kind of representing relation in the experiment is another of the relevant considerations. Real (material) experiments

TABLE 11.3. Types of Experiment: All Three with Representing Relations

	Ideal lab experiment	Hybrid experiments		Mathematical model experiment
		Virtually	*Virtual*	
Controls on:				
Inputs	experimental	experimental on	assumed	assumed
Intervention	experimental	inputs; assumed	assumed	assumed
Environment	experimental	on intervention and environment	assumed	assumed
Demonstration method	experimental in laboratory	simulation: experimental/ mathematical using model object		deductive in model
Degree of materiality of:				
Inputs	material	semimaterial	nonmaterial	mathematical
Intervention	material	nonmaterial	nonmaterial	mathematical
Outputs	material	nonmaterial	non- or pseudo-material	mathematical
Representing and Inference Relations	representa*tive of* . . . . . . to *same* in world representa*tive for* . . . . . . to *similar* in world		representa*tion of* . . . . . . back to *other* kinds of things in world	

may allow inference to the same kinds of things in the world if they can be considered representative of them and to similar things if they are representative for them. We have, in statistical reasoning and in analogical case-based reasoning, arguments and procedures for helping to judge such representative qualities. Mathematical model experimental results, such as those in economics, present much greater difficulties: they may only be inferred to other kinds of things—the things they are representations of—in the world, but we have less agreement on procedures for deciding when a representation is a good one.[21] In the case of the different hybrids of experiment and model treated in this chapter, we need to pay attention to the details in each case. To the extent that the hybrid experiments are on well-judged *representative* objects, inference may be limited in range but may really have the possibility to tell one something about the specific objects represented; to the extent that inference relies on *representations of* objects, experimental results may be suggestive but less able to teach one about the specifics of the object.

As we get further into the details of how modern science uses models and experiments together, we may find that, in many cases, the kind of representing relation is mixed—namely, that the objects of experiment embody qualities of being both representatives and representations of things in the world. We can see this in another model organism example that comes from Rachel Ankeny's analysis of the worm *c. elegans*. She recounts how a small sample of worms was used to produce a "wiring diagram" of that species of worm, a *representation of* the worm that she labels a "descriptive model" (see Ankeny 1999, 2000). The description is an average and slightly simplified diagram of the worm's nervous system that is used in the community of worm people to clarify the nature of a "normal worm," that is, what is typical, average, or generic to this worm strain. It is used as a *representative model of* the "normal" worm to assess particular "abnormal" worms by a process of comparison and as *representative for* other worms or model organisms to explore similarity relations in terms of case-based comparisons and inference. In its construction, it embodies qualities that suggest it lies between Tony Keaveny's two bone examples; and while it does not, as yet, seem to have become an experimental object, it is an object on which inferences are based and on which causal explanations may be constructed. Hughes (1999) discusses another case, the Ising model in physics (a regular array of points in geometrical space) and suggests that this model, too, can be understood both as a representative and a representation. His case provides an example of what has become a wide class of computer experiments upon particular mathematical objects. Each mathematical model used in the experiment can be understood as a *representation of* something in the natural or social world. At the same time, by virtue of the fact that each model is a member of a class of mathematical objects, it also acts as *representative of* that class of model. In addition, it may act as *representative for* some class of other objects or processes in the world that share similar features.

4. Conclusion: Virtual Experiments and Virtually Experiments

The impressive science studies literature of the past twenty years has taught us how difficult it is to cut cleanly, in any practical way, between the philosopher's categories of theory, experiment, and evidence. Modern scientists, in a different way, also refuse these cuts: in experimenting on and with models, running simulations and the like, they talk of using such wholly nonmaterial methods to produce observations or to effect experiments. Even if we were to ignore the former literature and dismiss the latter's terminology, we would still be left with the presence of these vicarious experiments and a need to understand them.

In this account, I have suggested that some of these vicarious experiments should be treated as hybrids, and I have pictured them lying between the limiting cases of mathematical model-experiments, at one end, and material laboratory experiments, at the other (see tables). By analyzing how these different kinds of hybrid experiments work, we can suggest a taxonomy of hybrid things "in between" that include virtual experiments (entirely nonmaterial in object of study and in intervention but which may involve the mimicking of observations) and virtually experiments (almost a material experiment by virtue of the virtually material object of input).

The qualities of these in-between hybrids turn out to run along several dimensions. They involve a mix of models and experiments. They mix mathematical and experimental modes of demonstration. They mix types of control by assumption with experimental controls. They mix nonmaterial and material elements. They represent the world via mixed modes of representation and representativeness. Exploring the case materials on hybrids and comparing them to the limiting cases shows that these mixtures are not independent of each other. And, since modern science is busy multiplying the number of hybrids on our epistemological map, the situation may be growing daily more difficult. I have taken materiality to be one of the prime characteristics of interest in these hybrids and suggest that when faced with such a hybrid experiment, we need to look carefully at where and how much materiality is involved, and where it is located, before we can say much more about its validity.

NOTES

This chapter was prepared for the workshop "Towards a More Developed Philosophy of Scientific Experimentation" (Amsterdam, June 2000) at the invitation of Hans Radder. The content was prompted by two events: Tony Keaveny's seminar on bone experiments at the University of California, Berkeley, in the fall of 1999 and the Princeton Workshops on "Model Systems, Cases and Exemplary Narratives" during 1999–2000. My thanks go to Tony Keaveny and his colleague, Michael Liebschner, for spending time discussing their work with me and the Princeton workshop organizers and members for a stimulating series of workshops that gave me the opportunity to learn about model organisms. I thank Marcel Boumans, Francesco Guala, Arthur Petersen, Hans Radder, Norton Wise, Angela Creager, Rachel Ankeny, Margaret Morrison, Dan Hausman, Rom Harré, and the participants at the Amsterdam workshop for many helpful comments. I thank the British Academy for supporting my research during this period.

1. I take Franklin (1986) and Gooding, Pinch, and Schaffer (1989) as paradigmatic.

2. The "new" view is taken to date from Hacking (1983a). Heidelberger and Steinle (1998) provide an excellent selection of papers on current views about experiments.

3. For example, Rohrlich (1991) suggests that simulations are a new methodology and prompt us to provide a new syntax for theory. Galison (1997), particularly in his chapter 8, writes about computer simulations in physics as a new method, one which is not quite theory and not quite experiment. This line is taken up by Dowling (1999), one of several papers relevant for the relations between simulations and models in the special issue of *Science in Context* on "Modelling and Simulation" (referenced under Sismondo [1999]; see also his introduction and papers by Winsberg and Merz). Dahan Dalmedico (2000) explores the issue

of a new epistemology for simulation. Hughes (1999) provides the most thoughtful philosophical analysis of simulations, models, and experiments relevant for the mixed cases raised in this chapter.

Two important and useful collections of papers on simulations in social science, providing both commentary and cases for exploration, are Guetzkow, Kotler, and Schultz (1972) (which, because of its date, discusses simulations of many sorts, not only computer-based ones) and Hegselmann, Mueller, and Troitzsch (1996). Particularly relevant for the combination of models and experiments are chapters in the former collection by Robert Schultz and Edward Sullivan and by Thomas Naylor, and in the latter collection by Hartmut Kliemt, Karl Muller, and Stephan Hartmann.

4. The label comes from Schultz and Sullivan (1972, 9) who, in writing about simulations, say that "[e]xperiments can take place in a vicarious way."

5. For a more complete account and comparison of the use of models, laboratory experiments, and hybrids in economics, see Morgan (2002). This latter paper contains some additional material on economics, found in the original Amsterdam workshop paper, but omitted here, that was subsequently extended for the Model-Based Reasoning Conference held at Pavia in 2001.

6. "Models," more recently than "experiments," have become a category of interest for science studies. In Morrison and Morgan (1999), among other characteristics, we treated them as embodying representing properties. This quality is explored further in section 3 of this chapter.

7. Boumans's work on this problem connects to Hasok Chang's (2001) discussion of the development of thermometers. Both works are outcomes of the joint "Measurement in Physics and Economics" research project at the London School of Economics (Centre for Philosophy of Natural and Social Science) and University of Amsterdam's History and Philosophy of Economics group.

8. Further discussion of the control issues for material and mathematical model experiments can be found in Boumans and Morgan (2001).

9. Of course, if the material is mathematical, then a material experiment is a mathematical one. I seek to distinguish in this chapter between experiments on mathematical models (and, later on, computer models) of natural or social systems and experiments directly in the material of the natural or social worlds.

10. The seminar, entitled "Micromechanics of Trabecular Bone—A Look at the Future of Virtual Experiments," was given in the Neyman Seminar Series, October 20, 1999.

11. Note that the experimenters use cow bones because of the difficulty of getting human bones for experiment, though the context of the seminar made clear that the ultimate aim of the lab's research is to understand the strength factors in human bones.

12. The translation from bones to computer images makes it difficult to penetrate further into the exact means of demonstration involved and how far we should regard the demonstration method as involving experimental or mathematical or some other kinds of resources. Not all simulations involve computer models or mathematical models. Oreskes (2000) discusses how geologists used to simulate geological process at a different scale and in different media but using material resources. Mimetic experiments, as in Jevons's or Wilson's original cloud chamber experiments, both designed to mimic cloud creation (see Maas 1999; Galison and Assmus 1989, respectively), used material resources to produce effects that mimic natural effects but in an analogical media. Hackmann (1989) calls these "model experiments," though it is not clear that he separates these from material analogical model experiments, such as those conducted on the Phillip's Machine (see Morgan and Boumans 2002). Mimetic experiments, analogical experiments, and material simulations provide fertile ground for

philosophical analysis of experiments but are not examined here, as they do not share the hybrid material/nonmaterial qualities of interest.

13. Note that the issue is not one of complexity here: either of the two kinds of bone model may be more or less simple or complex than the other (though exactly what that means is ambiguous). The issue is how the object is constructed and whether or not, at the point of interest, there are elements that accurately reflect material qualities. I thank Arthur Petersen for questions that helped me clarify this and several other points.

14. These experiments were observed during a visit to the lab in October 1996 as part of the Models in Physics and Economics Research Project at the London School of Economics.

15. Francesco Guala, another member of the "Modelling" and "Measurement in Physics and Economics" projects at the London School of Economics has conducted a serious analysis of the "parallelism" problem and its literature in experimental economics (see Guala 1999b). Although not primarily concerned with material/nonmaterial hybrids (he reserves the term "hybrids" for experiments-cum-simulations), he treats experiments as playing the same kind of mediating function as models (see Guala 1999a, 2002). I thank Francesco Guala for help in clarifying my thinking about this and other aspects of experiments.

16. I stress the term *relatively* because of course it is never easy to establish laboratory results into the field, as the science studies literature makes abundantly clear. I assume all such difficulties to exist; I point here to additional ones.

17. The terminology is specific but easily confused with other usages. My use of the terminology "representative of" and "representative for" is designed to suggest different kinds of relations between the model object and the object of inference comparison. It is not concerned with indicating the purpose of modeling as in the positive/normative distinction and terminology: a model of the economy versus a model for policy purposes. Nor is it to be confused with the ways in which different models represent different objects within simulations in the same domain, as in Evelyn Fox Keller's chapter in this volume.

18. I summarize the argument in Cartwright (2000) and omit her discussion of external validity but note that what is a typical shared process across a wide range of circumstances may not count for enough if the level of variation between materials that share the same process remains high. For example, the general process of change from solid to liquid to gas in chemical substances can be easily isolated. But knowing this process tells the chemist very little, for variation in that process means that working with these substances requires knowing the details of the individual cases.

19. Although Morrison and Morgan (1999) did not distinguish between representations and representatives, we argued that models can be intended to represent both theory and world at the same time; the term "representation" as used here has similar possibilities.

20. The fact that each bone was translated into a computer representation of that hipbone before it could be used in a computer experiment complicates the case but does not invalidate the label of a representative hipbone, as we shall see.

21. Two points: first, of course the fact that we have some procedures for some kinds of inference arguments in the "representative of and for" cases does not make it at all easy to make valid inferences in practice. Second, in the "representation of" case it is generally agreed that some kind of outside evidence or additional information is required to establish external validity in the case of mathematical model results as well as for these hybrid experimental results and simulation procedures (see Oreskes, Shrader-Frechette, and Belitz 1994; Oreskes 2000). What kind of evidence is less agreed: for example, economists take the accuracy of the representation to depend on the realism of the model assumptions; Cartwright (2000) suggests we need outside knowledge of causes; and Hartmann (1996) argues that we need independent reasons for believing the model used.

Daniel Rothbart

12 / Designing Instruments and the Design of Nature

Until recently, scant attention was paid by twentieth-century philosophers of science to laboratory instruments, even to devices that produced stunning results. According to empiricist-oriented philosophers of the modern era, the validity of findings rests on the same kind of quasi-transparent standard that determined a genuine experience with the naked senses. We now know better from our understanding of current research practices. The old aspirations for transparent methods have been discredited by philosophers on both sides of the Atlantic Ocean. In a resurgence of interest in experimental technologies, we read how instrumental techniques provide a source of philosophical insight into experimentation. Recent work includes writings by R. Ackermann (1985), D. Baird and T. Faust (1990), J. Bogen and J. Woodward (1988), A. Franklin (1986), P. Galison (1997), D. Gooding (1990), R. Harré (1998), B. Latour (1987), A. Pickering (1989), H. Radder (1996), J. Ramsey (1992), and S. Shapin and S. Schaffer (1985). Ian Hacking's work is particularly influential. Resurrecting the pragmatists' theme that knowing is achieved by doing, Hacking gives priority to the engineering of instrumentation over validation of theories, proclaiming that the use of a laboratory instrument is dense with philosophical meaning (Hacking 1983a).

But none of these authors examines the philosophical commitments associated with a central mission of engineers, which is design. Even in Hacking's writings, design is clearly subordinated to the technician's skilled manipulation of instruments as a basis for philosophy. Design plans for an instrument provide experimenters with a prescription for engaging a segment of the environment, prescriptions that are driven by various philosophical commitments. In this chapter, I argue that instrumental design provides a robust expression of an experimenter's philosophical commitments about inquiry. The visual models of a designer's plans for instruments replicate a specimen's capacities, an instrument's powers, and an experimenter's abilities. My entry into experimentation is a study of visual modeling in design.

1. Thought and Vision

The history of philosophy includes frequent references to cognition as a form of vision. For some of the ancients, syllogistic reasoning was analogized to a cognitive vision. Aristotle identifies intuition as a mental observation of thought, and error as visual illusion. The intellect is said to perceive an object through a metaphysical light. In this way, to know is to see. Vision and intellect establish a relation between the soul and the object looked upon (Pomian 1998, 211). Descartes' insistence on clear and distinct ideas rests on his conception of intuition as "the natural light of reason." Intellectual vision is an entry to metaphysics, providing knowledge of a body's primary qualities of extension and motion. For Locke, intuition is a power to see relationships between ideas without need for justification.

Charles Peirce extends this tradition with the bold thesis that cognition is a form of seeing. All rational thought is thoroughly blended with a kind of mental perception; all reasoning is diagrammatic. The mind forms visual diagrams in the imagination, producing skeletal patterns of segments of an environment. He writes: "By diagrammatic reasoning, I mean reasoning which constructs a diagram according to a precept expressed in general terms, performs experiments upon this diagram, notes their results, assures itself that similar experiments performed upon any diagram constructed according to the same precept would have the same results, and expresses this in general terms" (Peirce 1976, 4:47–48). The mind forms diagrams in the imagination for the purpose of representing a solution to a problem. Such mental diagrams are then subject to a kind of mental "experiment." Such experiments are performed upon a diagrammatic representation, not through the use of laboratory apparatus. In this way, the proposed solution can be tested vicariously through active experimentation in the imagination.

For Peirce, diagrammatic reasoning requires (1) a hypothetical introduction of new elements not previously given in the definition of the problem, (2) the use of such elements as general concepts, and (3) the creation of new hypotheses for future tests (Fernández 1993, 236). Such an experiment requires an active and imaginative search for physical circumstances that test the proposed solutions (Fernández 1993, 237).

Of course, visualizations are ubiquitous in the theoretical sciences. For centuries investigators have relied on conceptual models for visualizing real-world processes. Through modeling, scientists can simulate processes of the world in an abstract realm of ideas.[1] The categories of truth and falsity cannot determine the success or failure of a theoretical model. At best, a model offers simplification, approximation, and abstraction. A theoretical model functions as a conceptual replication of certain aspects of worldly events. Ludwig Boltzmann argued that Maxwell's theory requires more than a system of equations. The theory is defined by its visualizable models, resulting in the formation of mental images, or "thought-pictures," which presumably represent phenomena (Nyhof 1988). The point is not that every scientific theory must be defined by its visualizable models, but that visualizations are commonplace in the theoretical sciences, as I argue elsewhere (Rothbart 1997).

One major goal of engineering is the design of artifacts (Layton 1991, 66). The result is an idealized vision, a quasi-pictorial model of function that is conveyed through spatial media of drawings, illustrations, and schemata. Such media are well suited to convert a mental vision into information. The mind's eye requires a synthesis of past experiences and a projection of future contrivances (Ferguson 1992, 42).

Designers of instruments provide a graphic simulation of experimental technique. Designers often construct a nested series of drawings of increasing depth. As a reader unfolds layers of drawings, information becomes progressively more detailed. Certain pictorial symbols serve as interrogatives about one process and invite further inspection at a deeper level. In a design plan, lines of one schematic illustration are given specificity by appeal to other drawings representing deeper processes. The visual model of an instrument entices vicarious participation in detecting a transition from microscopic to macroscopic events. Such a sequence mimics the lawful character of embeddedness.

Visualizations associated with instrumental design are diagrammatic in a Peircian sense. Through diagrammatic reasoning, engineers establish a design space for simulating the function of a machine. A design space is an abstract representational space used to replicate movement, where certain conceptual elements are imagined and the effects are anticipated. Rather than providing a media for visual copies of actual machines, a design space defines a range of possible movements of "objects," based on principles of engineering and the physical sciences. Architects, artists, and mapmakers work in such spaces.

Peirce's requirements for diagrammatic reasoning can be extended to the construction of design plans for instruments as follows: (1) new elements must be introduced hypothetically into a design space; (2) such elements must be used as general concepts for the detection of possible structures; and (3) new hypotheses are created for testing. A designer imagines how an instrument should be used, what changes will occur to the specimen, and whether desired results can be achieved. I explore these three requirements in the context of engineering design.

2. New Elements of a Design Space

The alleged supremacy attributed to discursive language over visual media has held a monolithic grip over most contemporary philosophical discussions about knowledge. For logical empiricists, pictures, illustrations, schematic drawings, and diagrams serve, at best, as heuristic aids and, at worst, as sources of illusion for purposes of conveying information. Pictures are often denigrated, in contrast to the superiority of mathematical equations, for example. To be sure, some scholars recently challenged this conviction through rich historical and philosophical studies (Baigrie 1996). But the familiar dismissal of schematic illustrations in science remains a dominant position and is tied to the conviction that scientific reason is only expressible in discursive form.

But this conviction cannot account for the work of instrumental designers, who marshal an array of pictographic inscriptions for the purpose of conveying relationships between components. For centuries, engineers have supplemented discursive descriptions with line drawings to show how an instrument will operate under various conditions. Illustrations rarely offer a camera-ready mirror of the actual performance of a physically manifested machine and do not refer directly to this or that feature of a particular device. Experiments in contemporary research are often depicted through visual language, based on schematic drawings associated with an instrument's design. Visual media are often employed to distill practice and prescribe a course of action. Chemical engineers use flow sheets, electrical engineers use circuit diagrams, and many engineers use block diagrams as tools in design plans (Mitcham 1994, chap. 8).

Visual literacy requires deciphering a code, as if translating a language. This is well known to engineers: "The knack of reading a schematic [illustration] is somewhat similar to, although very much simpler than, translating from a foreign language" (Mann, Vickers, and Gulick 1974, 45). Engineers are expected to comply with rules of meaning and are "reprimanded" if they do not. Pictorial images of unique shapes and lines are often included in a designer's vocabulary. A designer's vocabulary is a set of unique shapes, and a shape is a finite collection of lines. Design plans are often expressed through the arrangement of such shapes. To read visual information about instruments from such drawings, a

FIG. 12.1. Alphabet of lines and line weights

THE ALPHABET OF LINES

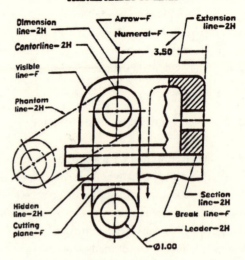

LINE WEIGHTS

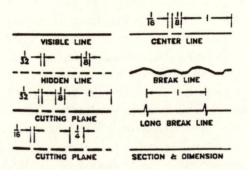

literate observer must have some understanding of the grammar of pictorial symbols. *Rules for a designer's language include a grammar of points, lines, and shapes.* In this respect, the language of design is defined by a shape-grammar familiar to the vocabulary of geometry.

A designer's vocabulary includes lines of various thicknesses, with intermediate cuts at different angles and different colors. Computer graphics plotters use pens with points of varying widths, ranging from 0.3 mm to 0.7 mm for the purpose of drawing lines of different thicknesses. Consider the example of "an alphabet of lines," according to one analysis of engineering graphics in figure 12.1 (Earle 1994, 187). Information is conveyed in the foreground by the dashed edges and pencil thickness. Notice in this figure that the lengths of dashes in the hidden lines and center lines are drawn longer as a drawing's size increases. In general, schematic illustrations are not drawn to scale, and the geometrical relationships between components are grossly distorted.

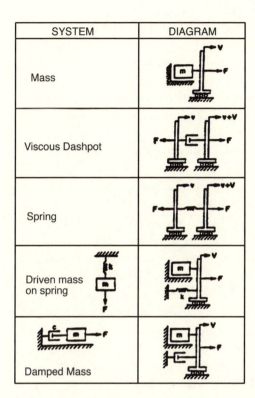

SYSTEM	DIAGRAM
Mass	
Viscous Dashpot	
Spring	
Driven mass on spring	
Damped Mass	

FIG. 12.2. Pictorial symbols for mechanical properties

The symbols are used in a conceptual model of a machine's anticipated operation. Consider, for example, a table from a technical handbook on mechanical engineering showing some of the basic elements of a mechanical system, characterized by masses, springs, and forces. Figure 12.2 introduces concepts associated with these elements and their impedances (Ungar 1996, 5.35). For example, mass is depicted graphically through its relationship to velocity and force. Rather than corresponding directly to a particular machine, such pictorial symbols convey information about abstract properties that define a mechanical system. The figure functions as a kind of translation manual for the proper use of an engineer's pictorial symbols. Of course, such information must conform to theoretical principles of the physical sciences as well as operational principles of engineering.

3. Conception from Visual Images

In a design plan, new elements must be used as general concepts. But how can concepts arise through visual imagery of design without resorting to abstract theories? The answer can be found in a mode of experience, familiar in everyday contact with material bodies, which unites sensation with conception. This is revealed in our experience of occlusion.

From the earliest sketches on cave walls to the current design of technological apparatus, line drawings of corners, edges, cracks, and contours have captured our imagination. These drawings exploit certain perceptual skills that one expects others to have. Based on such skills, an observer associates lines with surfaces. The visualizations associated with a design plan show striking similarities to certain sensory perceptions associated with occlusion. The experimental psychologist James J. Gibson studied how infants learn that toys and parents have a continual existence, even though they temporarily go out of sight (1986). From experience one learns that surfaces go out of, and come back into the visual frame. Surfaces disappear when the light is turned off, when another object blocks the line of sight, or when an observer moves his or her head in certain ways. Sometimes, when one surface disappears, other surfaces are exposed.

Gibson discovered an important clue in one's ability to discriminate between different sources of disappearance. He has demonstrated empirically that surfaces are perceived through the experience of occluding edges. An occluding edge has a double life: it hides some surfaces and exposes others. One's perception of an exposed surface is conjoined with awareness of hidden ones, linking actualities with possibilities. In figure 12.3 the line segment is not occluding, because our perception of this line provides no information about surfaces. But each line segment in figure 12.4 depicts an occluding edge. Each edge runs along an exposed surface as well as a hidden one that we imagine to be located behind the page. The lines in this figure are perceived as surrounding a surface that appears in the foreground. An observer tends to "fill in the space" in ways that produce perceptions of a surface, prompting awareness of possible perceptions. Of course, hidden surfaces are revealed as a result of a change in our reference frame. Can we imagine what surfaces would be exposed if we were to move to the left or right? In such cases, some aspects of the scene would change, such as the angles between line segments. Other aspects would remain the same, such as the verticality of two line segments.

Occlusion is not limited to the experience of an exposed surface but invites attention to a realm of possible, but hidden, surfaces. An illustration of an apple's contours reveals a range of possible shapes and outlines, enticing an observer to envision how certain lines can be extended around corners and how surfaces can appear from different perspectives. Our fascination with geographical maps is not explained by the triviality that maps reveal actual regions. This value of maps can be understood through the perception of occluding lines. We are drawn by the lure of the next frontier. A map tempts us to peer over that line dividing the known and the unknown, raising curiosity for vicarious travel.

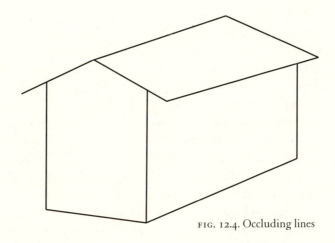

FIG. 12.3. Non-occluding line

FIG. 12.4. Occluding lines

An engineer employs lines to depict features from possible movements of machinery. To an untrained observer, schematic illustrations seem static, providing a still-life vision of a bulky device. But literate viewers can apprehend movement from such diagrams. An attentive observer anticipates how our perceptions would change from the possible movement of objects that clutter an environment. The lines of an engineer's design plan lie between exposed surfaces and hidden ones, in relation to a reader's visual perspective. A history of occlusion for a particular material object reveals how hidden surfaces can be exposed and revealed ones concealed by a change in perspective. Whatever goes out of sight can return by changing the layout of the environment (Gibson 1986, 92). Such awareness is enhanced by perception of depth. To apprehend contours of a house, edges of a fence, and outline of poles from this illustration, we resort to our ability to pick out certain features of our environment that remain invariant from one scene to another.

Occlusion draws upon a history of past perceptions and invites exploration of future possibilities, imagining surfaces from different perspectives. Whether perceiving apples or the metal and wires of a spectrometer, an observer recognizes how attributes would change as a result of movement around corners, under or even through a material body to recover "hidden" surfaces. Again, the lines of figure 12.4 depict edges that were there before I gazed upon it and will remain when I move out of the line of sight. In ordinary experiences of middle-sized objects, movement is reversible. Surfaces that are temporarily blocked in relation to my line of sight can be exposed as a result of my movement to the left or right, up or down. Of course, such movement is reversible in most settings (Gibson 1986, 193).

In addition to the presence of occluding edges, occlusion can also be attributed to surfaces. A surface can be occluding in relation to exposed and hidden volumes. In figure 12.5, we identify one of the foreground surfaces as occluding, because it reveals one region and hides other regions in relation to one's line of sight. An occluding surface is identified as a wall between one region that we perceive immediately and another region that could be perceived as a result of changing our reference frame. The hidden region is not completely unknown to a skilled viewer. In fact, in some pictures, more geometrical information is acquired about the hidden volume than the exposed ones. In figure 12.5, information about the shape of the exposed volume of a building's interior can be gained by an experience of the occluding surface. An occluding surface entices a viewer to peer through a wall, as it were, into a hidden realm. Our experience of an occluding surface requires that we visualize the results of our own movement in relation to a scene. New information is gained and old information is stored from a change in our relative position. So, occlusion underlies our experiences with edges and with surfaces.

Returning to the engineering of instruments, the grammar of pictorial symbols in a design space exploits our ability to perceive occluding edges. In figure 12.6, a simple schematic illustration of an absorption spectrometer draws attention to an instrument's bulky (material) components (Parsons 1997, 262). Although this illustration lacks detail, it provides information about the functional relationships between components. Absorption spectroscopy is commonly used in chemistry for identification, structure elucidation, and quantification of organic compounds. The material realm includes a radiation source, sample, monochromator, detector, and readout. A beam of electromagnetic radiation is emitted from a source and then passes through a monochromator. The monochromator isolates the radiation from a broad band of wavelengths to a continuous selection of narrow-band wavelengths. Radiation then impinges on the sample. Depending on the molecular structure of the sample, various wavelengths of radiation are absorbed, reflected, or transmitted. That part of the radiation that passes through the sample is detected and converted to an electrical signal, comprising an event of the phenomenal realm.

Designers of analytical instruments often use line segments to depict a sequence of experimental phenomena. When an absorption spectrometer is used, photons from an artificial energy source impinge on a sample, producing detectable reactions. Signals are produced, leading to inscriptions at the readout

FIG. 12.5. Occluding surfaces

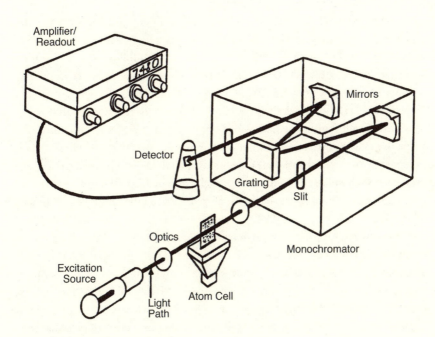

FIG. 12.6. Absorption spectrometer

device. In this context a source of energy that produces a narrow band of radiation is called a "line source" (Parsons 1997, 290). The term "line source" refers to a physical process associated with the progression of energy states. In figure 12.7, lines depict changes in the beam path, from energy source to detector (Coates 1997, 442). Obviously, the surfaces of the material components, such as the detector, are grossly distorted and not true to scale.

In figure 12.8, a radiation beam is conveyed by two shapes, not just a single line (Coates 1997, 445). The pie-shaped segment depicts the progression of radiation from light source to holographic grating. The pie-shape exploits our perceptual skills of occlusion. The direction of the beam's motion is obvious, going from light to grating, and then from grating to array detector. The shape of the radiation beam that is projected from the light source can change with appropriate changes in the energy source and the shape of the grating. The shading from holographic grating to detector maps out another area.

4. Thought Experiments

A third requirement of visualization in design is construction of hypotheses for testing. A designer's plan is often scrutinized by fellow researchers, prospective customers, and potential critics. Designers of instruments are often charged the task of explaining their work to manufacturers, justifying their research to funding agencies, or persuading reluctant experimenters to follow new techniques. Visualization of a proposed technique is a common rhetorical device of persuasion. We often read that thought experiments are confined to the realm of physics (Brown 1991, 31). But the design of an instrument can be tested through thought experiments without immediately deploying metals, wires, and apparatus in a particular laboratory. Such an experiment occurs in a design space where certain processes are conceived and their effects anticipated. Designers invite critical appraisal through the reader's vicarious participation in a hypothetical experiment. A thought experiment can be performed by visualizing the effects of instrumental manipulations.

One criterion for judging a particular design plan centers on the need for reproducible results. Can the design plan offer manufacturers and experimenters assurance that similar (reliable) results can be attained under similar laboratory conditions? Has the designer anticipated most of the interfering factors that might contaminate the results? In this respect, reproducibility of results becomes a norm of experimental research, a hallmark of success. A design plan that cannot warrant trust in the reproducibility of results is a sure sign of failure. Of

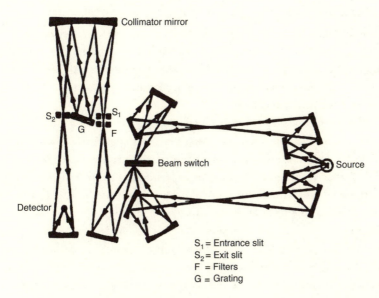

Collimator mirror

S_2 S_1
G   F

Beam switch

Source

Detector

S_1 = Entrance slit
S_2 = Exit slit
F = Filters
G = Grating

FIG. 12.7. An example beam path of a commercial dispersive IR instrument

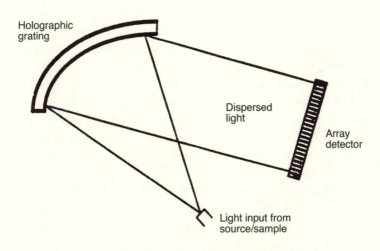

Holographic grating

Dispersed light

Array detector

Light input from source/sample

FIG. 12.8. Idealized layout for a simple diode array-based spectrograph

course, experimenters often discuss whether some empirical findings are reproducible. Talk of measuring dials, sources of radiation, and performance of apparatus is replaced by talk of reproducible results. Researchers assume that certain aspects of an experiment are reproducible in other laboratories. An event can have its status raised, as it were, from something localized in a specific laboratory setting to a potentially recurring state.

Some commentators argue that reproducibility is too demanding a standard because actual reproductions of an experiment rarely occur. Financial, institutional, and technological pressures frequently impose insurmountable obstacles on the ability of scientists to reproduce an experiment. Occasionally, experiments are repeated when the community of experimenters is faced with stunning results or when the need arises to improve apparatus, augment data, or refine instrumental techniques. Typically, however, no one actually repeats an experiment (Hacking 1983a, 231). After the results of a study are published in journals or disseminated through conferences, acceptance is achieved without reproducing the findings. This raises serious questions about research practices: Can the experimental findings of a particular study be validated without actually reproducing an experiment? If so, at what point in the process of validation does the community of scientists participate in the public hearing of these results? Can public participation be removed from the process of validation?

Hans Radder argues convincingly for the necessity of reproducibility, which he defines in terms of three types (1996, chap. 2). First, we can speak about the reproducibility of the material realization of an experiment. A material realization requires that experimenters conform to instructions from another experimenter, retaining the material properties of an "original" experiment. In this case, the same material realization can be reproduced under different interpretations. Second, in another sense, reproducibility applies to an experiment under a fixed theoretical interpretation, when researchers resort to a theoretical description of an original experiment as a guide for an actual performance. A theoretical description is used to identify the significant categories, distinctions, and relations in terms of the repeatable properties. Third, reproducibility may also apply to the results of an experiment. The experimental results, q, remain stable while allowing for alternative procedures, $p, p', p,'' \ldots, p^n$, for realizing q (Radder 1996, 11–18).

Additional insight into the centrality of reproducibility can be found in a process called virtual witnessing that Shapin and Schaffer examine in the context of Robert Boyle's experiments (1985). Boyle needed to secure testimonies about the true results of his experiments and tried to perform experiments in a social space, a common practice in seventeenth- and eighteenth-century England.

Such experiments could not reach a wide audience because of the need for large equipment. He then commissioned engravers to create visual images of the experimental scene, such as the schematized line drawings used to imitate Boyle's experiments with the air pump. The viewer of such observable impressions was encouraged to generate mental images, as a kind of conceptual simulation of the experiment, in order to critically evaluate the experiment. Through virtual witnessing, readers could endorse the methodology and accept the findings without actually reproducing the experiment (Shapin and Schaffer 1985, 60–62).

Returning to the contemporary scene, readers of design reports typically become virtual witnesses through their critical assessment of design plans. The reader rehearses, at least privately, the kind of malfunctions of apparatus, mistakes of implementation, and interfering effects that plagued past experiments. Are such dangers relevant to the present experiment? If so, are they avoidable? The answers to these questions require a command of the subject matter that is usually limited to expert witnesses. A reader vicariously participates in an experiment for the purpose of evaluating the original plan of action. When designers construct thought experiments for an instrument, they visually represent the anticipated function of a device. A reader of schematic illustrations anticipates how the device would perform under experimental conditions. A thought experiment is conceived, inviting critique of the experimental design. As a tool of persuasion, the plans provide readers with a cognitive vision of the laboratory events in ways that recommend endorsement. Based on virtual (vicarious) witnessing, readers are often persuaded that they could reproduce the same processes and would get the same correspondence of concepts to perceptions (Gooding 1990, 167). Even if the reader never performs the experiment described in the report, the author typically appeals to direct participation in experiments that are similar to those described. The reader is expected to follow a narrative that selects and idealizes certain steps of a procedure (Gooding 1990, 205). This narrative transports the reader from the actual to the possible by vicariously reenacting the significant features of the experiment, focusing on the instrument's design, material apparatus, and microscopic phenomena.

So, a design plan functions as an analytical tool for producing a hypothetical research technique. Cognition is diagrammatic in the Peircian sense. First, lines, shapes, and pictorial figures are introduced and manipulated in a design space, according to a grammar for the proper use of symbols. Second, these geometrical symbols represent general concepts, based on our perceptual skills of occlusion. Third, the design plan yields a thought experiment that entices knowledgeable participants to evaluate the designer's instructions to manufacturers, prescriptions to researchers, and predictions to everyone about the prospects for success.

5. Toward a Grammar of Pictorial Symbols

Ludwig Wittgenstein becomes a valuable ally in a campaign to link pictorial language with visual thinking. He never abandons his fascination with pictures as a basis of thought. A sentence in a story gives us a picture in the mind's eye. "[One] makes the story pass before me like pictures" (Wittgenstein 1974, para. 121). The use of a term calls before our mind the picture of what is named (1974, para. 37). What exactly does this mean? Wittgenstein is suggestive but not explicit. A clue is given by his analogy of language to instruments. Language functions as an instrument for the way we experience the world. "Look at the sentence as an instrument, and at its sense as its employment" (Wittgenstein 1958, para. 421, 569). His example of an engineer's drawings is quite telling. "What we call *'descriptions'* are instruments for particular uses. Think of a machine-drawing, a cross-section, an elevation with measurements, which an engineer has before him" (Wittgenstein 1958, para. 291). Descriptions of machinery, for example, rely on our experiences with such devices. The grammatical rules for any language are guided by practice. We follow a grammatical rule by engaging our environment, just as we use an instrument by picking out, and reacting to, certain salient features of the world. The visual character of language use is given depth through a study of how we engage nature by the use of tools.

But we should not read Wittgenstein as claiming that all rules are arbitrary in relation to our experiences. (This reading is advanced by Hacker 1986, 192–195.) A grammar is not completely free-floating in relation to the character of worldly events. The commitment to a rule is inseparable from the ability to follow that rule, which in turn rests on how we experience a world. Rule compliance requires skills for identifying known regularities and the ability to perform the same kinds of actions under the "appropriate" conditions. The concept of a rule is closely tied to the concept of doing the same thing (Wittgenstein 1958, para. 225). Compliance requires an expectation of a continuity of experiences: certain kinds of experiences are rendered salient and thus recommend specific responses. Of course, a speaker's appeal to a rule is no mirror of reality. But a speaker follows a rule by visualizing how we can engage a portion of our environment through certain kinds of experiences. Such an engagement is associated with the prescribed use of a term.

By reflecting on the grammar of schematic illustrations familiar to engineers, we can identify important aspects of an experiment. The shape grammar in engineering design is not arbitrary in relation to our experiences but presupposes an idealized scene for the ways in which we acquire experiences. A shape grammar exploits past successes and failures with similar devices and prescribes how experimenters should engage nature through proper use of the apparatus. The design plan places significant constraints on the proper means for investigating the world. By identifying the shape grammar of a design plan for instruments, we reveal idealized standards for proper experimental inquiry.

6. The Demand for Philosophical Instruments

In contemporary engineering, designers work with ideas, communicate in a visual language, and prepare for action. Design plans can be read in many different ways, depending upon a reader's purpose and degree of literacy. Manufacturers read such plans as prescriptions for construction, advertisers read them as marketing tools, and experimenters read them as signposts for research. By bringing instrumentation to the forefront of attention, we can identify epistemic commitments about the interaction between experimenters and specimens. One way for researchers to locate themselves in relation to microscopic processes is by reading design plans as epistemic maps. A design plan provides landmarks and paths for knowledge-seeking practices by identifying the opportunities for, and constraints upon, inquiry. In this respect, a design plan for laboratory technologies can be read as a channel of epistemic ideas about knowledge inquiry, dense with meaning about the idealized relationship between a skilled agent (experimenter) and a segment of the world (specimen). Such maps produce symbols for an experimenter's orientation, marking out the experimenter's place in the (detectable) world through the skillful use of instruments. A researcher's general orientation to the microworld is revealed in the plans for knowing by doing. Of course, no map is perfect: details are hidden, features are skewed, and landmarks are exaggerated. But an epistemic map offers a graphic agenda for further exploration.

The celebrated devices of sixteenth and seventeenth centuries were largely responsible for the glorious empirical discoveries of the day. Telescopes, microscopes, air pumps, and thermometers were hailed for their powers to reveal otherwise inaccessible regions of the world. Although some natural philosophers (Hobbes) found in such devices sources for deception and illusion, many others applauded their use for the insight they provided into regions of grand astronomical distances or microscopic scale. We read from researchers of the day how philosophical instruments enabled experimenters to discover the true identity of material bodies with respect to their mechanistic properties. Robert Hooke enjoyed a reputation as the foremost experimentalist in seventeenth-century England. His experimental skills were nurtured in machine shops as he immersed himself in the glass industry of Holland. By learning the machinists' trade, and improving on the production of lenses for his compound microscope,

Hooke was brought closer to the Creator, revealing the universal principles of Nature that determine the construction and refinement of any material body, whether produced from the machine's tools or from God's hands (Hooke 1961).

The stock in trade for Hooke and other designers of philosophical instruments of the seventeenth century were ideas and tools. I extend the seventeenth-century notion of a philosophical instrument to the realm of contemporary design, identifying the philosophical commitments about inquiry that underlie the discovery of new techniques. Inspired by Hooke's prescription that experimenters, including natural philosophers, should learn the material crafts of research, I believe that certain notions about contemporary inquiry are revealed in the designer's discoveries of instrumental techniques. Underlying such discoveries are commitments about the way in which experiments should be performed. At times, the elation of experimental discovery is inseparable from the achievement of creating an innovative technique. When philosophical instruments are deployed, the categories of human skill, instrumental power, and a specimen's capacities are put to good use, commissioned for service, as it were, in the construction and assessment of design plans. A commitment to such plans carries conviction about the possibility of empirical knowledge. Not composed of metals, wires, and plastic, a philosophical instrument is an analytical tool, distilled from past successes and failures, and used for discovering, or refining, detection methods. As laboratory events come and go, agency endures, as if lying in wait for another appearance. Undergirding each experiment are the capacities of apparatus, the abilities of experimenters, and the properties of raw materials.[2]

Since the time of the ancients, the metaphor of nature as machine has had a powerful influence on inquiry, motivated by the belief that a tamed segment of the world is more readily detected than a chaotic world. In the modern era, machines are responsible for putting something into motion. The machines of the seventeenth century were characterized by their powers to transmit, or modify, actions initiated from some other (mechanical) source. Mechanistic philosophers of the modern era found in nature's machines capacities to transmit, or modify, action that is initiated from an external source. According to one definition at that time, a machine functions to raise weights, move heavy bodies, and overcome resistances (Toulmin 1993, 141). Robert Hooke's mechanistic philosophy was born from the stunning advances in the construction of lenses, magnifying glasses, and microscopes. In his mechanistic philosophy, Hooke declares that the Grand Watchmaker created the cosmos as a cosmic machine, in which smaller machines are embedded in larger ones. The smallest machines comprise wheels, engines, and springs. In the eighteenth century, the mechanical, clocklike images were superseded by such technical advances as the steam engine.

The conception of nature as a generative mechanism continues to command attention in the sciences. A specimen functions as one of nature's machines, with capacities to generate movement when sufficiently agitated by a mediating technology called "instruments."[3] As such changes occur, capacities endure. A specimen's capacities are revealed through the new beings (states, events, or products) that are brought to occur during an experiment. Such capacities are tendencies that are activated by releasing a causal mechanism and by suppressing potential blocking agents. A cannonball has a tendency to fall toward the earth, just as an electron has a tendency to accelerate toward a positively charged plate, for example. Any tendency is manifested under the appropriate releasing conditions and the suppression of obstructing influences (Harré 1986, 284).

Nancy Cartwright's notion of a nomological machine offers guidance. Today, experimenters monitor the real, substantial changes that result from the operations of nature's nomological machines. For Cartwright, the physical world is a world of nomological machines. Underlying each machine are stable capacities that can be assembled and reassembled into different nomological machines, endless in their variety. A machine's capacities give rise to the kind of regular behavior that we represent in our scientific laws (Cartwright 1999, 49–52). In classical mechanics, attraction, repulsion, resistance, pressure, and stress are capacities that are exercised when a machine is running properly. Each law obtains due to the capacities of the system. Regularity is maintained when the machine runs properly (Cartwright 1999, 59). In Newton's law of universal gravitation, for example, the term "force" does not refer to yet another occurrent property, like mass or distance. "Force" refers to a capacity of one body to move another toward it, a capacity that can be used in different settings to produce a variety of different kinds of motions.

The revolutionary advances in biochemistry of the 1950s perpetuated the allure of an organism as a machine. DNA is characterized as an information processor. Molecular biology continues such a reliance on causal mechanisms (Bechtel and Richardson 1993; Brandon 1985; Burian 1996). In a provocative and well-researched article, "Thinking about Mechanisms," Peter Machamer, Lindley Darden, and Carl F. Craver (2000) examine the nature of mechanisms in neurobiology. In DNA replication, for example, the DNA double helix unwinds, exposing slightly charged bases to which complementary bases bond, eventually yielding two duplicate helices. In this case, protein synthesis is a mechanism comprising a process (activity) of hydrogen bonding of a DNA base and a complementary base (entities with their properties) (Machamer, Darden, and Craver 2000, 3).

Recent challenges to the machine/nonmachine duality are commonplace. For example, the brain/machine distinction became blurred with the advent of the Turing machine. In another area of research, the exploration of nanostructures may provide hints about both the emergence of life and the fabrication of new materials (Mainzer 1997). Such challenges reflect the following philosophical commitment about inquiry: the world can be known by exploiting the kind of mechanistic properties that we attribute to our own creations, properties that are exploited in the design of instrumental technologies.

NOTES

I would like to thank Hans Radder for his rigorous evaluation of an earlier draft of this chapter. Also, the substance of this chapter reflects constructive suggestions by participants at the workshop "Towards a More Developed Philosophy of Scientific Experimentation," June 15–17, 2000, Vrije Universiteit, Amsterdam. Marcel Boumans's commentary on my presentation at this workshop was particularly insightful.

1. The most compelling argument on the centrality of modeling in science appears in Rom Harré's *Principles of Thinking* (1970). Margaret Morrison and Mary S. Morgan have recently contributed to this subject in their "Models as Mediating Instruments," appearing in their coedited volume *Models as Mediators: Perspectives in Natural and Social Science* (1999).

2. Although the term "philosophical instrument" is used in original ways in this chapter, the notion of an experiment as a dynamic association of agents, both material and immaterial, has been established in the literature. In my opinion, the most compelling arguments on behalf of the priority of agency of research is found in David Gooding's *Experiment and the Making of Meaning* (1990). Based on his detailed study of Faraday's experiments, Gooding argues that an experimenter, as agent, engages nature through both mental processes and material manipulations (1990, xiv). In the present work, I extend this position to contemporary research.

3. A causal conception of substance is advanced in the important work by Eva Zielonacka-Lis (1998), and I thank her for valuable and inspiring discussions on this topic.

David Gooding

13 / Varying the Cognitive Span

EXPERIMENTATION, VISUALIZATION, AND COMPUTATION

1. Introduction

It is a commonplace that humans use technologies to extend their ability to see, remember, calculate, and reason. Science has always depended on such tools. I am interested in the interaction of ordinary human modes of perception with the technologies that enhance the modes of perception of empirical science. This dynamic has three features important to a developed philosophy of experimentation. First, there is a shift from ordinary human perception, construed in terms of contemporary commonsense notions, to the real (or super-real) objects of scientific investigation such as the primary qualities of Galileo and Newton. This first shift is often described as abstracting from the sensory sources of perceptual experience to mathematical properties that are beyond the reach of unmediated human sense perception. Although we perceive abstraction in retrospect as an intellectual process, this shift was engendered by the special technologies devised to access the measurable qualities of the new sciences. So the process whereby science selects, refines, and changes human perception is as much a history of machines and techniques as it is a history of ideas, discoveries, theories, and institutions. The second aspect of this dynamic is that technologies that expand the range of what scientists can observe also alter the evidential basis of the sciences. The overt function of the machines is to create new phenomena and data, yet these same machines constantly change how humans do science. A third aspect of the dynamic is that contemporary notions of what counts as ordinary or unmediated perception are constantly changed by the activities of scientists. As we shall see, scientists often disagree about the epistemic status of newly imaged objects. This last aspect is one of the key problems that science generates for epistemology because the technologies of scientific observation constantly change what counts as observation and, therefore, what counts as knowledge of nature. The very possibility of observing (as a kind of direct seeing) remains controversial in fields such as quantum chemistry (Humphreys 1999).

A philosophy of experiment needs to address the implications of this dynamic. How can a philosophy of experimentation be developed to accommodate the changes engendered by the advent of computation and data processing? In fields such as high-energy physics, geophysics, and meteorology, computing enables simulation, including (cheap) simulation of experiments "seeded" by ever-decreasing quantities of (expensive) "real" data. Visualization of data and of mathematical models of data has become an important means of restoring graphical representations to the processes of interpretation and communication. The drive to make biology more theoretical is largely based on mathematical modeling enabled by information technologies. Information-based programs such as the human genome project were created by the belief that decoding the genome is essentially a matter of information processing. Whereas these sciences predated cheap, powerful computation and have been transformed by its availability, new fields such as brain imaging are "born digital": the design of their digital detection and recording devices anticipates the need to re-present output in graphical form (Beaulieu 2000, 2001).

Of course the machines are changing, too. As their capabilities increase, so more of the processes of scientific investigation can be delegated to them. In the near future, data analysis, data mining, and robotics will produce investigative devices that are largely autonomous. Consider a science conducted primarily or even entirely by machines. Would such a science be science as we know it? I suggest that it would be only if it involved the same kind of knowing that humans have. However, the history of science shows that "our kind of knowing" is changed by science. So to answer this question we need to consider which cognitive features of humans have persisted (albeit enhanced through the technologies of observation and experiment) and those which have been marginalized or suppressed by science.

The new observational technologies of the sixteenth and seventeenth centuries shifted perception from variable, personal sensation (secondary qualities) to inherent and real properties (the primary qualities) by selecting certain aspects of ordinary experience and redefining them in terms of qualities accessed by mathematicians through their instruments and procedures. Although Aristotle and other pre-Renaissance philosophers understood the distinction between primary qualities (such as motion, rest, shape, size, and number) and secondary or perceived properties, Galileo understood that nature could not be mathematized on the basis of an Aristotelian physics. Galileo redefined the distinction between primary qualities (those properties of things that can be measured—quantities) and secondary qualities (which are properties as perceived by humans). Feyerabend characterized this as a process of "changing the sensory core" of experience, from an Aristotelian to a Galilean view of things.[1] Similarly, the advanced techno-science of the twentieth century raises a new question about the epistemic status of knowledge produced almost entirely by machines. This implies a further question: Do we understand well enough how science is done by humans to re-create science in machines? The answer is probably "no," and the consequence is that machine-based science will differ increasingly from science conducted by humans.

The question highlights the importance of studying different types of representation and argumentation in human activity, particularly in the sciences. Experimentation is central because it involves both representation of experience and argumentation about the theoretical meaning of that experience. A developed view of experimentation would recognize not only that the changing character of experiment is an indicator of the changing character of science but would recognize also the wider implications for epistemology. As the locus of knowledge production shifts from humans to machines and from laboratories to global networks of knowledge factories, the epistemic project of Locke, Berkeley, Hume, Kant, and many others will lose its relevance.

In this chapter I want to consider cognitive aspects of the agency of machines in order to understand how abstraction works in the context of material technologies designed to extend our senses and enhance our ability to interpret new information. I reserve judgment about whether such abilities could be re-created in machines (the failed dream of traditional artificial intelligence). I hope to unpack an apparent paradox arising from two aspects of science that are often seen to be opposed. What makes the sciences so successful is their highly focused approach to abstracted and simplified bits of the world. The sciences are very good at reducing problems to a form that can be solved by currently available methods. These constitute what Kuhn called the disciplinary matrix that defines a scientific discipline. Yet the selection of attributes and abstraction to tractable representations of them are also regarded as an obstacle to producing understanding that is more widely meaningful to humans. The distance between scientific discipline and lay understanding is lamented by those who seek greater public appreciation of science but is celebrated by those for whom it shows the superiority of scientific perceptions over commonsense ones (see Wolpert 1992 for this argument). This distance also gives rise to the charge that science is an alienating, dehumanizing force in culture. To some this alienation consists of the opposition of objective knowledge to the subjective, experiential knowledge of laypersons: "in its pursuit of increasingly refined observation of the material world, of the greatest agreement among observers, and of the highest degree of repeatability, science has resorted to methods which present a picture of that world alien in many ways to immediate experience" (Morison 1965, 273). Despite the alien appearance of the abstracted, technologically mediated picturings of science, we shall see that immediate experience remains important to science.

Many scientists have argued that science essentially and necessarily abstracts and reduces in order to achieve the goal of quantitative descriptions of the world. As we will see later, such arguments suggest that the qualitative, sensory world will be completely replaced by quantitative description. Nevertheless, qualitative, sense-based ways of thinking persist. Computations on digital representations do not wholly supplant informal reasoning about analogue representations. I shall argue that while abstraction is powerful, its technological embodiment in scientific instruments requires a narrowing both of sensory modalities and of the range of cognitive capacities invoked by scientific investigation. This function of cognitive narrowing is what makes an instrument scientific. Observational instruments appear to expand the sensory basis of human observation, say into the domain of the very small or the very distant. However, in many cases this narrowing has given primacy to visual perception over other sensory information. The newly quantified chemistry of Lavoisier, for example, systematically sought to eliminate smell, taste, and touch in favor of visually apprehended numerical measurements. In sociological terms the purpose of detecting, recording, and measuring devices is to decontextualize observation by transforming it into standardized, portable information or data (inscriptions). In psychological terms the purpose of instruments is to reduce and standardize the range of cognitive processes needed to make and interpret visual observations. The more successful this process of reduction is, the more readily the processing of information (as data) can be designed around established mathematical or other problem-solving procedures, most of which can be delegated to a context-independent world of machine processes.

Reductive practices and technologies and the criteria used to evaluate and validate them are peculiar to particular disciplines and will change as disciplines develop.[2] Moreover, this reduction is only the first stage of the process. Michael Heidelberger (this volume, chap. 7) describes this as the transposition of aspects of human experience to instruments as proxy perceivers and processors. He asks whether these surrogate "perceivers" are "material scapegoats" or are epistemologically superior to humans. This begs the question whether data-producing machines are "perceivers" in any humanly meaningful sense. Their output,

whether in numerical, digital, or some other encoded form, must be expanded
—that is, translated back into something that is meaningful to humans.[3] Cognitive reduction therefore has an important counterpart in the methods and
processes whereby the results of machine manipulation are re-presented and
recontextualized so that humans can assimilate, interpret, and reason with them,
typically through words, symbols, and images. We can therefore characterize
scientific "abstraction" as a process of reduction, computation, and expansion.
Since this sequence will have many iterations, it is a cyclical process. I shall argue
that, like most observation and measurement in science, digitalization involves
reduction (to enable computation) and expansion, which is the translation of
computed information into a meaningful context. This provides a scheme for
structuring accounts of the way that intellectual and material technologies
work in the sciences. Specifically, it identifies clearly the context of qualitative,
analogue-based work by which quantification and digitalization are achieved
so that computation can be used, and the further work that is subsequently done
to reintegrate the output into the analogue world of human cognitions, representations, and discourse. On this scheme, there is less difference between pre-
and postdigital technologies than we might expect. Digitalization is an extension involving further cognitive narrowing of what experimental technologies
have always done.

2. From Techno-Science to Cyborg Science

The model proposed is a very simple one. I hope to show that, suitably enhanced,
it enables us to expose a misunderstanding that supports the widely held belief
that quantification based on sorting, counting, and computation is more appropriate for scientific work than modes of cognition based on words and images. It
is tempting to draw this distinction in terms of analogue-pictorial and numerical-
digital modes of representation and reasoning and to argue that the increasingly mathematical character of the sciences shows that the numerical-digital is
the more primary mode of describing nature. This is a technologically updated
version of the arguments of Galileo, Newton, and others that reemerged in
nineteenth-century optimism about the finality of the alliance of mathematical
physics with precision measurement. It underlies confidence in the mathematical cast of theorizing in contemporary biology and futuristic views of the transformation of twentieth-century techno-science into cyborg science. I hope to
show that the analogue-digital distinction does not get to the heart of what makes
science different from other kinds of human activity and what, in particular,
alienates it from ordinary or lay perceptions and concerns.

It could be argued that my approach is far too cautious. Rather, we could celebrate techno-science as an activity whereby humans continually transform themselves. Thus science epitomizes the cyborgian nature of human beings as "bio-technical hybrids." Briefly, the argument invokes evolutionary explanations of the brain, tool making, science, and culture to conclude that culture is an externalized system of cognition. Information technologies are an integral part of this culture, in which the plasticity of the brain has the biological function of exploiting nonbiological knowledge resources. This shifts the locus of knowledge from the brain to culture, which is both the product of and the producer of embodied brains: "More than any other creature on the planet, we humans are indeed natural-born cyborgs, factory tweaked and primed so as to be ready to participate in cognitive and computational architectures whose bounds far exceed those of skin and skull" (Clark 2001, 23). On this view, the material, technological environment acts continually to change our brains and, therefore, what humans are. In this adventure all is process: there is neither nature nor human nature.

One can accept certain premises of this argument without accepting that the transforming effects of twentieth-century techno-science are so thoroughgoing or inevitably beneficial.[4] Human nature may not be fixed, but does it follow that there is no reference point from which we might evaluate the impact of techno-science on humans? Can we seek a reference point without invoking modernist certainties, natures, and essences? According to the cyborg vision of a continuous dynamic of biologically enabled, culture-producing, human nature–transforming technological activity, the persistence of some feature that makes us human is an illusion created by our limited vision of the future. However, I contend that our view of the past is just as relevant. History shows that humans have always needed to develop a view and that science has been an increasingly important source of ideas about this. It is therefore worth exploring the cognitive basis of the sort of science that humans have developed.

2.1. Two Cultures

Consider the fact that for much of its history science has been disseminated through documents that have been authored. Historians and philosophers have argued that the experimental narrative was a new, literary form of technology and that it was crucial to the development of modern, *empirical* science (Shapin and Schaffer 1985; Bazerman 1988). The main stratagem of empirical science has been to weave descriptions of objects, features, events, and processes into evidential arguments. This reference-making is grounded both in the world and in the activity of writing about work in the laboratory or field. Scientists make knowledge by relocating it, moving it from the personal and local context to the larger domain of publicly reproducible phenomena, proofs, or processes. Knowledge is constituted by all of the activities described by literary technologies, not just by discourse about them. These activities tell us something about our cognitive makeup as well as the social processes whereby knowledge is defined and redefined.

This focus on process enables us to transcend the modernist binary distinction between art and science as wholly distinct and fundamentally opposed in character. Modernist views draw distinctions based on differences between the essential nature of science and other activities. The position glossed earlier identifies nature with numbers and science with numerical and digitizable processes, thus distinguishing both nature and science from the arts and culture. It goes on to ground the primacy of the digital over the analogue in scientific, computational models of brain processes. Another example is Wolpert's widely disseminated claim about the inferiority of ordinary, lay, or "commonsense" perceptions to the counterintuitive truths of science.

To counter such essentialist views of science and its products, historians point out that science did not emerge from an homogeneous static culture; rather, the distinction emerged during the Renaissance as both the mathematical-experimental sciences and the representational arts developed. Science became the domain of fact, while the arts became the domain of expression and interpretation (Jones and Galison 1998, 1–4).[5] Their argument is that each of these activities needed to emphasize their differences from the other. Yet many of the technologies of art influenced representational practices in the sciences. It follows that the "two cultures" of science and art, of the numerical-digital and of the experiential, do not express fundamentally different aspects of human cognitive capability, such as the primacy of pictorial-analogue modes of perception over numerical-digital ones. Rather, they express differences between the cultures in which particular cognitive capacities are applied.

Most historians are constructivists. Most constructivists now take for granted that both the givenness of the experience and phenomena presented in experimental narratives and the aboutness of the language that describes them can be explained as the products of human agency—writing and theorizing as well as manipulating (and designing the manipulation of) material objects and agents (see Gooding 1990; Latour and Woolgar 1979). The authors of experimental narratives have been intimately involved with the worlds that their accounts enable others to explore vicariously. To be involved with the world in this way is to find out about some particular world and to invent ways of describing it, rather than to demonstrate facts or general truths about it. Features of the phenomenological world are selected, shaped into representations that are criticized, and changed through the practices that, eventually, successfully link them to those aspects of the world that they purport to be about. This involves learning. It is not so different in aim or process from those of artists.

Finally, a point about the scope of a philosophy of experiment. Sociologists such as Latour and Collins reject cognitive and developmental processes insofar as they involve the intentions of individuals (Latour 1983, 145).[6] In highlighting the several "translations" by which Pasteur moved anthrax from the field to the laboratory and back to the field, Latour remarks several times that from the field condition Pasteur "learned" how to create connections between the disparate worlds of the farmyard and the laboratory. But the learning is passed over. This neglect reflects a preference for a kind of explanation that black-boxes what it does not know how to deal with. A developed philosophy of experiment has a broader remit. It should address the personal and cognitive dimension of experimentation rather than subsume them to the consensual.[7]

3. Abstraction and the Progress of Science

Here is another view of the progress of science that is widely held, especially among scientists. Since the seventeenth century, measurable quantities such as mass, dimension, position, and temperature were thought to be more reliable means of apprehending the natural world than the properties that individual humans perceive (such as weight or heaviness, size, place, and heat). Objectivity became attached to numerical descriptions rather than images or other qualitative descriptions. So, during the seventeenth century the locus of authoritative description of nature moved from the realm of ordinary, shared experience into the realm of observational technology, the scientific laboratory.

This development has also been described as a process of abstraction, creating from the particulars of experience, descriptions and concepts that are general— indicative of the features of natural laws and of natural kinds. A corollary is that the success of a particular scientific endeavor depends on, among other things, achieving and working at the right level of abstraction. The "right" level is one that allows problems to be described in terms amenable to the application of preferred mathematical or computational techniques. This presupposes a view of the relationship of representations (images, symbols, concepts, theories) to what (in the world of experimentation) they purport to be about—namely, that the former are approximations of the latter. There is an implied contrast between abstractions and exact replications or simulacra. As the now extensive literature on modeling and simulation demonstrates, there is always a trade-off between simplicity (associated with greater abstraction from the complexities of

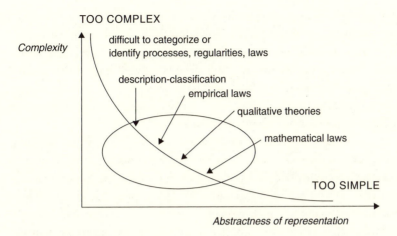

TOO COMPLEX

Complexity

difficult to categorize or
identify processes, regularities, laws

description-classification

empirical laws

qualitative theories

mathematical laws

TOO SIMPLE

Abstractness of representation

FIG. 13.1. The tension between the abstractness of a representation and its ability to describe real-world phenomena

the world) and realism (associated with the range of phenomena and relationships incorporated into the model). Abstraction gives us simplicity, transparency, clarity, and the ability to apply well-tried methods of analysis. Realism usually means empirical adequacy, which usually calls for predictive success.[8]

This trade-off can be illustrated by an image in which the vertical axis indicates complexity (where the phenomenological world is taken to be complex) and the horizontal axis, abstractness or artificiality (see fig. 13.1). This figure should not be read as suggesting that there is an "optimum" path or trajectory. The curve expresses an ever-present tension in science between simplification and realism. There is a continual movement back and forth along a trajectory of abstractedness. The view is sometimes put forward by physicists and mathematicians that science progresses as sense experience is eliminated by translation into abstractions that can be manipulated formally. By contrast to Galileo's recognition of the need to attend to the specifics of sensory experience, Max Planck described this linear view in terms of the renunciation—even suppression—of sensory experience: "The whole development of theoretical physics until now has been marked by a unification achieved by emancipating the system from its anthropomorphous elements, in particular from specific sense impressions. But if we reflect . . . that these impressions are the recognized starting point for all physical research, this conscious turning away from basic presuppositions must appear astonishing and even paradoxical" (quoted in Toulmin 1970, 6). Early in the last century Robert Millikan (1917, 4) argued, as he claimed Pythagoras had, that "any real scientific treatment of physical phenomena . . . is possible" only "on the basis of exact, quantitative measurement." He went on to claim that "the problem for *all* natural philosophy is to *drive out qualitative conceptions* and to replace them by quantitative relations," citing William Thomson's assertion of the identity of knowledge of something with an exact, numerical determination of it. (Thomson had remarked that "when you cannot express it in numbers, your knowledge is of a meagre and unsatisfactory kind.")

Does the increasingly mathematical, data-based character of contemporary science show that such claims about the mathematical nature of nature are true? Not at all. Notice the circularity of this sort of vindication. This view of science influences our ways of seeing and of conceptualizing things. It selects and emphasizes quantities (which can be handled through the manipulation of symbols) and numerical (and now digital) technologies. It is argued that these rely less on other human capabilities, such as qualitative forms of interpretation (for example, of patterns) and construction (for example, of structures), so that the rightness of quantitative measures of abstracted qualities appears inevitable. On this view it seems inevitable that, over time, visual and verbal modes of representation must be supplanted by numerical and digital ones.[9] This accomplishment is used to argue that the essentially mathematical character of science reflects the mathematical nature of reality. A similar case for the fundamental status of digital representations could be made: just as the new science of the seventeenth century reconstructed the world as primary qualities and humans as machines, so a wholly digitized science would present reality to us as nothing other than processable digital bit streams. This view of science and the associated view of human cognition fails to recognize that nonnumerical capabilities are —and continue to be—essential to the conduct of science, at least by humans.

On closer inspection the historical trend to abstraction has two aspects. The first is abstraction as the representation of qualities in terms of quantities. This is not a unidirectional process. Rather, in the history of any scientific field there is a searching back and forth to achieve an *appropriate* level of abstraction—not oversimple, capturing enough to be representative or "valid," yet not too complex to be mathematically or computationally tractable. Sir Robert May has described this process succinctly: "The world is complicated but not all the complexities are important all the time. . . . The real trick is to intuit . . . the subset of things that are important, and on this basis to formulate a *tentative* understanding in *un-ambiguous* terms. You then pursue where that tentative understanding leads, test it against the facts, and usually circle back to refine the original assumptions. This basic process holds whether you are trying to understand superconductivity, or the causes and consequences of biological diversity, or how best to translate Britain's excellent science base into industrial strength."[10] We could just as well label the axes in figure 13.1 to plot complexity against, say, solvability according to analytical methods or against computational tractability.

Abstraction is part of the process of preparing material for the application of established techniques, of *reducing* something unformed and unruly to a solvable form.

The second aspect of the trend to abstraction is that this dynamic sometimes involves reshaping a scientific field to exploit more powerful methods, such as computational models and techniques. This often leads to controversy. Consider current concerns about the way in which imported mathematical techniques are dictating biological approaches to organisms or the reduction of genetics to the study of genes as carriers of information (see Goodwin 1994). Ian Stewart argues that science needs to counter this quantitative turn. Noting that "one of the most striking features of recent mathematics has been its emphasis on general principles and abstract structures—on the qualitative rather than the quantitative," he argues that scientists need to develop another method that complements the quantitative. Number, he says, is "just one of an enormous variety of mathematical qualities that can help us understand and describe nature. We will never understand the growth of a tree or the dunes in the desert if we try to reduce all of nature's freedom to restrictive numerical schemes" (Stewart 1998, 169; see also 1999). Stewart proposes a new mathematical science of form that would be capable of explaining "why the organic and the inorganic worlds share so many mathematical patterns."

For a more historical view we should introduce a third axis against which to plot changes in the criteria invoked to evaluate the acceptability (veridicality, truthfulness) of a representation. A third, time-dependent axis would allow us to represent the historical fact that what counts as a good abstraction also changes. For example, in the seventeenth century abstraction from sensory experience—a move from concrete and particular to generalized representations (images, symbols, formulae) had a particular form: the proponents of scientific methods discredited personal, sensory-based experience as unreliable, in favor of quantities. In replacing secondary qualities, primary qualities such as mass, position, or temperature nevertheless retained some perceptual base in ordinary, lay perceptions of weight, place, or position on a scale. As we would now say, there had to be a phenomenology and procedures for relating abstract concepts and their tokens to events and experiences in the world.[11]

Nevertheless, it could be argued that even if visualization had been central to the process of operationalizing abstractions, this is no longer the case. In some areas of physics even the perceptual basis has become irrelevant. Arthur Miller has argued that quantum physicists abstracted not from experience but from mathematical formalisms. This is important because it shows that there is more than one kind of abstractive process and that to substitute images or verbal tokens for aspects of experience is no longer the only sort of abstraction that matters.[12]

Moreover, as the contested observability of electron orbitals shows, the operational path that links concepts, visualizations, percepts, and objects is increasingly tenuous.[13]

Was the Galilean mathematical reduction inevitable? The reasons emphasized by historians are sociopolitical as well as intellectual. There was a need to establish *nature* (rather than human artifice) as the object of study and as an authoritative source of knowledge. An important intellectual factor was the conviction that nature is inherently *mathematical*.[14] These intellectual and socioeconomic explanations do not invite us to consider reasons of a cognitive sort. Yet it is important that we do.

Let us reconsider Galileo's distinction. This was fundamental to the new science, which has made progress by selecting familiar, qualitative aspects of experience as the basis for quantifiable modes of representation. By *qualitative* I mean representations of experience that are *amenable* to ordinary or human modes of perception, as Miller (1986, 128) puts it. But what are "ordinary human modes of perception"? Typically these emphasize form, pattern, structure, and also number. These are features of the world that experience, cultural conditioning, or an evolved neurophysiology make us predisposed to recognize and to favor in our representation-making work.

3.1. Evolution and Cognitive Capacities

Before proceeding we need to note two misconceptions: one is about the cognitive status of common sense; the second is the alleged evolutionary pedigree for human knowledge-making capabilities. Given the displacement of secondary by primary qualities, the "ordinary modes of perception" are apparently not adequate for science. As noted earlier, Lewis Wolpert, a distinguished embryologist, celebrates this unnatural aspect of science, emphasizing the differences between ordinary, lay, or common sense and (superior) scientific sense (Wolpert 1992).[15] Wolpert offers an evolutionary explanation—namely that our brains have been selected for dealing with immediate experience and behavior rather than the remote, abstracted world of mathematical science to explain the alleged fact that "the way the universe works is not the way common sense works: the two are not congruent" (Wolpert 1992, 11). As an example of the conflict between common sense and scientific knowledge of the universe (showing the superiority of the latter), he appeals to the conflict between our interpretation of the stability of the earth we stand on (as sun, moon, and stars move by) and the scientific knowledge that the earth is rotating and moving through space. Another, school-science example is the variability of the perception of "temperature" induced by placing each hand into either hot or cold water. Our ordinary perceptions are too unreliable for science. Ironically, the sort of evolutionary explanation to which Wolpert alludes is also invoked to support the opposite view, that science is wholly commensurate with ordinary human capabilities. Robin Dunbar argues that the success of science is explained by the fact that the methods of science "are, at root, the mechanisms of everyday survival," while Alison Gopnik claims that "science gets it right because it uses psychological devices that were designed by evolution precisely to get things right" (Dunbar 1995, 96; Gopnik 1996, 489). This is reminiscent of Ernst Mach's appeal to experience-based intuitions to explain the power of thought experiments. Mach's explanation implies that different thinkers will have different abilities due to their different degrees of experience of the world. This is implied by any explanation that appeals to the benefit of sensory experience, whether that has been assimilated in the very brief span of cognitive processes or through the slow processes of natural selection. Mach believed that those with greater experience would produce better thought experiments because they have a more intimate knowledge of the environment or better empirical intuitions.[16]

These arguments fail to recognize that the "environment" of an embodied brain must be represented before it can be thought about and that, as a basis for such representations, ordinary or commonsense understanding is no more fixed than are the categories of science. As argued earlier, the sciences have long been an important agent for changing what we take to be "natural" or a matter of common sense. This indicates the naïveté of the view that biological evolution has conferred on humans a set of truth-seeking or "truth-tropic" cognitive capabilities, which, it is argued, are the basis of science and which therefore explain its success.

I am skeptical about the evolutionary explanations, because it is so difficult to disentangle different sources of the capabilities and predispositions that enable us to order experience in terms of shapes, patterns, and meaning. For example, it is most likely an evolved predisposition that makes most people see a face in certain rock formations on Mars (the "face of Cydonia"). As for the truth-seeking stratagems that psychologists such as Dunbar and Gopnik posit as the basis for science, these require considerable enculturated, contextual knowledge in addition to what is biologically "given." Past experience will affect, if it does not determine, whether we see ducks, rabbits, or antelopes in the gestalt images made familiar by Wittgenstein and Hanson (1958). Even more specific, culture-dependent experience will determine whether we can interpret Hanson's drawing of an X-ray tube as a made object, let alone identify it correctly.

Do any of these indicate basic cognitive functions endowed by evolution and which science exploits more effectively than lay practices do? If so, can scientific developments alter these cognitive functions? How would we know? One approach to an answer is to look for features of representations that scientists have continued to use, notwithstanding considerable advances in the reach, scale, and complexity of experimental technologies. Whereas historians and sociologists emphasize change in science, from a cognitive standpoint we need to consider also the continuities identified by a historical approach.

3.2. Abstraction Revised

The "abstraction" thesis glossed earlier asserts that science moves from ordinary or commonsense representations to abstracted or scientific ones. This is too simple a statement of how science changes both our experience and the status of that experience. Its more extreme form (as expressed by Millikan or Planck) plays down important cognitive continuities. Abstraction from experience does not require the elimination of features drawn from ordinary human sensory experience (though this may be a side effect of introducing another mode of representation). For example, the effectiveness of thought experiments depends on the creation of mental scenarios, worlds in which much is familiar and in which phenomena are sensed through the usual embodied sensory modes. This is no less true today than it was in Galileo's time. The phenomenology of this type of argument shows that ordinary modes of perception continue to be essential to the articulation and communication of arguments (Gooding 1992). So in some areas of science, at least, the historical process of displacement alleged by Millikan, Planck, and the cyborg visionaries is neither complete nor a one-way street. The cruciality of experimental evidence often turns on its being translated from an abstracted (usually numerical) form into objects or shapes that can be interpreted or manipulated. Visualizations of data may display patterns, and new properties such as symmetry may emerge as new features of the representation. In the case of the phenomenon of sea-floor spreading discussed below, such new features were central to geophysical explanations of the visualized data and to subsequent acceptance of the continental drift hypothesis (Le Grand 1990; Gooding 1998, 306–312). Far from being visible, such features are not even present in the data unless the appropriate visualization algorithms are applied. In other words, development within a particular domain requires movements (translations) in both directions—that is, between "primary" and "secondary," abstract and experiential, or between numerical-digital and analogue-pictorial.

There has not been a thoroughgoing assimilation of sensory-qualitative to digital-technological.[17] My next examples suggest that human experiential capabilities are resistant to displacement by the symbol-processing technologies developed to support the data-intensive, industrial-scale experiments of twentieth-century physics.[18] On the contrary, far from displacing qualitative experience, the quantity and character of numerical data demands new ways of generating it. Predominantly quantitative sciences do not deal solely in numerical data: visual and other qualitative descriptions precede and guide the reductions involved in abstracting from phenomena to data, and they are needed to aid expansion as the interpretation of large quantities of data through visualization. For example, in the 1960s, when Vine and Matthews were advancing a new mechanism to explain the spreading of ocean floors, they translated large quantities of numerical data (recording the magnetic field strengths measured across the northeast Pacific Ocean) into grayscale maps. This was digitalized output of numerical data representing measurements of field intensity above basaltic rock over thousands of square kilometers of the northeast Pacific Ocean. It is a highly condensed representation of recorded numerical data. Nevertheless, in the terms of my cyclical model of digitalization, it is an *expansion*. The tension between the demands of measurement and calculation, on the one hand, and the quite different sort of thinking involved in interpretation, on the other, often leads to disagreement. Some geophysicists objected to this expansion on the grounds that images move science away from the hard, that is, numerical, data. Nevertheless, certain images became "crucial" when it was realized that they displayed new properties, in particular, symmetries for which only one explanation —sea floor spreading—seemed plausible. Much effort goes into translating numbers into plots, maps, and diagrams because it is easier to think about and with the data in this form. Plots such as these made it possible to see (in analogue mode) what would not have been visible in the numerical data.[19] Certain features, such as the mirroring of field-strength patterns either side of the Juan de Fuca ridge became significant. As remarked earlier, these features of the plots were not observable "in nature." They suggested the physical mechanism that was eventually accepted as the true explanation of the expansion of the ocean floor and were subsequently written into the textbooks (and the histories) as "crucial" observations that changed the course of the science (see Le Grand 1990).

4. Analogue and Digital

My final example is the long-standing tension between numerical (digital) and graphical (analogue) modes of representation in high-energy physics. Galison (1997) has described how during the twentieth century two distinct approaches to the description and interpretation of information about subatomic particles developed. These distinct ways of establishing the existence and properties of elementary particles remained largely independent for several decades. As in geophysics, the distinct traditions in high-energy physics each emphasized a different feature: complexity of detail of events, on the one hand, and quantities of repeatable (but low-resolution) events, on the other. Each rejected the quality of the other's data because of its character. A mimetic or *image* tradition emerged from Charles Wilson's cloud chambers that mimic atmospheric processes in the laboratory. These had been resisted at first by critics who argued that one could not reproduce real cloud phenomena in a box in a laboratory. This is an objection to a "reduction." The cloud chamber led to the experimental technology of the particle chamber. These detectors called for new techniques of recording and displaying the tracks of elementary particles, culminating in nuclear emulsions and the bubble chamber. The mimetic approach prized detailed, high-quality visual information about very few particle events. In contrast, what Galison calls the *logic* approach involved detecting, identifying, classifying, and counting very large numbers of particle events. There was no richness in the description of these events: large quantities of low-resolution data were analyzed for certain features. If the required set could be detected, an event was counted, otherwise not. The approach was implicitly digital and amenable to machine-based processing. During the 1950s and 1960s, the process of making discriminations was mechanized and automated. Alvarez and others first recruited armies of women to interpret traces of events, converting these traces into large quantities of numerical data. Then, in what appears to be a typical process of automating a human capability (pattern recognition), he developed computing devices to replace the armies of human computers. Eventually, automatic counters produced statistical information about whole populations of events. These were based on electronic devices that operated according to logic circuits, where the design specified the conditions under which an event would be recognized as, and therefore counted as, an instance of the decay of a particular type of particle—rather than an artifact. The digital processor became an integral part of the experimental system.

Digitalization is based on the ability to make discriminations and to preserve the identity of the tokens that denote objects or attributes of objects. What Alvarez mechanized is the process of establishing and preserving the "identity" of each particle-indicating event.[20] So, in the logic tradition, a reduction was made for each of the steps needed to display—as histograms—the effects of particle interactions inside large complex machines. Scanning, measuring, reconstructing tracks, kinematic analysis of tracks, and experiment analysis were digitalized so as to be carried out by a complex but unified experimental system.[21]

The tension between the two experimental traditions was resolved, finally, through wholesale digitalization of their distinct methods so that their modes of representation could be combined (Galison 1997). They are interesting for two reasons. First, each tradition drew upon a different aspect of human cognition—

the visual, pattern-seeking or *analogue* capability, on the one hand, and the *digital* operations of classification, counting, and calculation, on the other. This shows that capacities that are fundamental to human cognition remain an important influence on the very medium of scientific thought—notwithstanding the mediation of sophisticated observational technologies. Second, as with geophysics in the 1960s, the numerical/digital and the image-based traditions of high-energy physics converged. During the 1960s the time projection chamber (TPC), with the use of pulse counting and computer analysis, could generate three-dimensional visual images from data gathered from large numbers of events. This is the technology of *expansion* that moves information from a computationally tractable form back into the area of representations that are "amenable" to human understanding and reasoning. Expansion allows humans to work with more familiar and cognitively more tractable representations. In this case, expansion involved combining high-quality imaging with large quantities of data. This second feature of microphysical practice illustrates what we already know about human cognition: it is most effective when different modalities are combined.

This technology enabled the convergence of sensory-analogue and numerical-digital modes of representing nature that high-technology physics had formerly kept quite distinct. Where the physics of the 1950s and 1960s has led, the availability of cheap computing power allows many other disciplines to follow.[22] As we have seen, technologies and methods for visualizing counts were readily applied to the analysis of numerical data in other fields. These methods produced images and plots—the representations that scientists actually worked with.

4.1. The Analogue-Digital Distinction in Context

Technology has made the analogue-digital distinction ubiquitous. The purchase of Time-Warner by America Online in January 2000 was described as the acquisition of a vast source of analogue content by a digital technology for access and distribution.[23] Does this distinction capture anything as important as, say, the Galilean distinction between primary (quantifiable) and secondary (qualitative) properties? In the twentieth century the rise of the image and logic traditions reflects two modes of human experience (sensory experience and counting) and two basic modes of thinking (visualization and classification/symbol manipulation), while their merger indicates that they are complementary and that both are necessary. This is in keeping with an interpretation of the history of human civilization as showing that humans have basic or innate capacities for verbal and written languages, for the production and interpretation of images, and for counting. However, for some the existence of counting raises a modernist question about fundamentals: is counting a basic human aptitude or is it a derivative capacity conferred through culture? After all, counting depends on the capability of making discriminations in a systematic way; this in turn requires a scheme for representing these as differences and for manipulating symbolic representations (such as letters into words, words into sentences, and so on). These are needed to develop and disseminate the discriminations that underlie counting and measuring (Johnson-Laird 1988; Barrow 1992, 102). Others highlight the fact that humans are much better at interpreting experience in terms of patterns and at using text and images than at performing digital (numerical) operations. This suggests that the brain is basically an analogue device rather than a digital one (Gregory 1981).[24] The analogue-digital distinction is not without problems, to which I'll return.

Suppose that visual perception and verbal expression are the prescientific or "default" modes of cognition and expression.[25] Let us express this supposition in the following way: humans are naturally analogue devices, capable of interpreting experience without recourse to sorting and counting. Because of our technological prowess we are living in an increasingly digital (digitalized) environment. The sciences, particularly the physical and mathematical sciences, have been the main source of this change. In such a digitalized environment it makes sense to ask: Are we analogue all the way down, or digital? This cannot be answered by historical methods, for example, by looking at the emergence and persistence or disappearance of different modes of representation or different uses of symbolic representations. Nevertheless, looking at how science has changed and yet remained the same does have something to offer. Studies of the development of scientific fields show a dialectical play of qualitative and quantitative modes of description and argumentation, as illustrated in figures 13.1 and 13.2. The examples from geophysics and particle physics show that the reductions that translate qualitative experience into to quantitative and digital representations, however contested they may be at particular times, indicate neither the cognitive primacy nor the epistemic superiority of the numerical-digital mode. As Galison (1999, 394) has remarked, "the status of visual and anti-visual reasoning has continually been in flux across the whole of the last 150 years. Neither side has 'won,' whether in the theoretical reaches of mathematics or physics or in the domain of experiment or simulation."

Yet the aspiration to redescribe all cognitive processes in terms of one type of representation remains. The driving hypothesis of traditional artificial intelligence, construed as an experimental research program, is that all important and interesting aspects of human thought are reducible to symbol manipulation. In its strong form, the physical symbol system hypothesis presupposes that our analogue capabilities are ephemeral and that we are symbol manipulators (but not necessarily digital processors) all the way down. Advocates of "strong" artificial intelligence have argued that a hypothesis of this type could be tested by the attempt to reproduce all human capabilities in artificial devices. However, to test this hypothesis properly we would have to reproduce far more than the semblance of discrete, context-specific capabilities such as pattern recognition, induction, or abductive inference. We would have to reproduce human capabilities such as those cognitive and social abilities that humans bring together to make science.

From the standpoint of some future historian of cyborg techno-science, the analogue-digital distinction may turn out to have the same epistemic status and function as the primary-secondary qualities distinction had for Galilean and Newtonian science. But from the context of modern science, it does not appear as deep as other distinctions we can draw between broader categories of representation, such as verbal-textual, visual-graphical, and symbolic. Number systems are just one type of symbolic system, and digital or binary notation is a further subset of this. As applied to representations rather than computers, the analogue-digital distinction is not even well drawn.[26] The digital is well defined in engineering and mathematics, but it is very difficult to capture the essence of an analogue representation, except by reference to examples or definitions that appeal to notions of sameness that beg the criteria of similarity.[27] An analogue representation is supposed to bear some resemblance of form or structure to what it denotes. The resemblance depends on an analogy between two different types of experience. For example, in a notational system for time, analogue symbols for a quantity of time elapsed would be a portion of a clock face or a portion of a line (though a clock face bearing no marks at all, apart from an hour hand, does not make use of a symbol system).[28] Here the "analogue" or underlying similarity is the individuation of objects (tally sticks), the perception of space, the experience of time passing. These are rudimentary, experiential—the stuff of human awareness unmediated by scientific instruments.[29]

By contrast, digital representations bear no "likeness" to what they represent. The character "5" bears no more of a resemblance or "analogy" to five objects than a digital encoding of this symbol does. Here, internal resemblance is everything: preservation of the identity of such symbols (and of what they denote) depends on a system of rules and procedures. When implemented in machines these maintain the required degree of similarity through successive manipulations. As the following example illustrates, the achievement of a sufficient degree of similarity requires technologies whose emergence is closely associated with major scientific developments (clocks and other measuring devices, machine-based repeatability, mechanized calculation, programmable computers, the Internet).

Galileo and Newton showed that the apparently irregular motions we find in the terrestrial world can be described in the same terms as the orderly, uniform motions of the celestial world. As the clockmakers realized, if one cannot find the same precision on earth that is found in the heavens, then it can be engineered. Precision and repeatability do not have to be found in nature; they can be manufactured. Variability can then be defined in terms of uniformities measured out in seconds, centimeters, and degrees. As the units become smaller, so the degree of uniformity can be more precisely defined. When even greater uniformity is required, technologies are sought that can produce or measure even smaller units. The failure of Charles Babbage's attempts to mechanize calculation in the 1830s and 1840s illustrates the need for greater degrees of precision. Babbage observed that there was too much variability in man-made methods and devices for calculating the logarithms used by navigators. Babbage's project failed, not because his design was flawed but because he could not produce parts engineered with sufficient precision (Swade 2000). Engineers would say that the tolerances were too great—the point is, parts could not be made "exactly" identical.[30] Twentieth-century atomic science provides a further illustration: cesium clocks show that the celestial, "solar" clock is not uniform. This required a "small" adjustment to the length of the solar year at the end of 1998.

4.2. Technology and the Nature of Thought

If the numerical-digital mode of cognition were primary and if science were the realization of this fundamental cognitive capacity, then we would expect science to represent the world in ways that are alien to nonscientific experience. In particular, we would expect digitalization to involve only reduction and computation, without expansions of the sort that I have identified. We can now restate the problem of alienation as deriving from the allegedly opposed character of scientific knowledge and ordinary human experience. Let us say that we are analogue devices. Numerical and symbolic representations are difficult abstractions even to some educated people; digital representations appear alien to most of us. How and why have scientists created an increasingly digitalized world?

To answer this let's return to the example of time. We experience time as continuous—in an *analogue* mode—except when focusing on a measure, such as the ticking of a clock. In science, however, time is treated both as a mathematically continuous variable and as something that can be measured out in discrete quantities. It has been quantized and, more recently, digitalized. Newton achieved the generality of mathematical description that Galileo had foreseen, according to which *everything* in the universe obeys a set of mathematical expressions. These expressions describe the world, but only insofar as it has *already* been reduced—that is, redescribed in the same terms as a machine. That redescription is achieved by finding ways of making ever more repeatable, reliable, and accurate experiments. In a Newtonian experiment, the world is made to repeat itself with the same precision as the currently available machine technology. But such experiments are designed to produce quantities (like position, elapsed time, and temperature) rather than experiences. Experimental technologies are central to the process of abstracting from and quantizing nature into units of ever-decreasing size.

Where Babbage applied mechanically ordered repeatability to mathematical calculations, Turing saw that you could take this far beyond arithmetic. He argued that in principle any process that could be represented by symbols and rules of symbol manipulation could be made computable. Thus, Turing enlarged the whole conception of what could be done with precisely specified, repeatable procedures, extending this to human thinking processes. This insight has had the same transforming effect as Newton's earlier insight that earthly and heavenly motions obey the same laws. *Human* thought and *machine* processes appear to us to be very different. Turing showed that these can be treated in the same way. Unifying insights such as those of Newton and Turing depend on whole traditions of practical problem solving, experimental and mathematical techniques, and their associated linguistic practices (see Crosby 1997).

The physical realization of powerful computing procedures during and after World War II and the advent of cheap computing in recent decades has transformed the sciences (see Keller, this volume, chap. 10). Yet the total translation of qualitative into quantitative that Galileo anticipated, Millikan advocated, and Planck celebrated has not occurred. In order to show why, we need a better account of the transition from experiential modes to symbolic and technological modes in the representation and dissemination of scientific results. As outlined here, such an account will show that although the sense-extending capabilities of scientific instruments appear to subvert the sensory modalities that humans favor, displacing them by numerical, digitally encodable information, this reductive activity is neither the whole of the process nor its end. Sense-based experience is woven back into results in order to facilitate thinking, argumentation, and dissemination.

5. Digitalization as Reduction, Computation, and Expansion

Humans are much less competent than their machines at logical and mathematical operations—the sorts of tasks that can be reduced to algorithms—than they are at perceiving pattern and structure. The methods and technologies of science are designed to overcome our incompetence as reasoners and our unreliability as observers. What is the significance of this observation for a philosophical theory about the knowledge produced by experimentation? Increased dependence on instruments to access primary, abstracted features of the physical world marks an important change. Although this change clearly displaces ordinary human modes of perception and cognition, it does not eliminate them. We can characterize this schematically as follows. Aspects of the world are selectively redescribed to make them amenable to manipulation according to rules that in many cases are now implemented in machines. The effect is that certain human modes of cognition and certain skills apparently cease to be relevant. They are replaced, to a greater or lesser extent, by the very limited modes of cognition of a machine. An episode in the history of applied statistical computation provides a way of conceptualizing this.

In the 1950s Paul Meehl and others applied statistical methods to problems of identifying certain psychiatric disorders such as schizophrenia and predicting which people on probation were most likely to reoffend (Meehl 1954). Meehl used statistical regression methods for a variety of clinical judgments. Later programs such as INTERNIST also used clustering methods, pattern matching, and decision and production rules. These applications performed as well as and sometimes better than experienced humans. The simulation of informed, expert judgments was embarrassing both to clinicians and cognitive scientists (who generally rejected statistical methods as a basis for models of inference). Nevertheless, it was widely supposed that the superiority of diagnostic machines consists in their speed, ability to handle large amounts of data, and their accuracy. Once again it appeared that machine-based science could improve upon human performance.

5.1. Narrowing the Cognitive Span

A clinician, Marsden Blois, countered this supposition as a misunderstanding. Blois drew attention to a very different aspect of the diagnostic process—the range of skills, competences and types of knowledge needed to complete a diagnosis. He called the totality of this set of competences the "cognitive span." At its widest, our cognitive capacities must confront the world as experienced from moment to moment. At its narrowest, only a few highly specialized, skilled, and context-specific capacities are needed. A typical process of diagnosis would begin with a wide "span" or range invoked when a patient first enters the consulting room: there is a preliminary conversation, reading of body language, taking a history, conducting a physical examination, and so on. Each of these involves many cognitive and social skills—choosing the next question during the

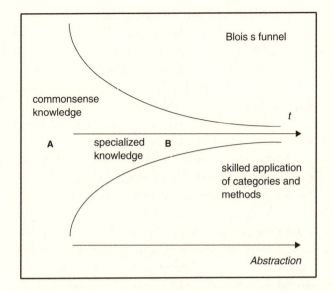

FIG. 13.2.
Reduction as a
narrowing of the
cognitive span

interview, criteria invoked in searching for particular symptoms, the order in which tests should be done, interpretation of their results, a choice of alternative diagnoses, and so on. Together these make up the human ability of making "clinical judgements" about possible causes of the symptoms identified (Blois 1980, 192). Later in the process, results of laboratory tests or X rays reduce the number of possible diagnoses to a small set of most probable conditions, for which a set of interventions may be specified.

Blois's central point is that "at the start, the physician's breadth of comprehension must extend to the totality of the everyday world." As the clinician approaches the point of delivering a diagnosis, the range of competences is far narrower. Blois visualizes this process as a funnel (see fig. 13.2).

The horizontal axis represents time, while the vertical axis indicates "cognitive span"—indicating the range and number of skills and context-sensitive knowledge; so, narrowing of the funnel shows the actual reduction in this range, in the quantity of factual knowledge needed, the number of judgments to be made, and so on. In this figure, narrowness indicates both fewer cognitive, social, and other types of competence, greater specificity of the inferences made, and specificity of the task being performed. Nevertheless, as the progression from left to right through time indicates, specialized, often formalized inferences at point B depend directly upon the larger base of ordinary human judgments required at A—the area of abstraction, formalization, or digitization. This visualizes the cognitive effect of processes of *reduction*.[31] The effectiveness of diagnostic computation at the narrowest point of the "span" is due to the fact that the "programs operate in small, well-structured task domains that were initially organized (formalized) by a human judge" at point A. In the design of expert systems, knowledge elicitation is a form of reduction.

The analogy to experimentation is that the ability to replicate an experiment presupposes that its methods have been articulated sufficiently to be expressed as a set of rule-governed procedures (as at point B). But this would not be possible without the wide range of cognitive capacities needed (at point A) for the exploratory, learning phase of experimentation that typically precedes the making of a discovery and the development of skilled practices needed to effect and disseminate an experimental demonstration. The introduction of instruments into a science involves identifying quantities, operationalizing them by specifying ways of measuring them, integrating procedures, scales, and standards for calibration, recording, and autonomous instrumentation. At point A, the widest span of the funnel, scientists confront the chaotic world on the same terms (phenomenologically speaking) as other humans. There is an enormous distance and a great deal of work between the widest span and the narrowest point at which scientific instruments produce the quantified world in which measurement and calculation are possible.

5.2. Expansion

There is a tension here, but the model shows that it is not a paradox. Observing instruments like telescopes and microscopes extend by several orders of magnitude the range of what could be "seen."[32] However, most instruments do not preserve the "analogue" quality of images in this way (Galison 1997 is an excellent source of examples). At the narrow point of the funnel the information-bearing symbols they produce have been stripped of any sensory or other meaning. Blois argued, as others such as Dreyfus and Collins have since, that much of the effectiveness of the diagnostic system is due to the human interpretation and information processing that takes place in order that the model(s) implemented as the statistical or other programs can operate. But Blois's analysis is incomplete. Having reduced some aspect of the world to a form that can be processed according to rules, the output of the computation needs to be reintroduced into the world of meaningful, human action. To be put to work theoretically, information has to be reintegrated as meaningful and relevant evidence into a system of concepts, assumptions, and hypotheses. This involves translating the output into a familiar notational system and, in some cases, restoring more basic sensory modes of apprehension, as in the case of data visualization or the phenomenology of a thought experiment. I call this *expansion*. Just as the possibility of using an instrument depends on reduction, so the interpretation and use of its

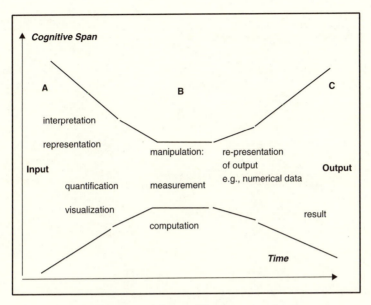

FIG. 13.3. Experimentation and the cognitive span

output requires expansion. This scheme, as applied to experimentation, is shown in figure 13.3. The skills needed for expansion (or their machine routines) in area C may of course differ from those needed for reduction. The point is that an increasing array of skills or routines is needed as we get closer to the world as interpreted and ordered by human cognitive capacities.

Quantification in science is a form of reduction. Quantification involves the introduction of procedures and technologies of detection, counting, calibration, and measuring. These do two things: they highlight features and processes that we cannot personally observe (such as temperature or the collisions of subatomic particles) and they introduce uniformity and unambiguous meaning into the description of nature. We finally come to digitalization. To digitalize is to represent features of the world, including relationships between them, in a manner that establishes and fixes unambiguous meaning.[33] Number systems have always done this, but some lend themselves better to computation than others.[34]

Digitalization involves reduction (to enable computation) and the expansion or translation of computed information into a meaningful context (see fig. 13.4). This scheme of the way that intellectual and material technologies work in the sciences improves upon the "abstractive" process visualized in figure 13.1 because it readily expresses the never-ending, cyclical character of the process. Quantification depends on technologies that produce and regulate uniformity. This is achieved by abstraction (from qualities to quantities). The abstractive process

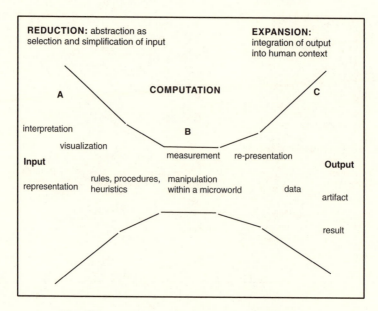

FIG. 13.4. Digitalization as reduction, computation, and expansion

involves *reduction*—the narrowing both of cognition and of ontology—moving from a teeming, complex world to one containing just those objects and quantities which can be regarded as uniform, identical, and stable—subjected to repeatable procedures. These procedures need not be algorithms; for example, they may be informal, intuitive rules for the manipulation of shapes or images of objects.[35]

6. Conclusion

I have explored the analogue-digital distinction, supposing that humans are naturally analogue devices, capable of interpreting experience without recourse to sorting and counting. Quantification is a process of extending human counting capabilities through the creation of concepts, categories, and the procedures that enable classification, manipulation, and data analysis. Digitalization is an extension of this process. It is inherently practical and technological, depending on machines as well as abstract concepts such as number and rigor. I have characterized digitalization as a process of reduction that makes information into a computationally tractable form and the expansion of output into cognitively appropriate forms.[36] I have drawn an analogy between digitalization and the processes of experimentation that involve (1) displacing human (analogue) in-terpretations and error-prone procedures by numerical description and exactly repeatable (digital) procedures, and (2) a shift from relatively unmediated in-teraction with a world of natural, elemental forces (via the development, for ex-ample, of particle- and bubble-chamber science as a means of re-creating nature in laboratory experiments), to a world in which "natural" phenomena are com-pletely replaced by representations of a very particular kind. The analogy be-tween experimentation and the reduction of a domain to a computably tractable form is shown in figures 13.3 and 13.4. It suggests that achieving uniformity of measurement in different circumstances and for different observers involves the same sort of process as that which produces the apparent, digitalized expert-ise of artificial experimental systems. The history of empirical, experimental science can be described as a process of reduction, which often leads to quanti-fication and this, in turn, to digitalization. Nevertheless, I have argued, this his-torical development does not support the opinion of Kelvin, Millikan, and Rutherford that "qualitative is just poor quantitative": there do appear to be modes of understanding that are inherently analogue rather than numerical or digital. These are inherent in the practices even of the most technologically so-phisticated experimentation.

If humans are analogue devices, and it is humans that continue to do science, then we would expect the cyborg vision of the onward march of digitalization to need qualification. Instead of looking for cognitive capacities of the sort re-quired by an algorithmic view of science as rule-based reasoning about an in-herently digitizable world, we should investigate those cognitive capacities that enable practitioners from different cultures to exchange meanings and methods.

NOTES

I wish to acknowledge the very helpful commentary on this chapter by Maarten Franssen at the Amsterdam conference "Towards a More Developed Philosophy of Scientific Experi-mentation" in June 2000; the guidance of the editor and an anonymous referee; comments on versions presented to the British Society for the History of Mathematics Conference at Birk-beck College, London, in January 2000; and the History and Philosophy of Science collo-quium at Cambridge in June 2001.

1. See Feyerabend (1975). His example was Galileo; for an application of this concept to Faraday's novel characterizations of electromagnetic phenomena, see Gooding (1982).

2. Examples from twentieth-century science include the contrasting approaches to phe-nomenology in biology and physics (Knorr-Cetina 1996) and the changing status of visual-ized data in high-energy physics (Galison 1998).

3. Historically, this has always been the case. In large experimental systems where "re-sults" are processed by other machines such translation would not be necessary.

4. On techno-science, see Latour (1987) and Keller (this volume, chap. 10).

5. Technologies also shaped the very character of vision in painting; see, for example, David Hockney's recent studies of the use of mirrors in Flemish art (Hockney 2001).

6. Latour proclaimed that "a sociology of science is crippled from the start if it believes in the results of one science, namely sociology, to explain the others" (Latour 1983, 144); by 1987 the exclusion of psychology had been elevated to the status of methodological precept (Latour 1987, 228, 258). This is because cognitive factors are not observable in the way that social interactions are (Latour 1990, 21–22).

7. Galison challenges the tacit knowledge gambit that sociologists use to support a view of science as an array of distinct, incommensurable cultures. He treats inarticulateness about rules and judgments as an invitation to attend to those pieces of practice that do not yet fit the public discourse of a field and to ask, "Why not?" (Galison 1997). As he points out (Galison 1999, 393), what appears as an elliptical expression of technique in one situation may well be articulated in another.

8. Note that this is achieved with experiments that are also abstractions, in that they eliminate most of the complexity in the world in order to achieve a degree of control.

9. In his exploration of the scientific data revolution Frederick Suppe notes that "the data that count scientifically are digitized data" (unpublished typescript 1993, lecture 3, 1).

10. Robert May, UK Government Chief Scientific Adviser (quoted in Kelly 1995, 6).

11. For the contrasting phenomenologies of modern biology and physics, see Knorr-Cetina (1996).

12. Miller (1986, 1991) argues that during the first half of the last century abstraction to representations drew not on sense experience (visualization as *anschauung*) but on mathematical formalism (the visualizability of mathematical forms, *anschaulichkeit*).

13. The controversy turns in part on disagreement about what an orbital can be a visualization of (see Humphreys 1999).

14. Galileo asserted that "philosophy is written in this grand book, the universe, which stands continually open to our gaze . . . [which] . . . cannot be understood unless one first learns to comprehend the language and read the letters in which it is composed." This language is "the language of mathematics, and its characters are triangles, circles, and other geometric figures, without which it is *humanly* impossible to understand a single word of it" (quoted in Drake 1957, 237–238).

15. For an opposite, evolutionary biological explanation of the congruence of human truth-seeking capacities and those of science, see Gopnik and Meltzoff (1997).

16. See "On Thought Experiments" (Mach 1976, 134–147) and the discussion in Sorensen (1992).

17. The failures of traditional artificial intelligence suggest that complete assimilation is not possible. See Dreyfus (1992), Collins (1992), Radder (1999).

18. These issues are discussed in DeMey (1992) and Smith (1998).

19. Note, however, that in terms of notation, the black and white maps are a reduced, binary representation. Grayscale images of this data contain more information but do not display these regularities. Similarly, facial features disappear from grayscale images of the face of Cydonia.

20. An adherent of this view remarked: "Anything can happen once"—in other words, to establish whether an event is typical, commonplace, and therefore statistically significant, many instances must be observed for comparison with many others (Galison 1997, 434).

21. In this case, reduction eliminates the need for variable and therefore unreliable human judgment. However, it is not always the case that observational technologies eliminate judgment in this way. See Jones and Galison (1998).

22. As for the displacement of hands-on, bench-top manipulation by experience of virtual or simulated worlds, and of experimental data by Monte Carlo simulations of data (see Keller, this volume, chap. 10), it is more likely that our concept of "how the world is" will change. This happened in the sixteenth and seventeenth centuries with thought experiments, when the convincingness of the mental enactment proved that reality is ultimately mathematical.

23. BBC Radio 4, "The Money Programme," broadcast January 15, 2000. According to a familiar distinction in the history of art, works can be classified as representational or non-representational. Since digital is notational but nonrepresentational, that makes it truly post-modern.

24. For the debate about the status of mental imagery in thought processes, see Block (1981) and Miller (1991).

25. To call them *natural* modes would imply that the emergence of counting techniques, measuring technologies, and the digital information culture has somehow made us unnatural. As remarked earlier, we can acknowledge the cyborg status of human beings as a consequence of human activity without accepting either the inevitability or the goodness of this particular vision of the future.

26. In the 1940s computing machines were labeled "analogy machines," but the term "analogue" became part of common parlance due to the introduction of numerical displays on consumer products such as clocks and watches. There was then a need to distinguish these from traditional watches that display the passage of time by the movement of the hands. This usage was, I believe, introduced by Seiko in the 1980s.

27. For a systematic treatment of the distinction, see Goodman (1968, 157–164).

28. Many devices combine analogue and digital modes of representation (e.g., as an analogue watch with numerals denoting hours and a graduated scale for minutes). Some familiar methods of reckoning are neither analogue nor digital.

29. As Lakoff and Johnson (1980) have demonstrated, metaphors often have a nonverbal, experiential basis and are meaningful without translation into a notational system. For an opposing view, see Pylyshyn (1981).

30. This is not an oxymoron. The "exactness" of identity is constantly being redefined. A notion of "identity" is useless unless it can be realized and regulated through a material technology having a greater degree of discrimination. These capabilities change. An analogous point can be made about the foundational concept of randomness as used in evolutionary explanations. Randomness is definable only with respect to the best digital pattern recognition system we have so far devised.

31. Although this process could be represented diagrammatically—e.g., by the pruning of a decision tree—this would not indicate as clearly the reducing demand for contextual or background knowledge.

32. On extending the scope of vision beyond the empirically observable, see Lowe and Schaffer (2000).

33. It is a method designed to achieve two things: to preserve the invariance of tokens in a symbol manipulation system and to make the value of the tokens unambiguous.

34. The superiority of the binary system for computation is largely a matter of computational efficiency, see Wiener (1965, 116ff).

35. For examples from geology and physics see Gooding (1998) and from paleontology Gould (1989, chap. 3). I suspect that the visualization of mathematical formalisms that Miller calls "visualizability" is another form of expansion.

36. This digitalization process is what most nonscientists have in mind when asked what distinguishes science from other human activities.

REFERENCES

Ackermann, R. (1985) *Data, Instruments and Theory*. Princeton: Princeton University Press.

Ankeny, R. A. (1999) "Model Organisms as Case-Based Reasoning: Worms in Contemporary Biomedical Research." Paper for Princeton Workshop on Model Systems, Cases and Exemplary Narratives, October 1999.

————. (2000) "Fashioning Descriptive Models in Biology: Of Worms and Wiring Diagrams." *Philosophy of Science* 67 (supplement to no. 3): S260–S272.

————. (2001) "Model Organisms as Cases: Understanding the 'Lingua Franca' at the Heart of the Human Genome Project." *Philosophy of Science* 68 (supplement to no. 3): S251–S261.

Aristotle (1970) "Physica." In *The Works of Aristotle*, edited by D. Ross. Vol. 2. Oxford: Clarendon Press.

Babbage, C. (1864) *Passages from the Life of a Philosopher.* London: Longmans and Green.

Bacon, F. (1859 [1620]) *Novum Organum*. Translated by A. Johnson. London: Bell & Daldy Fleet Street.

————. (1960 [1620]) *The New Organon*. Edited with an introduction by F. H. Anderson. Englewood Cliffs, N.J.: Prentice Hall.

————. (2000 [1620]) *The New Organon*. Edited by L. Jardine and M. Silverthorne. Cambridge: Cambridge University Press.

Baigrie, B. (ed.) (1996) *Picturing Knowledge: Historical and Philosophical Problems Concerning the Use of Art in Science*. Toronto: University of Toronto Press.

Baird, D. (1987) "Factor Analysis, Instruments, and the Logic of Discovery." *British Journal for the Philosophy of Science* 38:319–337.

————. (1988) "Five Theses on Instrumental Realism." In *PSA 1988*, edited by A. Fine and J. Leplin, 1:165–173. East Lansing: Philosophy of Science Association.

————. (1989) "Instruments on the Cusp of Science and Technology: The Indicator Diagram." *Knowledge and Society: Studies in the Sociology of Science, Past and Present* 8:107–122.

————. (1995) "Meaning in a Material Medium." In *PSA 1994*, edited by D. Hull, M. Forbes, and R. Burian, 2:441–451. East Lansing: Philosophy of Science Association.

————. (2000a) "Encapsulating Knowledge: The Direct Reading Spectrometer." *Foundations of Chemistry* 2:5–46.

————. (2000b) "The Thing-y-ness of Things: Materiality and Spectrochemical Instrumentation, 1937–1955." In *The Empirical Turn in the Philosophy of Technology*, edited by P. Kroes and A. Meijers, 99–117. Amsterdam: JAI/Elsevier.

————. (Forthcoming) *Thing Knowledge: A Philosophy of Scientific Instruments*. Berkeley: University of California Press.

Baird, D., and T. Faust (1990) "Scientific Instruments, Scientific Progress and the Cyclotron." *British Journal for the Philosophy of Science* 41:147–176.

Baird, D., and A. Nordmann (1994) "Facts-Well-Put." *British Journal for the Philosophy of Science* 45:37–77.

Barrow, J. (1992) *Pi in the Sky. Counting, Thinking and Being.* Oxford: Oxford University Press.

Bazerman, C. (1988) *Shaping Written Knowledge: The Genre and Activity of the Experimental Article in Science.* Madison: University of Wisconsin Press.

Beaulieu, A. (2000) "The Space Inside the Skull: Digital Representations, Brain Mapping and Cognitive Neuroscience in the Decade of the Brain." Ph.D. diss., University of Amsterdam.

———. (2001) "Voxels in the Brain." *Social Studies of Science* 31:635–680.

Bechtel, W., and R. Richardson (1993) *Discovering Complexity: Decomposition and Localization as Strategies in Scientific Research.* Princeton: Princeton University Press.

Beck, J. D., B. L. Canfield, S. M. Haddock, T. J. H. Chen, M. Kothari, and T. M. Keaveny (1997) "Three-Dimensional Imaging of Trabecular Bone Using the Computer Numerically Controlled Milling Technique." *Bone* 21:281–287.

Bhaskar, R. (1978) *A Realist Theory of Science.* Hassocks, U.K.: Harvester Press.

Biagioli, M. (ed.) (1998) *The Science Studies Reader.* New York: Routledge.

Bigelow, J., and R. Pargetter (1987) "Functions." *Journal of Philosophy* 84:181–196.

Birdsall, C. K. (1966) "Computer Experiments with Charged Particles, Charged Fluids and Plasmas: A Classification and Bibliography." Institute for Plasma Physics, Japan, no. 5, December (NTIS PB174351).

Block, N. (ed.) (1981) *Imagery.* Cambridge, Mass.: MIT Press.

Blois, M. (1980) "Clinical Judgement and Computers." *New England Journal of Medicine* 303: 192–197.

Bogen, J., and J. Woodward (1988) "Saving the Phenomena." *Philosophical Review* 97:303–352.

———. (1992) "Observations, Theories and the Evolution of the Human Spirit." *Philosophy of Science* 59:590–611.

Böhme, G., W. van den Daele, and W. Krohn (1983) "The Scientification of Technology." In *Finalization in Science,* edited by W. Schäfer, 173–205. Dordrecht, The Netherlands: Reidel.

Bohr, N. (1958) *Atomic Physics and Human Knowledge.* New York: Wiley.

———. (1963) *Essays 1958–1962 on Atomic Physics and Human Knowledge.* New York: Wiley.

Boumans, M. (1999) "Representation and Stability in Testing and Measuring Rational Expectations." *Journal of Economic Methodology* 6:381–401.

Boumans, M., and M. S. Morgan (2001) "*Ceteris Paribus* Conditions: Materiality and the Application of Economic Theories." *Journal of Economic Methodology* 8:11–26.

Brandon, R. (1985) "Grene on Mechanism and Reductionism: More Than Just a Side Issue." In *PSA 1984,* edited by P. Asquith and P. Kitcher, 2:315–353. East Lansing: Philosophy of Science Association.

Brock, W. A., and C. Hommes (1997) "Models of Complexity in Economics and Finance." In *System Dynamics in Economic and Financial Models,* edited by C. Heij, J. M. Schumacher, B. Hanson, and C. Praagman, 3–44. New York: Wiley.

Brown, H. R., and R. Harré (1988) *Philosophical Foundations of Quantum Field Theory.* Oxford: Oxford University Press.

Brown, J. R. (1991) *The Laboratory of the Mind.* New York: Routledge.

Bucciarelli, L. L. (1994) *Designing Engineers.* Cambridge, Mass.: MIT Press.

Buchwald, J. Z. (1993) "Design for Experimenting." In *World Changes: Thomas Kuhn and the Nature of Science,* edited by P. Horwich, 169–206. Cambridge, Mass.: MIT Press.

———. (1994) *The Creation of Scientific Effects.* Chicago: University of Chicago Press.

———. (1998) "Issues for the History of Experimentation." In Heidelberger and Steinle (1998), 374–391.

——— (ed.) (1995) *Scientific Practice. Theories and Stories of Doing Physics.* Chicago: University of Chicago Press.

Buneman, O., and D. A. Dunn (1965) *Computer Experiments in Plasma Physics*, Report No. SU-0254-2 (NASA January 1965).

Bunge, M. (1966) "Technology as Applied Science." *Technology and Culture* 7:329–347.

Burian, R. (1996) "Underappreciated Pathways Toward Molecular Genetics as Illustrated by Jean Brachet's Cytochemical Embryology." In *The Philosophy and History of Molecular Biology: New Perspectives,* edited by S. Sarkar, 671–685. Dordrecht, The Netherlands: Kluwer.

Cantor, G. (1989) "The Rhetoric of Experiment." In Gooding, Pinch, and Schaffer (1989), 159–180.

Cardwell, D. S. L. (1971) *From Watt to Clausius: The Rise of Thermodynamics in the Early Industrial Age.* London: Heinemann.

Carrier, M. (1998) "New Experimentalism and the Changing Significance of Experiments: On the Shortcomings of an Equipment-Centered Guide to History." In Heidelberger and Steinle (1998), 175–191.

Cartwright, N. (1983) *How the Laws of Physics Lie.* Oxford: Clarendon Press.

———. (1989) *Nature's Capacities and Their Measurement.* Oxford: Clarendon Press.

———. (1992) "Aristotelian Natures and the Modern Experimental Method." In *Inference, Explanation, and Other Frustrations,* edited by J. Earman, 44–71. Berkeley: University of California Press.

———. (1999) *The Dappled World: A Study of the Boundaries of Science.* Cambridge: Cambridge University Press.

———. (2000) "Laboratory Mice, Laboratory Electrons, and Fictional Laboratories." Paper for Princeton Workshop on Model Systems, Cases and Exemplary Narratives, January 2000.

Cartwright, N., and M. Jones (1991) "How to Hunt Quantum Causes." *Erkenntnis* 35: 205–231.

Chang, H. (2001) "Spirit, Air and Quicksilver: The Search for the 'Real' Scale of Temperature." *Historical Studies in the Physical and Biological Sciences* 31:249–284.

Clark, A. (2001) "Natural Born Cyborgs?" In *Cognitive Technology: Instruments of Mind,* edited by M. Beynon, C. L. Nehaniv, and K. Dautenhahn, 17–23. Berlin: Springer.

Coates, J. (1997) "Vibrational Spectroscopy: Instrumentation for Infrared and Raman Spectroscopy." In *Analytical Instrumentation Handbook,* 2d ed., edited by G. W. Ewing, 393–556. New York: Marcel Dekker.

Cohen, M., and D. Baird (1999) "Why Trade?" *Perspectives on Science* 7: 231–254.

Coleman, J., and T. Hoffer (1987) *Private and Public High Schools: The Impact of Communities*. New York: Basic Books.

Collins, H. M. (1985) *Changing Order: Replication and Induction in Scientific Practice*. London: Sage.

———. (1992) *Artificial Experts*. Cambridge, Mass.: MIT Press.

Collins, H., and M. Kusch (1998) *The Shape of Actions*. Cambridge, Mass.: MIT Press.

Considine, D. M. (ed.) (1983) *Van Nostrand' s Scientific Encyclopedia*. New York: Van Nostrand Reinhold.

Cook, T., and D. Campbell (1979) *Quasi-Experimentation: Design and Analysis Issues for Field Settings*. Boston: Houghton Mifflin Co.

Coquillette, D. R. (1992) *Francis Bacon*. Stanford, Calif.: Stanford University Press.

Craig, E. (ed.) (1998) *Routledge Encyclopedia of Philosophy*. London: Routledge.

Crick, F. (1988) *What Mad Pursuit: A Personal View of Scientific Discovery*. New York: Basic Books.

Crosby, A. W. (1997) *The Measure of Reality*. Cambridge: Cambridge University Press.

Cummins, R. (1975) "Functional Analysis." *Journal of Philosophy* 72:741–765.

Dahan Dalmedico, A. (2000) "Between Models as Structures and Models as Fictions: Computer Modeling Practices in Post–World War II." Paper for Princeton Workshop on Model Systems, Cases and Exemplary Narratives, February 2000.

Darrigol, O. (1999) "Baconian Bees in the Electromagnetic Fields: Experimenter-Theorists in Nineteenth-Century Electrodynamics." *Studies in History and Philosophy of Modern Physics* 30:307–345.

Davenport, W. R. (1929) *Biography of Thomas Davenport: The "Brandon Blacksmith." Inventor of the Electric Motor*. Montpellier: Vermont Historical Society.

Dehue, T. (1997) "Deception, Efficiency, and Random Groups: Psychology and the Gradual Origination of the Random Group Design." *Isis* 88:653–673.

DeMey, M. (1992) *The Cognitive Paradigm*. 2d ed. Chicago: University of Chicago Press.

Dingler, H. (1928) *Das Experiment: Sein Wesen und seine Geschichte*. München: Verlag Ernst Reinhardt.

Dipert, R. R. (1993) *Artifacts, Art Works and Agency*. Philadelphia: Temple University Press.

Dowling, D. (1999) "Experimenting on Theories." *Science in Context* 12:261–273.

Drake, S. (trans.) (1957) *Discoveries and Opinions of Galileo*. New York: Doubleday.

Dreyfus, H. L. (1992) *What Computers Still Can't Do*. Cambridge, Mass.: MIT Press.

Duhem, P. (1974 [1906]) *The Aim and Structure of Physical Theory*. New York: Atheneum.

Dunbar, R. (1995) *The Trouble with Science*. London: Faber.

Earle, J. H. (1994) *Engineering Design Graphics*. Reading, Mass.: Addison-Wesley.

Faraday, M. (1821a) "Historical Sketch of Electro-Magnetism." *Annals of Philosophy* 18:195–200, 274–290.

———. (1821b) "On Some New Electromagnetical Motions, and on the Theory of Magnetism." *Quarterly Journal of Science* 12:74–96.

————. (1822a) "Description of an Electro-Magnetical Apparatus for the Exhibition of Rotatory Motion." *Quarterly Journal of Science* 12:283–285.

————. (1822b) "Electro-Magnetic Rotations Apparatus." *Quarterly Journal of Science* 12:186.

————. (1822c) "Historical Sketch of Electro-Magnetism." *Annals of Philosophy* 19:107–121.

————. (1971) *The Selected Correspondence of Michael Faraday*. Vol. 1, *1812–1848*. Edited by L. P. Williams. Cambridge: Cambridge University Press.

Farmer, D., T. Toffoli, and S. Wolfram (1984) *Cellular Automata: Proceedings of an Interdisciplinary Workshop, Los Alamos, March 7–11, 1983*. Amsterdam: North-Holland Physics Publishers.

Feenberg, A. (1995) *Alternative Modernity*. Berkeley: University of California Press.

Fehér, M. (1993) "The Natural and the Artificial: An Attempt at Conceptual Clarification." *Periodica Polytechnica: Humanities and Social Sciences* (Budapest) 1 (1):67–76.

Ferguson, E. (1992) *Engineering and the Mind's Eye*. Cambridge, Mass.: MIT Press.

Fernández, E. (1993) "From Peirce to Bohr: Theorematic Reasoning and Idealization in Physics." In *Charles S. Peirce and the Philosophy of Science,* edited by E. C. Moore, 233–245. Tuscaloosa: University of Alabama Press.

Feyerabend, P. K. (1975) *Against Method*. London: New Left Books.

Feynman, R. (1967) *The Character of Physical Law*. Cambridge, Mass.: MIT Press.

Franklin, A. (1986) *The Neglect of Experiment*. Cambridge: Cambridge University Press.

————. (1989) "The Epistemology of Experiment." In Gooding, Pinch, and Schaffer (1989), 437–460.

————. (1990) *Experiment, Right or Wrong*. Cambridge: Cambridge University Press.

Freedman, D. (1997) "From Association to Causation via Regression." In *Causality in Crisis? Statistical Methods and the Search for Causal Knowledge in the Social Sciences,* edited by V. McKim and S. Turner, 113–161. Notre Dame: University of Notre Dame Press.

Galileo, G. (1974 [1632]) *Dialogue Concerning the Two Chief World Systems*. Translated by S. Drake. Berkeley: University of California Press.

Galison, P. (1987) *How Experiments End*. Chicago: University of Chicago Press.

————. (1996) "Computer Simulations and the Trading Zone." In *The Disunity of Science: Boundaries, Contexts, and Power,* edited by P. Galison and D. J. Stump, 118–157. Stanford: Stanford University Press.

————. (1997) *Image and Logic: A Material Culture of Microphysics*. Chicago: University of Chicago Press.

————. (1998) "Judgement against Objectivity." In Jones and Galison (1998), 327–358.

————. (1999) "Author's Response." *Metascience* 8:393–404.

Galison, P., and A. Assmus (1989) "Artificial Clouds, Real Particles." In Gooding, Pinch, and Schaffer (1989), 225–274.

Gardner, M. (1970) "Mathematical Games: The Fantastic Combinations of John Conway's New Solitaire Game 'Life.'" *Scientific American* 223 (October):220–223.

Garman, R. L. (1942) "AI-10 Trainer Simulation at I. F. Level." Report 105-1. MIT Radiation Laboratory, August 15.

Gasking, D. (1955) "Causation and Recipes." *Mind* 64:479–487.

Gee, B. (1991) "Electromagnetic Engines: Pre-technology and Development Immediately Following Faraday's Discovery of Electromagnetic Rotations." *History of Technology* 13:41–72.

Gibson, J. J. (1986) *The Ecological Approach to Visual Perception.* Hillsdale, N.J.: Lawrence Erlbaum Associates.

Giere, R. (1988) *Explaining Science: A Cognitive Approach.* Chicago: University of Chicago Press.

Glymour, C., R. Scheines, P. Spirtes, and K. Kelly (1987) *Discovering Causal Structure.* Orlando: Academic Press.

Golinski, J. (1998) *Making Natural Knowledge: Constructivism and the History of Science.* Cambridge: Cambridge University Press.

Gooding, D. (1982) "Empiricism in Practice: Teleology, Economy and Observation in Faraday's Physics." *Isis* 73:46–67.

———. (1990) *Experiment and the Making of Meaning.* Dordrecht, The Netherlands: Kluwer.

———. (1992) "Putting Agency Back into Experiment." In Pickering (1992), 65–112.

———. (1998) "Picturing Experimental Practice." In Heidelberger and Steinle (1998), 298–322.

Gooding, D., T. Pinch, and S. Schaffer (eds.) (1989) *The Uses of Experiment.* Cambridge: Cambridge University Press.

Goodman, N. (1955) *Fact, Fiction, and Forecast.* Cambridge, Mass.: Harvard University Press.

———. (1968) *Languages of Art: An Approach to a Theory of Symbols.* Indianapolis: Bobbs-Merrill.

Goodwin, B. (1994) *How the Leopard Changed Its Spots: The Evolution of Complexity.* London: Phoenix.

Gopnik, A. (1996) "The Scientist as Child." *Philosophy of Science* 63:485–514.

Gopnik, A., and A. Meltzoff (1997) *Words, Thoughts and Theories.* Cambridge, Mass.: MIT Press.

Gould, S. J. (1989) *Wonderful Life: The Burgess Shale and the Nature of History.* Harmondsworth, U.K.: Penguin.

Grand, S. (2002) *Creation: Life and How to Make It.* London: Weidenfeld and Nicolson.

Gregory, R. L. (1981) *Mind in Science.* Harmondsworth, U.K.: Penguin.

Guala, F. (1999a) "Economics and the Laboratory." Ph.D. diss., University of London.

———. (1999b) "The Problem of External Validity (or 'Parallelism') in Experimental Economics." *Social Science Information* 38:555–573.

———. (2002) "Models, Simulations, and Experiments." In *Model-Based Reasoning: Science, Technology, Values,* edited by L. Magnani and N. J. Nersessian, 59–74. New York: Kluwer/Plenum.

Guetzkow, H., P. Kotler, and R. L. Schultz (1972) *Simulation in Social and Administrative Science.* Englewood Cliffs, N.J.: Prentice-Hall.

H. (1822) "Letter to the Editor: Account of a Steam-Engine Indicator." *Quarterly Journal of Science, Literature and the Arts* 13:91–95.

Habermas, J. (1970) *Toward a Rational Society.* Boston: Beacon Press.

———. (1978) *Knowledge and Human Interests.* 2d ed. London: Heinemann.

Hacker, P. M. S. (1986) *Insight and Illusion: Themes in the Philosophy of Wittgenstein.* Rev. ed. Oxford: Clarendon Press.

Hacking, I. (1983a) *Representing and Intervening.* Cambridge: Cambridge University Press.

———. (1983b) "Was There a Probabilistic Revolution 1800–1930?" In *Probability since 1800: Interdisciplinary Studies of Scientific Development,* edited by M. Heidelberger, L. Krüger, and R. Rheinwald, 43–68. Bielefeld, Germany: B. Kleine Verlag.

———. (1984) "Experimentation and Scientific Realism." In *Scientific Realism,* edited by J. Leplin, 154–172. Berkeley: University of California Press.

———. (1989a) "Philosophers of Experiment." In *PSA 1988,* edited by A. Fine and J. Leplin, 2:147–156. East Lansing: Philosophy of Science Association.

———. (1989b) "Extragalactic Reality: The Case of Gravitational Lensing." *Philosophy of Science* 56:555–581.

———. (1992) "The Self-Vindication of the Laboratory Sciences." In Pickering (1992), 29–64.

Hackmann, W. D. (1989) "Scientific Instruments: Models of Brass and Aids to Discovery." In Gooding, Pinch, and Schaffer (1989), 31–65.

Hakfoort, C. (1986) *Optica in de eeuw van Euler.* Amsterdam: Rodopi.

———. (1995) *Optics in the Age of Euler, 1700–1795.* Cambridge: Cambridge University Press.

Hanson, N. R. (1958) *Patterns of Discovery.* Cambridge: Cambridge University Press.

Harré, R. (1970) *Principles of Scientific Thinking.* Chicago: University of Chicago Press.

———. (1986) *Varieties of Realism.* Oxford: Blackwell.

———. (1988) "Parsing the Amplitudes." In Brown and Harré (1988), 59–71.

———. (1998) "Recovering the Experiment." *Philosophy* 73:353–377.

Hartmann, D., and R. Lange (2000) "Epistemology Culturalized." *Journal for General Philosophy of Science* 31:75–107.

Hartmann, S. (1996) "The World as a Process." In Hegselmann, Mueller, and Troitzsch (1996), 77–100.

Hausman, D., and J. Woodward (1999) "Independence, Invariance and the Causal Markov Condition." *British Journal for the Philosophy of Science* 50:521–583.

Heelan, P. A. (1983) *Space-Perception and the Philosophy of Science.* Berkeley: University of California Press.

Hegselmann, R., U. Mueller, and K. G. Troitzsch (1996) *Modelling and Simulation in the Social Sciences from the Philosophy of Science Point of View.* Dordrecht, The Netherlands: Kluwer.

Heidelberger, M. (1979) "Über eine Methode der Bestimmung theoretischer Terme." In *Aspekte der physikalischen Begriffsbildung,* edited by W. Balzer and A. Kamlah, 37–48. Brunswick, Germany: Vieweg.

———. (1980) "Towards a Logical Reconstruction of Scientific Change: The Case of Ohm as an Example." *Studies in History and Philosophy of Science* 11:103–121.

———. (1983) "Zur logischen Rekonstruktion wissenschaftlichen Wandels am Beispiel der 'Ohm'schen Revolution.'" In *Zur Logik empirischer Theorien,* edited by W. Balzer and M. Heidelberger, 281–303. Berlin: Walter de Gruyter.

————. (1998) "Die Erweiterung der Wirklichkeit im Experiment." In Heidelberger and Steinle (1998), 71–92.

Heidelberger, M., and F. Steinle (eds.) (1998) *Experimental Essays—Versuche zum Experiment*. Baden-Baden: Nomos Verlagsgesellschaft.

Hentschel, K. (1998) "Feinstruktur und Dynamik von Experimentalsystemen." In Heidelberger and Steinle (1998), 325–354.

Hillis, W. D. (1984) "The Connection Machine: A Computer Architecture Based on Cellular Automata." *Physica* 10D:213–218.

Hills, R. L. (1989) *Power from Steam: A History of the Stationary Steam Engine*. Cambridge: Cambridge University Press.

Hintikka, J. (1988) "What Is the Logic of Experimental Inquiry?" *Synthese* 74:173–190.

Hockney, D. (2001) *Secret Knowledge: Rediscovering the Lost Techniques of the Old Masters*. London: Thames and Hudson.

Hon, G. (1989a) "Franck and Hertz versus Townsend: A Study of Two Types of Experimental Error." *Historical Studies in the Physical and Biological Sciences* 20:79–106.

————. (1989b) "Towards a Typology of Experimental Errors: An Epistemological View." *Studies in History and Philosophy of Science* 20:469–504.

————. (1995) "Going Wrong: To Make a Mistake, to Fall into an Error." *Review of Metaphysics* 49:3–20.

————. (1998a) "'If This Be Error': Probing Experiment with Error." In Heidelberger and Steinle (1998), 227–248.

————. (1998b) "Exploiting Errors." *Studies in History and Philosophy of Science* 29:465–479.

Hooke, R. (1961 [1665]) *Micrographia or Some Physiological Descriptions of Minute Bodies Made by Magnifying Glasses with Observations and Inquiries Thereupon*. New York: Dover Publications.

Hoover, K. (1988) *The New Classical Macroeconomics*. Oxford: Blackwell.

Horton, R. (2001) "Thalidomide Comes Back." *New York Review of Books* 48 (May 17): 12–15.

Hughes, R. I. G. (1997) "Models and Representation." *Philosophy of Science* 64 (supplement to no. 4): S325-S336.

————. (1999) "The Ising Model, Computer Simulation, and Universal Physics." In Morgan and Morrison (1999), 97–145.

Hughes, T. P. (1979) "The Electrification of America: The System Builders." *Technology and Culture* 20:124–161.

————. (1983) *Networks of Power: Electrification in Western Society, 1880–1930*. Baltimore: Johns Hopkins University Press.

————. (1998) *Rescuing Prometheus*. New York: Pantheon Books.

Humphreys, C. J. (1999) "Electrons Seen in Orbit." *Nature* 401:21–22.

Humphreys, P. (1991) "Computer Simulations." In *PSA 1990,* edited by A. Fine, M. Forbes and L. Wessels, 2:497–506. East Lansing: Philosophy of Science Association.

Ihde, D. (1990) *Technology and the Lifeworld*. Bloomington: Indiana University Press.

————. (1991) *Instrumental Realism: The Interface between Philosophy of Science and Philosophy of Technology*. Bloomington: Indiana University Press.

James, W. (1897) *The Will to Believe and Other Essays in Popular Philosophy*. New York: Longmans.

Janich, P. (1978) "Physics—Natural Science or Technology?" In *The Dynamics of Science and Technology*, edited by W. Krohn, E. T. Layton, and P. Weingart, 3–27. Dordrecht, The Netherlands: Reidel.

———. (1985) *Protophysics of Time*. Dordrecht, The Netherlands: Reidel.

———. (1996a) *Konstruktivismus und Naturerkenntnis*. Frankfurt /M.: Suhrkamp Verlag.

———. (1996b) "Kulturalistische Erkenntnistheorie statt Informationismus." In *Methodischer Kulturalismus*, edited by D. Hartmann and P. Janich, 115–156. Frankfurt/M.: Suhrkamp Verlag.

———. (1997) *Das Maß der Dinge*. Frankfurt/M.: Suhrkamp Verlag.

———. (1998) "Was macht experimentelle Resultate empiriehaltig? Die methodisch-kulturalistische Theorie des Experiments." In Heidelberger and Steinle (1998), 93–112.

Jardine, L. (1974) *Francis Bacon: Discovery and the Art of Discourse*. Cambridge: Cambridge University Press.

Joerges, B., and T. Shinn (eds.) (2001) *Instrumentation between Science, State and Industry*. Dordrecht, The Netherlands: Kluwer.

Johnson-Laird, P. (1988) *The Computer and the Mind*. London: Fontana.

Jones, C. A., and P. Galison (eds.) (1998) *Picturing Science, Producing Art*. London: Routledge.

Judson, H. F. (1979) *The Eighth Day of Creation*. New York: Simon and Schuster.

Kant, I. (1992) "Lectures on Logic." In *The Cambridge Edition of the Works of Immanuel Kant*, translated and edited by J. M. Young. Cambridge: Cambridge University Press.

Keaveny, T. M., E. F. Wachtel, X. E. Guo, and W. C. Hayes (1994) "Mechanical Behavior of Damaged Trabecular Bone." *Journal of Biomechanics* 27:1309–1318.

Keller, A. (1984) "Has Science Created Technology?" *Minerva* 22:160–182.

Keller, E. F. (2000) "Making Sense of Life: Explanation in Developmental Biology." In *Biology and Epistemology*, edited by R. Creath and J. Maienschein, 244–260. Cambridge: Cambridge University Press.

———. (2002) *Making Sense of Life: Explaining Biological Development with Models, Metaphors, and Machines*. Cambridge, Mass.: Harvard University Press.

Kelly, M. (1995) "Whitehall's Superconductor." *Times Higher Education Supplement*, September 29, 6.

King, W. J. (1963) "The Development of Electrical Technology in the Nineteenth Century." *Contributions from the Museum of History and Technology*, 29–30, 231–407. Washington D.C.: Smithsonian Institution.

Kirkwood, J. G. (1935) "Statistical Mechanics of Fluid Mixtures." *Journal of Chemistry and Physics* 3:300.

Knorr-Cetina, K. (1981) *The Manufacture of Knowledge*. Oxford: Pergamon Press.

———. (1996) "The Care of the Self and Blind Variation: The Disunity of Two Leading Sciences." In *The Disunity of Science*, edited by P. Galison and D. J. Stump, 287–309. Stanford: Stanford University Press.

Kohler, R. (1994) *Lords of the Fly*. Chicago: University of Chicago Press.

Kosso, P. (1989) *Observability and Observation in Physical Science*. Dordrecht, The Netherlands: Kluwer.

Kroes, P. A. (1991) "Teleologie, Wiskunde en Natuurkunde." In *Rationaliteit kan ook redelijk zijn; bijdragen over het probleem van de teleologie,* edited by G. Debrock, 125–128. Assen: Van Gorcum.

———. (1994) "Science, Technology and Experiments: The Natural versus the Artificial." In *PSA 1994,* edited by D. Hull, M. Forbes, and R.M. Burian, 2:431–440. East Lansing: Philosophy of Science Association.

———. (1996) "Technical and Contextual Constraints in Design: An Essay on Determinants of Technological Change." In *The Role of Design in the Shaping of Technology,* edited by J. Perrin and D. Vinck. *COST A4* 5:43–76.

———. (1998) "Technological Explanations: The Relation between Structure and Function of Technological Objects." *Techné: Journal of the Society for Philosophy and Technology* 3. Available at http://scholar.lib.vt.edu/ejournals/SPT/v3n3.

———. (2001) "Technical Functions as Dispositions: A Critical Assessment." *Techné: Journal of the Society for Philosophy and Technology* 5:1–16. Available at http://scholar.lib.vt.edu/ejournals/SPT/v5n3.

Kroes, P. A., and A. W. M. Meijers (2000) "Introduction: A Discipline in Search of Its Identity." In *The Empirical Turn in the Philosophy of Technology,* edited by P. A Kroes and A. W. M. Meijers, xvii–xxxv. Amsterdam: JAI/Elsevier.

Kuhn, T. S. (1970) *The Structure of Scientific Revolutions*. 2d ed. Chicago: University of Chicago Press.

Lakatos, I. (1970) "Falsification and the Methodology of Scientific Research Programmes." In *Criticism and the Growth of Knowledge,* edited by I. Lakatos and A. Musgrave, 91–197. Cambridge: Cambridge University Press.

Lakoff, G., and M. Johnson (1980) *Metaphors We Live By*. Chicago: University of Chicago Press.

Lange, R. (1996) "Vom Können zum Erkennen—Die Rolle des Experimentierens in den Wissenschaften." In *Methodischer Kulturalismus,* edited by D. Hartmann and P. Janich, 157–196. Frankfurt/M.: Suhrkamp Verlag.

———. (1999) *Experimentalwissenschaft Biologie*. Würzburg, Germany: Königshausen & Neumann.

Langton, C. G. (1984) "Self-Reproduction in Cellular Automata." *Physica* 10D:135–144.

———. (1986) "Studying Artificial Life with Cellular Automata." *Physica* 22D:120–149.

———. (ed.) (1989) *Santa Fe Studies in the Sciences of Complexity*. Vol. 6, *Artificial Life*. Reading, Mass.: Addison-Wesley.

Latour, B. (1983) "Give Me a Laboratory and I Will Raise the World." In *Science Observed,* edited by K. D. Knorr-Cetina and M. Mulkay, 141–170. London: Sage.

———. (1987) *Science in Action*. Milton Keynes, U.K.: Open University Press.

———. (1990) "Drawing Things Together." In *Representation in Scientific Practice,* edited by M. Lynch and S. Woolgar, 19–68. Cambridge, Mass.: MIT Press.

———. (1999) *Pandora's Hope: Essays on the Reality of Science Studies*. Cambridge, Mass.: Harvard University Press.

Latour, B., and S. Woolgar (1979) *Laboratory Life: The Social Construction of Scientific Facts*. Beverly Hills: Sage.

Layton, E. (1991) "A Historical Definition of Engineering." In *Critical Perspectives on Nonacademic Science and Engineering*. Vol. 4. Research in Philosophy and Technology Series, edited by P. Durbin, 60–79. Bethlehem, Pa.: Lehigh University Press.

Lee, K. (1999) *The Natural and the Artefactual: The Implications of Deep Science and Deep Technology for Environmental Philosophy*. Lanham, Md.: Lexington Books.

Le Grand, H. E. (1990) "Is a Picture Worth a Thousand Experiments?" In *Experimental Inquiries,* edited by H. E. Le Grand, 241–270. Dordrecht, The Netherlands: Kluwer.

Lelas, S. (1993) "Science as Technology." *British Journal for the Philosophy of Science* 44:423–442.

———. (2000) *Science and Modernity. Toward an Integral Theory of Science*. Dordrecht, The Netherlands: Kluwer.

Leplin, J. (ed.) (1984) *Scientific Realism*. Berkeley: University of California Press.

Lowe, A., and S. Schaffer (2000) *Noise: Universal Language, Pattern Recognition, Data Synaesthetics*. Cambridge, U.K.: Kettle's Yard.

Lüer, G. (1998) "Aus der Innenperspektive des Experimentierens: Etablierung und Wandlungen des psychologischen Experiments." In Heidelberger and Steinle (1998), 192–208.

Maas, H. (1999) "Of Clouds and Statistics: Inferring Causal Structures from the Data." Measurement in Physics and Economics Discussion Paper Series, no. 99-7, London and Amsterdam.

Mach, E. (1883) *The Analysis of Sensations*. Translated by T. J. McCormack. Chicago: Open Court.

———. (1896) "Kritik des Temperaturbegriffs." In E. Mach, *Die Prinzipien der Wärmelehre: Historisch-kritisch entwickelt,* 39–57. Leipzig: Barth.

———. (1976 [1905]) *Knowledge and Error: Sketches on the Psychology of Enquiry*. Edited by B. McGuinness. Dordrecht, The Netherlands: Reidel.

Machamer, P., L. Darden, and C. Craver (2000) "Thinking about Mechanisms." *Philosophy of Science* 67:1–25.

Mainzer, K. (1997) "Symmetry and Complexity." *HYLE: An International Journal for the Philosophy of Chemistry* 3:29–49.

Mann, C., T. Vickers, and W. Gulick (1974) *Basic Concepts in Electronic Instrumentation*. New York: Harper & Row.

Martin, J. (1992) *Francis Bacon, the State, and the Reform of Natural Philosophy*. Cambridge: Cambridge University Press.

Mayo, D. G. (1996) *Error and the Growth of Experimental Knowledge*. Chicago: University of Chicago Press.

McMullin, E. (1985) "Galilean Idealization." *Studies in History and Philosophy of Science* 16:247–273.

Meehl, P. E. (1954) *Clinical versus Statistical Prediction*. Minneapolis: University of Minnesota Press.

Mendoza, E. (ed.) (1960) *Reflections on the Motive Power of Fire by Sadi Carnot and Other Papers on the Second Law of Thermodynamics by E. Clapeyron and R. Clausius*. New York: Dover Publications.

Menzies, P., and H. Price (1993) "Causation as a Secondary Quality." *British Journal for the Philosophy of Science* 44:187–203.

Metropolis, N., and S. Ulam (1949) "The Monte Carlo Method." *Journal of the American Statistics Association* 44:335–341.

Mill, J. S. (1872) *A System of Logic*. London: Routledge.

Miller, A. (1986) *Imagery in Scientific Thought*. Cambridge, Mass.: MIT Press.

———. (1991) "Imagery and Meaning: The Cognitive Science Connection." *International Studies in the Philosophy of Science* 5:35–48.

Millikan, R. A. (1917) *The Electron*. Chicago: University of Chicago Press.

Mitcham, C. (1994) *Thinking through Technology*. Chicago: University of Chicago Press.

Mize, J. H., and J. G. Cox (1968) *Essentials of Simulation*. New York: Prentice-Hall.

Morgan, M. S. (2001) "Models, Stories and the Economic World." *Journal of Economic Methodology* 8:361–384.

———. (2002) "Model Experiments and Models in Experiments." In *Model-Based Reasoning: Science, Technology, Values,* edited by L. Magnani and N. J. Nersessian, 41–58. New York: Kluwer/Plenum.

Morgan, M., and M. Boumans (2002) "The Secrets Hidden by Two-Dimensionality: Modelling the Economy as a Hydraulic System." In *Displaying the Third Dimension: Models in the Sciences, Technology and Medicine,* edited by S. de Chadarevian and N. Hopwood. Stanford: Stanford University Press.

Morgan, M. S., and M. Morrison (eds.) (1999) *Models as Mediators: Perspectives on Natural and Social Science*. Cambridge: Cambridge University Press.

Morison, R. S. (1965) "Toward a Common Scale of Measurement." In *Science and Culture: A Study of Cohesive and Disjunctive Forces,* edited by G. Holton, 273–289. Boston: Beacon Press.

Morrison, M. (1990) "Theory, Intervention and Realism." *Synthese* 82:1–22.

Morrison, M., and M. S. Morgan (1999) "Models as Mediating Instruments." In Morgan and Morrison (1999), 10–37.

Müller, R. (1946) "Monthly Column: Instrumentation in Analysis." *Industrial and Engineering Chemistry, Analytical Edition* 18 (3):29A–30A.

Mumford, S. (1998) *Dispositions*. Oxford: Oxford University Press.

Newton-Smith, W. H. (ed.) (2000) *A Companion to the Philosophy of Science*. Oxford: Blackwell.

Niebur, G. L., M. J. Feldstein, J. C. Yuen, T. J. Chen, and T. M. Keaveny (2000) "High Resolution Finite Element Models with Tissue Strength Asymmetry Accurately Predict Failure of Trabecular Bone." *Journal of Biomechanics* 33:1575–1583.

Nyhof, J. (1988) "Philosophical Objections to the Kinetic Theory." *British Journal for the Philosophy of Science* 39:81–109.

Olby, R. (1974) *The Path to the Double Helix*. Seattle: University of Washington Press.

Oreskes, N. (2000) "Why Believe a Computer: Models, Measures and Meaning in the Natural World." In *The Earth Around Us: Maintaining a Liveable Planet,* edited by J. S. Schneiderman, 70–82. San Francisco: W. H. Freeman.

Oreskes, N., K. Shrader-Frechette, and K. Belitz (1994) "Verification, Validation, and Confirmation of Numerical Models in the Earth Sciences." *Science* 263 (February 4):641–646.

Parsons, M. (1997) "Atomic Absorption and Flame Emission Spectrometry." In *Analytical Instrumentation Handbook,* 2d ed., edited by G. W. Ewing, 257–326. New York: Marcel Dekker.

Pearl, J. (1995) "Causal Diagrams for Empirical Research." *Biometrika* 82:669–688.

———. (2000) *Causality*. Cambridge: Cambridge University Press.

Peirce, C. S. (1934) *The Collected Papers of Charles Sanders Peirce*. Cambridge, Mass.: Harvard University Press.

———. (1966) *Selected Writings*. Edited with an introduction and notes by P. P. Wiener. New York: Dover.

———. (1976) *The New Elements of Mathematics*. 4 Vols. The Hague: Mouton.

Pickering, A. (1981) "The Hunting of the Quark." *Isis* 72:216–236.

———. (1989) "Living in the Material World: On Realism and Experimental Practice." In Gooding, Pinch, and Schaffer (1989), 275–297.

———. (1995) *The Mangle of Practice: Time, Agency and Science*. Chicago: University of Chicago Press.

———. (ed.) (1992) *Science as Practice and Culture*. Chicago: University of Chicago Press.

Pinch, T. (1985) "Towards an Analysis of Scientific Observation: The Externality and Evidential Significance of Observational Reports in Physics." *Social Studies of Science* 15:3–36.

Pitt, J. (1999) *Thinking About Technology*. New York: Seven Bridges Press.

Planck, M. (1909) "Die Einheit des physikalischen Weltbildes." *Physikalisches Zeitschrift* 10 (2):62–75.

Pomian, K. (1998) "Vision and Cognition." In Jones and Galison (1998), 211–231.

Popper, K. R. (1965) *Conjectures and Refutations*. New York: Harper & Row.

———. (1972) *Objective Knowledge: An Evolutionary Approach*. Oxford: Oxford University Press.

Prausnitz, J. M., and B. E. Poling (1999) "Molecular Dynamics." *Encyclopedia Britannica Online* (www.britannica.com).

Preston, B. (1998) "Why Is a Wing Like a Spoon? A Pluralist Theory of Function." *Journal of Philosophy* 95:215–254.

Price, D. de Solla (1980) "Philosophical Mechanism and Mechanical Philosophy: Some Notes Toward a Philosophy of Scientific Instruments." *Annali dell' Istituto é Muséo di Storia Della Scienza di Firenze* 5:75–85.

———. (1984) "The Science/Technology Relationship, the Craft of Experimental Science, and Policy for the Improvement of High Technology Innovation." *Research Policy* 13:3–20.

Price, H. (1991) "Agency and Probabilistic Causality." *British Journal for the Philosophy of Science* 42:157–176.

Pylyshyn, Z. (1981) "The Imagery Debate: Analog Media versus Tacit Knowledge." In Block (1981), 151–206.

Radder, H. (1979) "Bohrs filosofie van de quantummechanica. Analyse en kritiek." *Kennis en Methode* 3:411–432.

———. (1986) "Experiment, Technology and the Intrinsic Connection between Knowledge and Power." *Social Studies of Science* 16:663–683.

————. (1987) "De relatie tussen natuurwetenschap en techniek." *Krisis* 7 (1):6–23.

————. (1988) *The Material Realization of Science*. Assen, The Netherlands: Van Gorcum. Originally published as H. Radder (1984) *De materiële realisering van wetenschap* (Amsterdam: VU-Uitgeverij).

————. (1995) "Experimenting in the Natural Sciences: A Philosophical Approach." In Buchwald (1995), 56–86.

————. (1996) *In and About the World: Philosophical Studies of Science and Technology*. Albany: State University of New York Press.

————. (1997) "Philosophy and History of Science: Beyond the Kuhnian Paradigm." *Studies in History and Philosophy of Science* 28:633–655.

————. (1998) "Issues for a Well-Developed Philosophy of Scientific Experimentation." In Heidelberger and Steinle (1998), 392–404.

————. (1999) "Conceptual and Connectionist Analyses of Observation: A Critical Evaluation." *Studies in History and Philosophy of Science* 30:455–477.

————. (2002) "How Concepts Both Structure the World and Abstract From It." *Review of Metaphysics* 55:581–613.

Ramsey, J. (1992) "On Refusing to Be an Epistemological Black Box: Instruments in Chemical Kinetics in the 1920s and 30s." *Studies in History and Philosophy of Science* 23:283–304.

Rheinberger, H.-J. (1997) *Toward a History of Epistemic Things. Synthesizing Proteins in the Test Tube*. Stanford: Stanford University Press.

Richards, E. (1991) *Vitamin C and Cancer: Medicine or Politics?* London: Macmillan; New York: St. Martin's Press.

Richtmyer, R. D., and J. Von Neumann (1947) "Statistical Methods in Neutron Diffusion." *Los Alamos Manuscripts* 551 (April 9).

Robison, J. (1797) "Steam Engine." *Encyclopedia Britannica*. Reprinted in *System of Mechanical Philosophy,* by J. Robison (1822). Edinburgh: J. Murray.

Rohrlich, F. (1991) "Computer Simulation in the Physical Sciences." In *PSA 1990,* edited by A. Fine, M. Forbes, and L. Wessels, 2:507–518. East Lansing: Philosophy of Science Association.

Roper, S. (1885) *Engineer's Handy-Book*. Bridgeport, Conn.: Frederick Keppy, Scientific Book Publisher.

Rose, M., and G. Lauder (eds.) (1996) *Adaptation*. San Diego: Academic Press.

Rossi, P., R. Berk, and K. Lenihan (1980) *Money, Work and Crime*. New York: Academic Press.

Rothbart, D. (1997) *Explaining the Growth of Scientific Knowledge*. Lewiston, N.Y.: Edwin Mellen Press.

Rothbart, D., and S. W. Slayden (1994) "The Epistemology of a Spectrometer." *Philosophy of Science* 61:25–38.

Rothman, D. H., and S. Zaleski (1997) *Lattice-Gas Cellular Automata: Simple Models of Complex Hydrodynamics*. Cambridge: Cambridge University Press.

Rothman, D. J. (2000) "The Shame of Medical Research." *New York Review of Books* 47 (November 30): 60–64.

Rouse, J. (1987) *Knowledge and Power: Toward a Political Philosophy of Science*. Ithaca: Cornell University Press.

Royal Society (1841) "Dedication of the Copley Medal." *Proceedings of the Royal Society* 4:336.

Schaffer, S. (1989) "Glass Works: Newton's Prisms and the Uses of Experiment." In Gooding, Pinch, and Schaffer (1989), 67–104.

———. (1995) "Where Experiments End: Tabletop Trials in Victorian Astronomy." In Buchwald (1995), 257–299.

Scheibe, E. (1973) *The Logical Analysis of Quantum Mechanics*. Oxford: Pergamon Press.

Schultz, R. L., and E. M. Sullivan (1972) "Developments in Simulation in Social and Administrative Science." In Guetzkow, Kotler, and Schultz (1972), 3–47.

Shapin, S., and S. Schaffer (1985) *Leviathan and the Air-Pump: Hobbes, Boyle, and the Experimental Life*. Princeton: Princeton University Press.

Sibum, O. (1994) "Working Experiments: Bodies, Machines and Heat Values. The Physics of Empire." In *The Physics of Empire,* edited by R. Staley, 29–56. Cambridge: Cambridge University Press.

———. (1995) "Reworking the Mechanical Value of Heat." *Studies in History and Philosophy of Science* 26:73–106.

Sismondo, S. (ed.) (1999) *Modeling and Simulation*, *Science in Context* (Special Issue) 12 (2).

Sklar, L. (ed.) (2000) *Philosophy of Science: Collected Papers*. 6 vols. Andover: Garland Publishing.

Skyrms, B. (1984) "EPR: Lessons for Metaphysics." *Midwest Studies in Philosophy* 9:245–255.

Smith, B. C. (1998) *On the Origin of Objects*. Cambridge, Mass.: MIT Press.

Sohn, D. (1999) "Experimental Effects: Are They Constant or Variable Across Individuals?" *Theory and Psychology* 9:625–638.

Solomon, J. R. (1998) *Objectivity in the Making: Francis Bacon and the Politics of Inquiry.* Baltimore: Johns Hopkins University Press.

Sorensen, R. (1992) *Thought Experiments*. Oxford: Oxford University Press.

Spirtes, P., C. Glymour, and R. Scheines (1993) *Causation, Prediction and Search*. New York: Springer-Verlag.

———. (2000) *Causation, Prediction and Search*. 2d ed. Cambridge, Mass.: MIT Press.

Staubermann, K. (1998) "Controlling Vision: The Photometry of Karl Friedrich Zöllner." Ph.D. diss., University of Cambridge.

Staudenmaier, J. M. (1985) *Technology's Storytellers*. Cambridge, Mass.: Society for the History of Technology and MIT Press.

Steinle, F. (1998) "Exploratives vs. theoriebestimmtes Experimentieren: Ampères erste Arbeiten zum Elektromagnetismus." In Heidelberger and Steinle (1998), 272–297.

Sterckx, S. (ed.) (2000) *Biotechnology, Patents and Morality*. 2d ed. Aldershot, UK: Ashgate.

Stewart, I. (1998) *Nature's Numbers. Discovering Order and Pattern in the Universe*. London: Phoenix.

———. (1999) *Life's Other Secret. The New Mathematics of the Living World*. Harmondsworth, U.K.: Penguin.

Suppes, P. (1961) "A Comparison of the Meaning and Use of Models in Mathematics and the Empirical Sciences." In *The Concept and Role of the Model in Mathematics and Natural and Social Science,* edited by H. Freudenthal, 163–177. Dordrecht, The Netherlands: Reidel.

———. (1962) "Models of Data." In *Logic, Methodology and Philosophy of Science: Proceedings of the 1960 International Congress,* edited by E. Nagel, P. Suppes, and A. Tarski, 252–261. Stanford: Stanford University Press.

———. (1967) "What Is a Scientific Theory?" In *Philosophy of Science Today,* edited by S. Morgenbesser, 55–67. New York: Basic Books.

Swade, D. (2000) *The Difference Engine: Charles Babbage and the Quest to Build the First Computer.* London: Little, Brown.

Thornhill, R. (1992) "Female Preference for the Pheromone of Males with Low Fluctuating Asymmetry in the Japanese Scorpionfly." *Behavioral Ecology* 3:277–283.

Tiles, J. E. (1992) "Experimental Evidence vs. Experimental Practice?" *British Journal for the Philosophy of Science* 43:99–109.

———. (1993) "Experiment as Intervention." *British Journal for the Philosophy of Science* 44:463–475.

Tiles, M., and H. Oberdiek (1995) *Living in a Technological Culture.* London: Routledge.

Toffoli, T. (1984) "Cellular Automata as an Alternative to (Rather Than an Approximation of) Differential-Equations in Modeling Physics." *Physica* 10D:117–127.

Toffoli, T., and N. Margolus (1987) *Cellular Automata Machines: A New Environment for Modeling.* Cambridge, Mass.: MIT Press.

Torretti, R. (1999) *The Philosophy of Physics.* Cambridge: Cambridge University Press.

Toulmin, S. (1967) *The Philosophy of Science.* London: Hutchinson.

———. (1993) "From Clocks to Chaos: Humanizing the Mechanistic World-View." In *The Machine as Metaphor and Tool,* edited by H. Haken, A. Karlqvist, and U. Svedin, 139–154. Berlin: Springer-Verlag.

———. (ed.) (1970) *Physical Reality.* New York: Harper & Row.

Ulam, S. (1952) "Random Processes and Transformations." *Proceedings of the International Congress of Mathematicians 1950* 2:264–275.

———. (1990) *Analogies between Analogies.* Berkeley: University of California Press.

Ungar, E. (1996) "Mechanical Vibrations." In *Mechanical Design Handbook,* edited by H. Rothbart, sect. 5. New York: McGraw-Hill.

Van Fraassen, B. C. (1980) *The Scientific Image.* Oxford: Clarendon Press.

Vance, E. R. (1949) "Melting Control with the Direct Reading Spectometer." *Journal of Metals* 1 (October): 28–30.

Verlet, L. (1967) "Computer 'Experiments' on Classical Fluids. I. Thermodynamical Properties of Lennard-Jones Molecules." *Physical Review* 159:98–103.

Vichniac, G. Y. (1984) "Simulating Physics with Cellular Automata." *Physica* 10D:96–116.

Von Wright, G. H. (1971) *Explanation and Understanding.* London: Routledge and Kegan Paul.

Wainwright, T., and B. J. Alder (1958) "Molecular Dynamics Computations for the Hard Sphere System." *Nuovo Cimento,* ser. 10 (supplement 9): 116–143.

Wallace, W. A. (1996) *The Modeling of Nature.* Washington, D.C.: Catholic University of America Press.

Watson, J. (1981) "The Double Helix: A Personal Account of the Discovery of the Structure of DNA." In *The Double Helix: A Personal Account of the Discovery of the Structure of DNA,* edited by G. Stent, 1–133. New York: W.W. Norton and Co.

Weinberg, R. (1985) "The Molecules of Life." *Scientific American* 253 (4): 48–57.

Whitehead, A. N. (1929) *Science and the Modern World*. Cambridge: Cambridge University Press.

Wiener, N. (1965) *Cybernetics or Control and Communication in the Animal and the Machine*. 2d ed. Cambridge, Mass.: MIT Press.

Winsberg, E. (1999) "Sanctioning Models: The Epistemology of Simulation." *Science in Context* 12:275–292.

Wittgenstein, L. (1958) *Philosophical Investigations*. New York: Macmillan.

———. (1974) *Philosophical Grammar*. Berkeley: University of California Press.

Wolfram, S. (ed.) (1986) *Theory and Applications of Cellular Automata*. Singapore: World Scientific Publishers.

Wolpert, L. (1992) *The Unnatural Nature of Science*. London: Faber.

Woodward, J. (1980) "Developmental Explanation." *Synthese* 44:443–466.

———. (1997) "Explanation, Invariance and Intervention." *Philosophy of Science* 64 (supplement to no. 4): S26-S41.

———. (1999) "Causal Interpretation in Systems of Equations." *Synthese* 121:199–247.

———. (2001) "Probabilistic Causality, Direct Causes, and Counterfactual Dependence." In *Stochastic Causality,* edited by M. Galavotti, P. Suppes, and D. Costantini, 39–63. Stanford: CSLI Publications.

———. (Forthcoming a) *Making Things Happen: A Theory of Causal Explanation*. Oxford: Oxford University Press.

———. (Forthcoming b) "What Is a Mechanism? A Counterfactual Account." *Philosophy of Science* 69 (supplement).

Woodward, J., and C. Hitchcock (Forthcoming) "Explanatory Generalizations. Part I: A Counterfactual Account." *Nous*.

Wright, L. (1973) "Functions." *Philosophical Review* 82:139–168.

Zeisel, H. (1982) "Disagreement Over the Evaluation of a Controlled Experiment." *American Journal of Sociology* 88:378–389.

Zielonacka-Lis, E. (1998) "Some Remarks on the Specificity of Scientific Explanation in Chemistry." Paper presented at the Second International Conference of the Society for the Philosophy of Chemistry, Sidney Sussex College, UK.

INDEX

abstraction, 257–259, 262–266, 268, 275, 277, 279–280; and experimental replicability, 10–11, 18n5, 157–160. *See also* theory, as abstract

Ackermann, R., 25, 71, 236

action, free, 6, 94, 111–112

affordances, 37–38

agency, 108–112, 127

air pump, the, 167–168, 171, 172, 248–249

Alder, Berni, 206

Alvarez, L., 270

Ampère, André-Marie, 161, 163, 164

animals, 17, 65–66, 106, 141, 222; *C. elegans* (worms), 229, 232; *Drosophila* (fruit flies), 27, 29, 133, 227–228; insects, 98–101; mice, 227–228. *See also* model organisms

Ankeny, Rachel, 228, 232

anomalies, 142, 143, 144, 151n1

anthropomorphism, 80, 87–89, 94, 126–127

apparatus, 5, 19, 153, 165–166; advanced, 169; calibration of, 180; classifications of, 14; fluctuations of, 183; history of, 251–252; malfunctioning of, 71; as models, 20, 26–32, 34; and nature, 22–25; neglect of, 22; and ontology, 25, 34; as stage of experiment, 191, 194; states of, 170; theories of, 3; use of, 250. *See also* instruments; *names of specific apparatus*

apparatus-world complexes, 20, 25–26, 28–31, 34–35, 37–38

applied science, 130

apprenticeship, 131, 135

Archimedes, 7

Aristotle, 22, 31, 70, 72–73, 85–86n2, 86nn4–5, 102, 117n14, 237; and errors, 184, 186; and qualities, 256–257

artifacts, 70–71

artificial. *See* natural and artificial, distinction between

artificial intelligence, 213, 272–273, 282n17

artificial life, 11, 208–213, 214n11

Assmus, A., 234n12

astro-photometers, 63

astrophysics, 15

Atwood's machine, 27, 28

Babbage, Charles, 198, 274, 275

back inference, 27–28, 31–32, 33–34, 37

Bacon, Francis, 9, 174, 184–190, 195

Baigrie, B., 239

Baird, Davis, 3, 5, 8, 12, 14, 18n4, 118n22, 147, 157, 178, 236

Barlow, Peter, 47–49

barometers, 90–92, 98, 109, 116n3

Barricelli, 214n12

Barrow, J., 272

Bazerman, C., 260

Beaulieu, A., 256

Bechtel, W., 253

Beck, J. D., 222

behavioral science, 88, 96

Belitz, K., 235n21

Berk, R., 117n10

Berkeley, George, 257

Bernard, Claude, 141

Bhaskar, R., 18n3

Biagioli, M., 2

Bigelow, J., 86n6

biology, 65, 98–101, 117n13, 133, 147, 212–213, 213n3; molecular, 106, 253; physiology, 141; reproduction, 209, 210. *See also* DNA; genetics

biomedicine, 88, 96

Birdsall, Charles, 207

black box, 24

black boxing, 58–60

Block, N., 283n24

Blois, Marsden, 276–278

Blondlot, P., 191

Bogen, J., 115, 236

Böhme, G., 153

Bohr, Niels, 6, 26, 28–30

Bohrian apparatus, 26, 38

Bohrian phenomena, 31

Bohrian theory of experimentation, 34–38

Boltzmann, Ludwig, 238
Boscovich, 35, 36
Boumans, Marcel, 14, 173*n9*, 219, 234*nn7–8, 12*
Boyle, Robert, 167–168, 171, 172, 248–249
Brandon, R., 253
Brock, William, 226
Brown, H. R., 31
Brown, J. R., 246
Bucciarelli, Larry, 43–44
Buchwald, Jed, 2, 56, 67*n9,* 148, 167, 175–176,
    214*n10*
Bunge, M., 153
Buneman, O., 205, 207–208
Burian, R., 253

Campbell, D., 88, 117*n13*
Cantor, G., 190
Cardwell, D. S. L., 67*n6*
Carrier, M., 164, 173*n13*
Cartwright, Nancy, 32, 91, 92, 116*n4,* 117*n10,*
    196, 228–229, 235*nn18, 21,* 253
causal claims, 6–7, 105, 107, 109, 115, 127;
    and interventions, 89–92
causal experiments, 145–146, 150
causal inferences, 15, 87, 232; from passive
    observation, 100–101
causality, 6–7, 140, 146; interventionist ac-
    count of, 124–128; and theory-laden-
    ness, 139–140, 142–143
causation, 144; agency theories of, 108–112,
    127; anthropomorphic accounts of,
    87–89, 126–127; subjectivist accounts of,
    108–109, 126–127
causation, manipulability theory of, 87–116;
    anthropomorphic, 87–88, 94; basic ideas
    of, 88–89; and circularity, 93–95,
    116–117*n6*; and correlations, 87, 99, 101,
    103–105, 109–111; and experimenta-
    tion, 97–103; and hypothetical experi-
    ments, 95–96, 105, 106–107, 109; and
    inference problems, 103–105; and inter-
    ventions, 89–100, 104–105, 109, 111;
    and mechanism-preserving require-
    ment, 92–94, 100, 116*n3*; and prediction
    of outcomes, 103–107; and realism,
    108–110; subjectivist, 94, 111
cellular automata (CA), 209–213,
    214–215*nn14–16*
Cern laboratories, 30
Chang, Hasok, 234*n7*
chemistry, 34, 56–57, 133, 256, 258
circularity, 88, 93–94, 111, 116–117*n6,* 126,
    135, 264; as consequence of theory-
    ladenness, 170–171

Clapeyron, É., 51
Clark, A., 260
clocks, 274
cloud chambers, 23–24, 28, 270
Coates, J., 246
cognition, 237, 249, 256, 261, 262, 267–268;
    analogue, 270–271, 272; as digital, 264,
    271, 274; and reduction, 259, 266, 277;
    and technology, 257–258, 260
cognitive autonomy, 53–54
cognitive span, 276–280
Cohen, M., 59
Coleman, J., 89
Collins, Harry, 10, 63, 135, 157, 262, 278,
    282*n17*
color, 166–167, 171
computation, 201–204, 206, 207, 259, 274, 275
computer experimentation, 11–13, 200,
    205–208
computer images, 212, 222–223, 230
computer modeling and simulation, 11–13,
    199–213, 223, 232, 262–263, 265; ana-
    logue, 200; artificial life, 11, 209–210,
    213, 214*n11*; cellular automata (CA),
    209–213, 214–215*nn14–16*; definition of,
    214*n5*; discourse of, 214*n7*; and experi-
    ments in theory, 204, 205; history of,
    199–202; meaning of, 198–199; numeri-
    cal computation, 201–204, 206, 207;
    speed of, 204
computers, 210, 240, 256; rise of, 199, 271, 275
Considine, D. M., 58
constructionism, 19, 22–24
constructivism, 4–5, 138, 150, 261; methodi-
    cal, 154, 156
controls, 219, 223, 225
Conway, John, 210
Cook, T., 88, 117*n13*
Coquillette, D. R., 187
correlations, 115, 141; and causation, 87, 99,
    101, 103–105, 109–111; between objects
    and apparatus, 153, 165, 168, 171–172
Coulomb, C. A., 149
counting, 271–272
Cox, J. G., 203
Craig, E., 16
Craver, F., 253
Crick, Francis, 5, 45–47, 52, 53, 65, 66*n1*
Crosby, A. W., 275
Cummins, R., 86*n6*
cyborg science, 259–260, 268, 273, 283*n25*

Dalmedico Dahan, A., 233–234*n3*
Darden, L., 253

Darrigol, O., 179

data, 70–71, 122, 194; and computers, 202, 256; digitized, 282*n*9; economic, 225; numerical, 269; random, 203; reduction of, 192; visualization of, 256, 268, 278

Davenport, Thomas, 5, 40–42, 47, 49, 53, 65

Davy, Humphrey, 30, 31

Dehue, T., 18*n*9

De La Warr, 23, 38*n*1

DeMey, M., 282*n*18

Descartes, René, 34, 35, 166, 237

descriptions: functional, 74–78, 81–82, 83–84, 86*n*8; qualitative, 258, 259, 264, 269, 272; quantitative, 258, 259, 264, 272; structural, 74–78, 81, 84

design, 69, 76, 83. *See also* instruments, design of

Diesel, 24

digitalization, 259, 269, 270, 274, 276–277, 279. *See also* cognition, as digital

Dingler, Hugo, 8, 136*n*3, 154, 156

Dipert, R. R., 85*n*1

dispositions, 36–38, 38*n*2

DNA, 45–47, 52, 253

dogmas, 186

Dowling, Deborah, 214*n*6, 233*n*3

Drake, S., 282*n*14

Dreyfus, H. L., 278, 282*n*17

Duhem, Pierre, 9, 80, 140–141, 142, 143–146, 149, 150, 177, 189, 193–194

Dunbar, Robin, 267

Dunn, D. A., 205, 207–208

dynamics, molecular, 205–208

Earle, J. H., 240

economics, 88, 218–219, 225, 226, 229, 231

Edison, Thomas, 11, 158–160

education, specialist, 131, 132, 133, 136

Einstein, Albert, 28–29, 164

electricity, 147, 148–150, 158–160, 163

electromagnetic motors, 40–42, 47–49, 67*n*5

electrometers, 147, 150

electroscope, 148

empiricism, 7, 18*n*8, 21, 150, 159; inadequacy of, 171

energy, 36

engineering, 81, 154, 213; and design, 236, 238–239, 241, 243, 244, 250–251

engines, 24, 50

entities: functions of, 79, 81, 84; theoretical, 78

epistemological objects, 62–63

epistemology, 5, 25, 80, 171, 183, 217; and computer simulation, 200–204, 210, 212;

empirical, 204; of experiment, 71, 187, 221; materialist, 39–40, 42, 60; objective, 60–64; and obstacle to philosophy of experiment, 177–178; strategies of, 180–181; subjective, 44, 63; and technology, 256

error, experimental, 9–10, 13, 176, 182; avoidance of, 187, 196; elimination of, 184–185, 189; of interpretation, 194; probabilities of, 191–193; random, 183; sources of, 183–184, 185–193, 195; systematic, 183, 193–194; taxonomies of, 185–191

ethical standards, 17

evolution, 209, 210, 266–267, 282*n*15

expansion, 259, 269, 271, 274, 278–280, 283*n*35

expectations, 138, 143

experience, 141, 142–143, 265, 267, 268–269, 274, 275

experientialism, 150

experiment, philosophy of, obstacles to, 176–181

experimental instructions, 122–124, 130

experimentalism, 4; new, 217

experimental laws, 124, 135, 136*n*3

experimental practice, 152–153, 178, 179; nonlocality of, 131–136

experimental processes, 130, 173*n*2; classification of, 129

experimental systems, 131, 153; closed, 14, 168

experimenters' regress, 135–136

experiments: and abstraction, 158–159, 282*n*8; and advanced theory, 141, 150; artificial, 281; on bone, 221–224, 230, 232; causal, 145–146, 150; cognitive aspect of, 262; as common experience, 141, 142–143; definition of, 152; and etc. list, 181–183, 194–195; epistemology of, 71, 187, 221; exploratory, 161, 222; extrascientific (Lebenswelt), 121, 128; historical studies of, 1–3, 174–176; hybrid, 224–225, 227, 230–232; hypothetical, 95–96, 106–107, 109, 246; and manipulability theory of causation, 97–103; material, 221, 228; natural, 94, 105; nonmaterial, 217, 221, 227; objects of, 194; and ontology, 80, 229; phases of, 121–122, 190, 191; qualitative, 169; reproduction of, 135, 157–158, 173*n*3, 278; semimaterial, 223, 226, 227; as tests of theories, 162–163; in theoretical context, 163–164, 191, 193–194; theoretical interpretations of, 142, 164–165, 166, 168–172;

experiments (*continued*):

in theory, 204, 205; theory-free, 9–10, 145, 150, 161–162, 165–169, 170; theory-guided, 144, 145, 163; theory-laden, 139, 145, 146, 147, 162–169; thought, 218, 246, 249, 267, 268, 278, 283*n*22; vicarious, 217, 227; virtual, 217, 224, 226–227; virtually, 217, 224, 226–227. *See also* error, experimental; knowledge, experimental

experiments, physical, 29, 68, 70, 178–179; functional descriptions in, 74–78, 81–82, 83–84, 86*n*8; and interpretation, 140; and observation, 140; structural descriptions in, 74–78, 81, 84

facts: concrete, 140–141; theoretical, 140
Faraday, Michael, 30, 192, 254*n*2, 281*n*1; and electromagnetic motor, 5, 47–49, 53, 56, 62–64, 67*n*5; and ontology, 35, 36, 179
Farmer, D., 210
Faust, T., 236
Feenberg, A., 18*n*10
Féher, M., 69
Ferguson, E., 238
Fermi, Enrico, 203
Fernández, E., 237–238
Feyerabend, P. K., 171, 281*n*1
Feynman, Richard, 215*n*17
field concept, 35–36
Fizeau, Hippolyte, 164
force, 163, 179, 222, 241
Franck, J., 191, 194
Franklin, Allan, 10, 71, 180–181, 182–183, 189, 192, 233*n*1, 236
Franklin, Rosalind, 65
Fredkin, Ed, 212
Freedman, David, 88

Galileo, 28, 147, 193, 102, 196, 200, 273–275, 281*n*1; and mathematics, 259, 282*n*14; and qualities, 255, 257, 268, 271, 273; and reduction, 266; and senses, 263; and the telescope, 42
Galison, Peter, 56, 203, 234*n*12, 236, 261, 272, 278, 281*n*2, 282*nn*7, 20–21; and particle physics, 269–270; and simulations, 199, 200, 202, 233*n*3
galvanometers, 148, 150
Gardner, M., 210
Garman, R. L., 213*n*2
Gasking, D., 124–125
Gee, B., 49
Geiger counters, 23–24

genetics, 133, 256
geophysics, 256, 268–269, 272
Gibson, James J., 15, 37, 192, 242–243
Giere, R., 66*n*2
Gilbert, 35, 186
Glymour, C., 91, 96, 97, 117*n*10, 118*n*18
Golinski, J., 138
Gooding, David, 10, 13, 15, 18*n*5, 56, 62, 67*nn*9, 11, 163, 180, 181, 189, 233*n*1, 236, 249; and agency, 254*n*2; and Faraday, 62, 67*n*5
Goodman, N., 103, 198, 283*n*27
Goodwin, B., 265
Gopnik, Alison, 267, 282*n*15
Gould, S. J., 283*n*35
Grand, Steve, 213
Gregory, R. L., 272
graphics, computer, 240
Guala, Francesco, 6, 173*n*11, 235*n*15
Guetzkow, H., 234*n*3
Gulick, W., 239

Habermas, J., 8, 156
Hacker, P. M. S., 250
Hacking, Ian, 1, 6, 9, 15, 56, 62, 67*n*9, 112, 123, 151*n*2, 157, 176, 184, 233*n*2, 248; and creation of phenomena, 65, 68, 78–82, 84, 101; and elements of experiment, 181–182, 183, 194–195; and instruments, 236; and relation of theory to experimentation, 161–162, 164, 165, 170, 171, 172, 173*n*7
Hackmann, W. D., 148, 234*n*12
Hakfoort, C., 166
Hall effect, 78, 191
Hanson, Norwood Russell, 9, 139–141, 143, 144, 145–146, 171, 267
Harré, Rom, 2, 3, 4, 5, 6, 12, 14, 18*n*3, 173*n*7, 236, 253, 254*n*1
Hartmann, D., 128
Hartmann, Stephan, 234*n*3, 235*n*21
Hausman, D., 118*n*17
Heelan, P. A., 18*nn*4, 7, 170
Hegselmann, R., 234*n*3
Heidelberger, Michael, 2, 3, 6, 14, 174, 233*n*2, 258; and relation between experiment and theory, 8–9, 162, 164, 165, 167–169, 170, 172
Helmholtz, H., 192
Henry, Joseph, 40–42, 47, 53
Hentschel, K., 15
hermeneutics, 16, 18*n*4, 52, 153
Hertz, H., 191, 192, 194
Hillis, W. D., 215*n*14

Hills, R. L., 50, 67n6
Hintikka, J., 196
history of science: and abstraction, 264, 265, 266, 283n22; and apparatus, 251–252; and experimentation, 1–3, 174–176; and light, 113, 161, 165–167; and objectivity, 262; and representation, 268; and technology, 153. *See also names of specific scientists*
Hitchcock, C., 116n2
Hobbes, Thomas, 167–168, 251
Hockney, David, 282n5
Hoffer, T., 89
Holland, 214n12
Hommes, Cars, 226
Hon, Giora, 2, 3, 9–10, 13
Hooke, Robert, 167, 251–252
Hoover, Kevin, 88
Horton, R., 17
Hughes, R. I. G., 45–46, 66n2, 214n9, 229, 232, 234n3
Hughes, Thomas, 67n12, 159–160
human nature, 185, 260, 272
Hume, David, 7, 21, 257
Humphreys, C. J., 256, 282n13
Humphreys, Paul, 199, 214n5
hypotheses, 239, 246. *See also* experiments, hypothetical

idols: of the script, stage, spectator and moral, 190–191, 193–195; of the tribe, cave, marketplace and theater, 185–186
Ihde, D., 18n4, 156
incandescent lamp, 158–160
Indicator Diagram, 50–52, 63–64, 67n7
induction, 21, 22, 103, 184, 185
inference, 103–105, 217, 221, 227–228, 230–232. *See also* back inference
instrumentarium, the, 25
instruments, 3, 8–9, 13–14, 22–23, 34, 39–40, 137n8; and abstraction, 258; causal power of, 144; classifications of, 14, 25–26, 33–34, 54, 146–148; constructive, 146–149, 162, 167; design of, 13, 236–244, 246, 249, 251, 254; development of, 130, 134; imitative, 147; measuring, 50–53, 71, 136n2, 149, 156; observational, 154–155, 278; philosophical, 251–252, 254n2; productive, 146–149, 162, 167; representative, 147, 148, 149; roles of, 29, 56, 57, 145, 216–217, 258, 278; states of, 32; theory of, 149. *See also* apparatus; *names of specific instruments*

interpretation, 10, 123, 140, 191, 193
interventions, 4–5, 7, 122, 156; and causal claims, 89–97; and causality, 124–128; destructive, 116n3, 230; and manipulability theory of causation, 89–100, 104–105, 109, 111; material, 217; non-material, 216, 223
invariance, 109–110

James, W., 196
Janich, Peter, 6, 7, 8, 9, 136n2, 137nn7, 9, 158, 161, 172, 173nn2, 6; and physics, 172n1; and technology, 154–156
Jardine, L., 185
Jevons, W. S., 234n12
Joerges, B., 8, 153
Johnson, M., 283n29
Johnson-Laird, P., 272
Jones, C. A., 261, 282n21
Jones, M., 91, 92, 116n4
Joule, James, 159–160
Judson, H. F., 66n1

Kant, I., 35, 36, 183, 257
Kapitza, P. L., 192
Keynes, J. M., 218
Keaveny, Tony, 221–222, 224, 226, 230, 232
Keller, A., 153
Keller, E. F., 11–12, 117n11, 118n19, 235n17, 275, 282n4
Kelly, M., 282n10
King, W. J., 49
Kirchhoff, G. R., 148
Kirkwood, J. G., 205–206
Kliemt, Hartmut, 234n3
Knorr-Cetina, K., 281n2, 282n11
knowledge, 155, 261; detached, 58–60; experimental, 178, 191–192; material, 52, 54, 57, 58, 66, 178; model, 45–47, 50, 52, 54; objective, 44–45, 60–63, 258; obstruction of, 187; propositional, 52, 58, 177; representational, 45; skillful, 66; subjective, 44, 258; tacit, 53, 62, 66, 217; theoretical, 5, 9–11, 66; thing, 39–40, 41, 43, 54, 60, 64–66; working, 47–49, 50, 52–57, 65, 66n4
Kohler, R., 133, 137n11
Kosso, Peter, 170, 171
Kotler, P., 234n3
Kroes, Peter, 3, 4, 6, 67n8
Krohn, W., 153
Kuhn, Thomas, 9, 139, 141–145, 147, 149, 150, 151n1, 171, 175, 257
Kusch, Martin, 63

Lakatos, I., 62
Lakoff, G., 283*n29*
Lange, Rainer, 3, 6, 7, 8, 9, 14, 116*n6*, 154, 156, 169, 172*n1*
Langton, Christopher, 209, 210, 212, 214*n11*
language, 186; pictorial, 250; scientific, 138, 142; sense-datum, 139
Latour, Bruno, 8, 42–43, 56, 58, 60, 67*n9*, 154, 158–160, 172, 173*n3*, 175, 236, 261; and cognition, 262, 282*n6*; and instruments, 3, 22–25; and operationalism, 173*n6*; and techno-science, 282*n4*
Lauder, G., 98
Lavoisier, Anton-Laurent, 142, 258
Lawrence, E. O., 42
Layton, E., 238
Lee, K., 8, 17, 156
Le Grand, H. E., 268–269
Leibniz, Gottfried Wilhelm, 35, 36
Lelas, S., 2, 8, 71, 72, 156
Lenihan, K., 117*n10*
light, 113, 155, 157, 161, 164–167, 173*n8*
Locke, John, 237, 257
logic, 21, 22, 195, 270
logicism, 19, 21–22, 23
Lorentz, Hendrik, 164
Lowe, A., 283*n32*
Lüer, Gerd, 163

Maas, H., 234*n12*
Mach, Ernst, 21–22, 29, 147, 177, 179–180, 182, 189, 267, 282*n16*
Machamer, P., 253
machines, agency of, 257
manipulability theory of causation. *See* causation, manipulability theory of
manipulation, of phenotype, 98–101
Mann, C., 239
Margolus, N., 211, 214*n14*
Martin, J., 195
materialism, 5–6, 39
material realization, 4–6, 122, 164–165, 166, 248
matériel, 181–182
mathematics, 159, 208, 265, 275, 282*n14*; differential equations (DE), 200–202, 203, 204, 210–211; and nature, 264, 266; and physics, 204, 259. *See also* abstraction; models, mathematical; reduction
Matthews, D., 269
Maxwell, J. C., 72, 238
May, Robert, 264, 282*n10*
Mayo, Deborah, 10, 191–193
McMullin, Ernan, 147

measurement, 10, 77, 136*n2*, 263, 272. *See also* instruments, measuring
Meehl, Paul, 276
Meijers, A. W. M., 69
Meltzoff, A., 282*n15*
Mendoza, E., 51
Menzies, P., 87, 108, 110–111
Merz, M., 233*n3*
metaphysics, 21, 32, 34–35, 64–66, 178
methodology, 6, 129, 249; of computer simulation, 200; and functions of objects, 74, 75, 77–78; and obstacle to philosophy of experiment, 176–181; and structures of objects, 77; and tension with ontology, 77, 86*n8*; transverse principles, 179
Metropolis, N., 203
microscopes, 147, 251–252
Mill, J. S., 21–22
Miller, Arthur, 265, 266, 282*n12*, 283*nn24, 35*
Millikan, Robert, 263, 268, 275, 281
Mitcham, C., 18, 239
Mize, J. H., 203
model organisms, 227–230, 232
models, 66*n3*, 130, 217, 238; apparatus as, 20, 26–32, 34; descriptive, 232; mathematical, 218–221, 222–227, 229, 231, 232, 256; meaning of, 198–199; as representative for, 229–232; as representative of, 229–232; theoretical, 204. *See also* knowledge, model
Morgan, Mary S., 11–12, 14, 16
Morison, R. S., 258
Morrison, M., 169, 173*n12*, 219, 234*n6*, 235*n19*, 254*n1*
Mueller, U., 234*n3*
Müller, Ralph, 59
Muller, Karl, 234*n3*
Mumford, S., 38*n2*, 86*n11*

natural and artificial, distinction between, 4–5, 17, 68–69, 79, 80–82, 102; traditional, 70–74, 82, 85
natural experiments, 94, 105
nature, 22, 77, 185; as composed of separate mechanisms, 102–103; conceptions of, 68, 70, 72, 75, 77, 81–82, 85–86*n2*; laws of, 21, 69, 72–73, 76, 130, 155, 196; as machine, 252–253; as mathematical, 264, 266; representations of, 24, 80; and scientists, 72–73
Naylor, Thomas, 234*n3*
neo-Aristotelianism, 32
neo-Popperianism, 61
networks, actor, 159–160

Newton, Isaac, 34, 72, 165–167, 171–172, 253, 255, 259, 273–275
Newton-Smith, W. H., 1
Neyman, J., 191
Niebur, G. L., 221
normativity, 3, 12–13, 16–18, 55, 74–75, 121, 129, 155–156, 176, 194, 246
nuclear weapons, 203, 204, 214*n*7
Nyhof, J., 238

Oberdiek, H., 8
objectivity, 44–45, 60–64, 258, 262
observations, 15, 21–22, 71, 95, 115, 140; causal inferences from, 15, 100–101, 143; and instruments, 154–155, 278; of natural phenomena, 154; that presuppose causality, 146; theory-ladenness of, 138–139, 144. *See also* scientists, as spectators
occlusion, 242–246, 249
Oersted, Hans Christian, 148, 163
Ohm, Georg Simon, 139, 148–150, 151*n*5, 158, 159–160
Olby, R., 66*n*1
ontology, 4–5; of apparatus, 25, 34; for Bohrian theory, 34–38; and computer simulation, 202, 212; development of claims about, 114; dynamicist, 36, 38*n*3; and experiments, 80, 229; field, 35–36; of modern physics, 73; and tension with methodology, 77, 86*n*8
operationalism, 121, 130, 136*n*2, 155–156, 173*n*6
optics, 165–167
Oreskes, N., 234*n*12, 235*n*21

Pargetter, R., 86*n*6
Parsons, M., 244, 246
particle accelerators, 169
Pasteur, Louis, 60, 192, 262
Pearl, J., 91, 92, 96, 97
Pearson, K., 191
Peirce, Charles Sanders, 49, 191, 192, 237–238, 249
perception, 138, 142, 143, 151*n*1; commonsense, 255, 261, 266–267, 268; numerical-digital, 261; pictorial-analogue, 261, 272
Petersen, Arthur, 235*n*13
phenomena, 65, 155, 281*n*2; created, 65, 68, 78–82, 84, 101; natural, 154, 155; natural *vs.* artificial, 72, 80–81
physics, 6, 11, 15, 73, 156, 203, 232; computational, 200, 212; digitally-based, 215*n*17; experimental, 29, 68, 70, 77–78, 178–179; high-energy, 29–31, 256, 269–271, 272;

laws of, 83; mathematical, 204, 259; metaphysics of, 32, 34–35; observational, 154; as technology, 154; theoretical, 200, 202, 214*n*10, 263. *See also* computer modeling and simulation; experiments, physical; geophysics; protophysics
Pickering, Andrew, 2, 56, 67*n*9, 131, 171, 177–178, 181, 236
Pinch, T., 13, 15, 180, 189, 233*n*1
Pitt, J., 67*n*11
Planck, Max, 86*n*14, 263, 268, 275
Plato, 186
Poincaré, H., 80
Poisson, Siméon, 163
Poling, Bruce E., 208
Pomerai, David de, 229
Pomian, K., 237
Popper, Karl R., 5, 22, 60–66, 161, 173*n*4, 171
Prausnitz, John M., 208
predictions, theoretical, 103–107, 204
Preston, B., 86*n*6
Price, Derek de Solla, 43, 137*n*8
Price, H., 87, 108, 110–111
Priestley, Joseph, 142
prisms, 165–166, 172
probability theory, 203
problem solving, 120–121
propositions, 21, 61
protophysics, 136*n*2, 156
psychiatry, 276
psychology, 163, 169–170, 267, 282*n*6
Pylyshyn, Z., 283*n*29
Pythagoras, 263

qualities: primary, 32–33, 255–257, 264–266, 271, 273; secondary, 32–33, 255–257, 271, 273
quantification, 10, 258, 259, 264–266, 269, 272, 279–280

Rabi, I. I., 192
Radder, Hans, 116*nn*3, 6, 119, 137*n*9, 175, 192, 236, 248; and artificial intelligence, 282*n*17; and material realization, 66*n*4, 122
Ramsey, J., 236
rationality, 22, 180
realism, 5, 108, 110, 115–116, 170–171, 177, 181, 263; instrumental, 5, 112–116; scientific, 15, 38*n*3, 78–79
reasoning, diagrammatic, 237–238
reduction, 89, 258–259, 265–266, 274–275, 278–280; cognitive, 259, 277
reductionism, 119–120, 121, 126, 156

Reichenbach, H., 80
relativism, 10, 78, 83
replicability, 10–11, 20, 156–160
representations, 13, 45, 52, 80, 149, 257,
   261–262; analogue, 259, 269–273; digital, 274; of nature, 24, 80; numerical
   (abstracted), 268, 269–276; qualitative,
   264–266; quantitative, 264–266, 269, 272
reproducibility, 122–124, 132–136, 157–158,
   246, 248
reproduction, 209, 210; of experiments, 135,
   155, 157–158, 173n3, 278
results: experimental, 72, 134, 157–158,
   180–182; derivation of, 220; interpretation of, 123, 184, 191, 193; prediction of,
   103–107; production of, 220; pseudo-
   material, 224–227; reproducibility of,
   157–158, 246, 248; validity of, 132, 217
Rheinberger, H. J., 131
Richards, E., 17
Richardson, R., 253
Richtmyer, R. D., 203
Robison, J., 67n6
Roentgen, Wilhelm, 139, 142–143, 144,
   146–147, 188–189
Rohrlich, Fritz, 199, 200, 233n3
Roper, S., 51
Rose, M., 98
Rossi, P., 117n10
Rothbart, Daniel, 3, 4, 13, 15, 169
Rothman, D. H., 214n13
Rothman, D. J., 17
Rouse, J., 16, 157, 173n13
Rowland, Henry, 42
Rutherford, Ernest, 192, 281

Saunderson, Jason, 55, 58
Schaffer, S., 13, 15, 164, 166–168, 180, 189,
   233n1, 236, 248–249, 260, 283n32
Scheibe, E., 6
Scheines, R., 91, 96, 97, 118n18
Schultz, R. L., 234nn3–4
science and technology, 7–8, 79, 119, 137n8,
   153
science as technology, 8, 119–120, 154–156,
   158
scientific experience, 15
scientific laws, 121
scientists: communication among, 131–133,
   156; communities of, 19, 25, 135–136,
   170, 248; cooperation among, 131–133,
   135; as creators, 79–80; as spectators,
   72–73, 78, 79–80, 81, 116, 156
Seaborg, Glenn, 58

selection, sexual, 98–101
sex discrimination, 106–107
Shapin, S., 164, 167–168, 236, 260, 248–249
Shinn, T., 8, 153
Shrader-Frechette, K., 235n21
Sibum, Otto, 67n10
simplification, 262–263
Sismondo, S., 233n3
Sklar, L., 1
Skyrms, Brian, 96
Slayden, S. W., 169
Smith, B. C., 282n18
social science, 15–16, 88, 89–90, 96, 106–107
sociology of science, 19, 119, 186, 282nn6–7
Sohn, David, 169–170
Solomon, J. R., 195
Sorensen, R., 282n16
Southern, John, 5, 50–52
spectrometers, 55, 56–60, 66, 169, 244–246
spectroscopes, 147
Spirtes, P., 91, 96, 97, 118n18
stability, 2–3
statistics, 180, 191–192, 203, 204, 205, 270, 276
Staubermann, Klaus, 63
Staudenmaier, J. M., 153
Steinle, Friedrich, 2, 9, 144, 161–163, 164,
   170, 171, 172, 174, 233n2
Sterckx, S., 17
Stern-Gerlach apparatus, 28
Stewart, Ian, 265
stock market, 225–227
Stokes, G. G., 191
subjectivism and subjectivity, 44, 63, 94,
   108–109, 111, 126–127, 258
subjects, 17, 122. See also model organisms
Sullivan, Edward, 234nn3–4
Suppe, Frederick, 282n9
Suppes, P., 66n2
Swade, D., 274
symbols, 240–241, 249, 259, 272, 273
systems dynamics, 202

technology, 122, 128, 255, 266; and abstraction, 257–258; artifacts, 69, 76, 79,
   83–84, 86n13; and cognition, 257–258,
   260; extrascientific (Lebenswelt), 120,
   121, 124, 132; literary, 261; and observation, 256; philosophy of, 18; pseudoartifacts, 69–70, 84–85; and science, 7–8,
   130, 153–160; and theory, 71–72; without theory, 161–162. See also apparatus;
   computer modeling and simulation;
   instruments
techno-science, 259–260, 273, 282n4

teleology, 68, 69, 73, 78
telephones, 43–44
telescopes, 42, 147
televisions, 79
Theodoric of Friborg, 26–27, 28, 29
theoretical interpretations, 142, 164–165, 166, 168–172
theory, 140, 182, 204; as abstract, 141, 241; of apparatus, 3, 194; approximate, 206–207; and causality, 139, 142–143. *See also* causation, manipulability theory of; definition of, 171; descriptive or classificatory, 105–106; as knowledge, 5, 9–11, 66, 155; meaning of, 208; non-locality of, 158, 160; operational, 160; as paradigm, 141–144; and prediction, 103–107, 204; qualitative, 207; structural, 76; and technology, 71–72; tests of, 162–163
theory and experiment, relationship between, 9–11, 161–162, 164, 165, 170, 171, 173*n7*. *See also* physics, theoretical
theory-free experiments, 9–10, 145, 150, 162, 165–169, 170
theory-guidance, 143–144, 145, 163
theory-ladenness, 138–146, 149, 162, 170–171
thermometers, 50, 75–77, 82–83, 147
Thomson, William (Lord Kelvin), 42, 59, 264, 281
Thornhill, R., 98–100, 117*n13*
thought experiments, 218, 246, 249, 267, 268, 278, 283*n22*
Tiles, J. E., 71, 156
Tiles, M., 8
time, 243–244, 274
Toffoli, T., 210, 211, 214*n14,* 215*n15*
tools, 79, 84–85, 252
Torretti, R., 117*n14*
Torricelli, Evangelista, 163–164
Toulmin, S., 20, 173*n8*, 252, 263
Troitzsch, K. G., 234*n3*
truth, 36, 61, 108, 127, 135, 153, 184, 196; material, 54–55, 56, 57
Turing, Alan, 253, 275

Ulam, Stanislaw, 202–204, 205, 209–210, 212, 214*n12*
Ungar, E., 241
unobservables, 115, 171

Vance, E. R., 57
Van den Daele, W., 153
Van Fraassen, B. C., 18*n8,* 66*n2*, 115
Verlet, Loup, 206–207, 214*nn7, 10*
Vichniac, G. Y., 212
Vickers, T., 239
Vine, F. J., 269
visualization and images, 237–244, 246, 249–250, 258; of data, 256, 268, 278
Volta, A., 148
Von Liebig, Justus, 161
Von Neumann, John, 203, 209–210, 212
Von Wright, G. H., 7, 87, 110, 124–127, 173*n2*

Wainwright, Ted, 206
Wallace, W. A., 31
Watson, James, 5, 45–47, 52, 53, 54, 65, 66*n1*
Watt, James, 50–52, 62, 63–64
Weinberg, Robert, 106, 107
Whitehead, A. N., 176
Wiener, N., 283*n34*
Williams, Maurice, 65
Wilson, Charles, 23–24, 28, 31, 234*n12,* 270
Winsberg, E., 199, 204, 214*nn8, 10,* 233*n3*
Wittgenstein, Ludwig, 250, 267
Wolfram, Stephen, 210, 211, 215*n17*
Wolpert, Lewis, 256, 261, 266–267
Woodward, Jim, 3, 5–7, 15, 16, 17, 126–128, 236
Woolgar, Steve, 42–43, 173*n3*, 261
W particles, 30–31
Wright, L., 86*n6*

X rays, 142–143, 144–145, 208

Zaleski, S., 214*n13*
Zeisel, H., 117*n10*
Zielonacka-Lis, Eva, 254*n3*
Zöllner, Karl Friedrich, 63